Pigeon Health and Disease

PIGEON HEALTH AND DISEASE

David C. Tudor, B.S., V.M.D.

Iowa State University Press, Ames, Iowa, USA

David C. Tudor, B.S., V.M.D., is research professor in animal science, retired, Poultry Diagnostic Laboratory, Cook College, Rutgers University, New Brunswick, New Jersey.

♾ Printed on acid-free paper in the United States of America

First edition, 1991

Library of Congress Cataloging-in-Publication Data
Tudor, David C.
Pigeon health and disease / David C. Tudor.—1st ed.
p. cm.
Includes index.
ISBN 0-8138-1244-5 (alk. paper)
1. Pigeons—Diseases. 2. Pigeons—Health. I. Title.
[DNLM: 1. Bird Diseases. 2. Pigeons. SF 994.6 T912]
SF994.6. T83 1991
636.5′96—dc20
DNLM/DLC
for Library of Congress 91-7020

Contents

Preface

Research conducted during the past several years has materially advanced our knowledge and understanding of pigeon health and disease. The object of this text is to provide pigeon fanciers, attendant veterinarians, and research-oriented groups with factual information concerning pigeon diseases. The text is organized to permit a clear understanding of each disease entity together with present-day methods of control and/or prevention. Other basic information on pigeon breeding is included to broaden the scope of the text.

The material presented is derived from the author's practical farm experience in earlier years, private practice, and day-to-day activities in the Avian Diagnostic Laboratory at Cook College, Rutgers University, where he served for 27 years as a research professor and teacher. Invaluable observations of other research personnel are included and documented to enhance the complete understanding of each subject. Additional uncited references have been included to provide the reader with related information.

The text is divided into five parts. The subheading format permits a complete discussion of each disease. References listed after each section enable further research on the part of interested investigators. Appendixes listing pathological and anatomical disease conditions, a glossary, and an index provide additional access to the terminology and disease information.

The author wishes to express his deepest gratitude for the unlimited help and encouragement given by his wife, Emily. Artwork has been contributed by his daughter, Diane Louis. Inspiration for the publication stems from the memories of untiring efforts of devoted and loving parents and from the kindly and instructive suggestions of mentors F. R. Beaudette and C. B. Hudson.

I

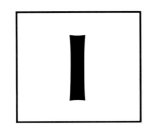

PIGEON STRUCTURE AND FUNCTION

Fig. 1.1. Exterior anatomy (adapted from Levi 1941).

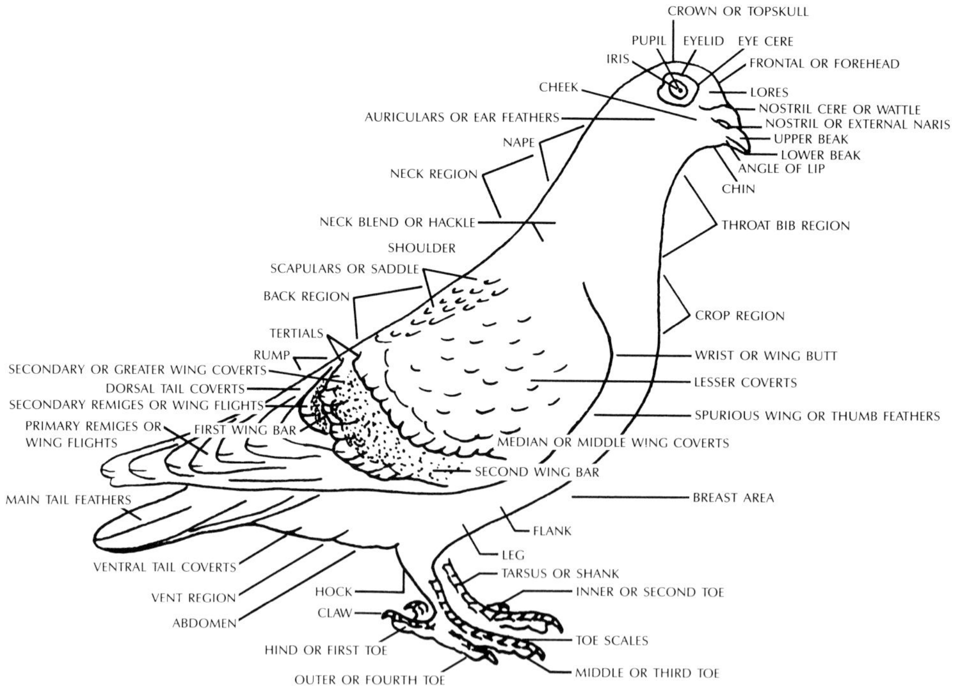

1

Anatomy and Physiology

Anatomy

Preventing pigeon diseases not only involves a planned disease control program; it also involves a knowledge of body structure and function. Without an understanding of the normal anatomy one is not conditioned to be aware of the abnormal or the disease condition.

Anatomy is the science of body structure (Fig. 1.1), both gross and microscopic. Following is a discussion of gross pigeon anatomy on a system basis.

Systemic anatomy includes (1) arthrology, or joint structure; (2) osteology, which deals with the skeleton; (3) myology, or muscles and their function; (4) the digestive system, which includes the alimentary tract and contributing glands; (5) the respiratory system, or the organs of breathing; (6) the excretory system; (7) the genital system, or the organs of reproduction; (8) the circulatory or blood vascular system; (9) the nervous system, which includes the brain, peripheral nerves, and sympathetic system; (10) the sensory system, which includes the sense organs, and (11) the endocrine system, which includes the secretory glands with regulatory function.

Arthrology. An *articulation* (joint) is formed when two or more bones or cartilage fit together either solidly or so as to permit motion. Movable joints are characterized by a joint cavity with a synovial membrane within a joint capsule. From the disease standpoint, the synovial fluid within the joint capsule occasionally becomes infected with mycoplasma or staphylococcus organisms. Joints become puffy and distended from excess turbid fluid.

Bursa, which provide synovial cushions over prominent bones, may also become involved. The sternal bursa at the lower point of the keel is a common site of infection.

Skeleton. The skeleton is the bony framework that protects and supports the soft tissues and organs of the body. The bones are identified in separate pictures. The pigeon, like other birds, has pneumatic bones that contain air spaces. These spaces are lined with mucous membrane and communicate with external air. The *foramen* (the hole in the proximal posterior notch of the humerus) permits air to enter the bone so that it will actually float in water.

The skull also has air spaces in its spongy bone. It provides space for a rounded brain compartment and large eye sockets. The openings to the ear canals may be observed below and behind the eye sockets. The long upper and lower mandibles form the beak. The lower mandible articulates posteriorly with the ventral skull. The hyoid bone provides support for the tongue. (See Fig. 1.2.)

The skull articulates posteriorly with the atlas, the first vertebrae, which in turn articulates with the axis. There are fourteen cervical vertebrae in all. Thoracic vertebrae are fused and articulate with the synsacrum. Five coccygeal vertebrae articulate with the vertical bladelike urostilus of the tail. (See Fig. 1.3.)

The wing bones consist of the humerus, which articulates proximally with the coracoid and scapula and distally with the radius and ulna, followed by the radial carpal and ulnar carpal bones. A divided, partly fused III and IV carpometacarpal articulates laterally with a tiny phalanx 1, digit II. It is tipped by phalanx

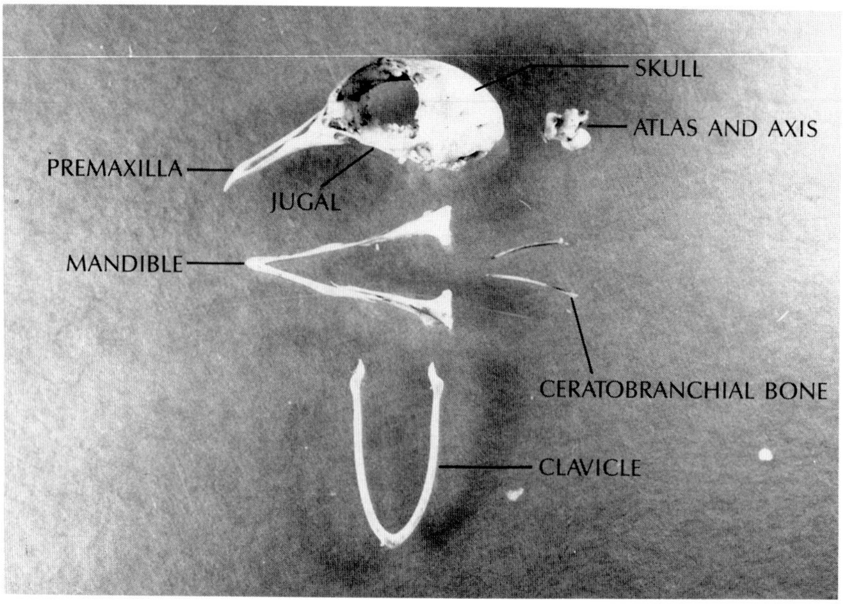

Fig. 1.2. Skull. A tiny bone can be observed in the ear canal. The ceratobranchial bone is part of the hyoid.

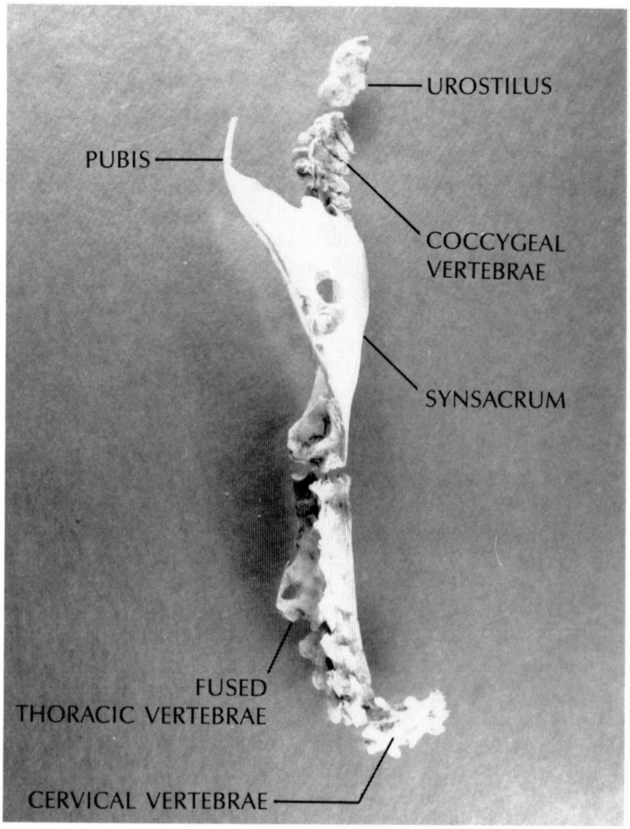

Fig. 1.3. Back bones. These are fused.

2, digit II. The distal end of the carpometacarpal is joined by two bones. The larger is phalanx 1, digit III. To it is attached phalanx 2, digit III. The smaller bone extending from the carpometacarpal is phalanx 1, digit IV (Chamberlain 1943). (See Figs. 1.4 and 1.5.)

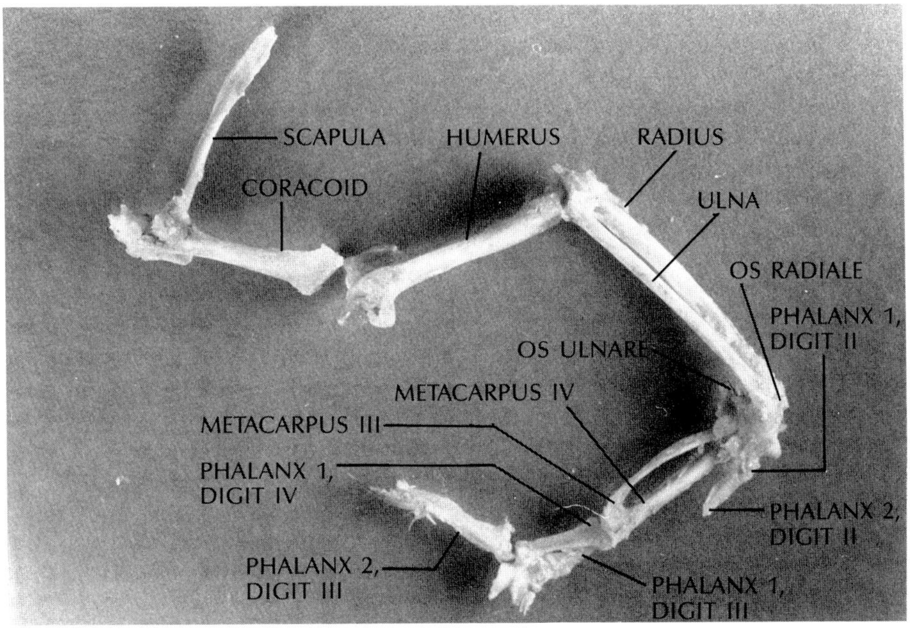

Fig. 1.4. Wing bones.

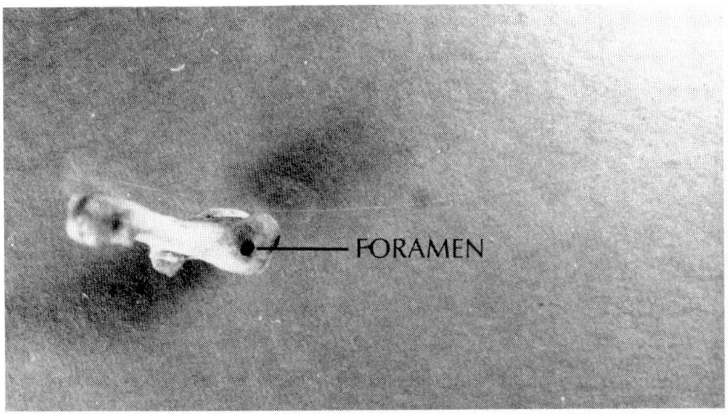

Fig. 1.5. Foramen in the humerus, which allows air movement. This permits the bone to float in water.

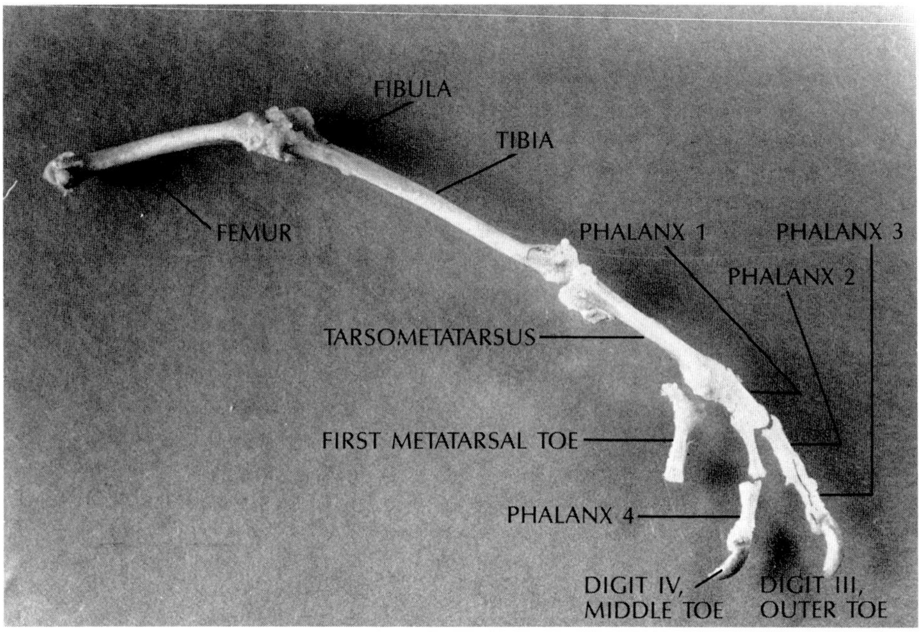

Fig. 1.6. Leg bones. These are hollow. Digit II (mediae toe) is not shown. The claw is missing on the first metatarsal toe.

The leg bones consist of the femur, which articulates proximally with the os innominatum of the synsacrum, and the tibia and fibula. The *patella* (kneecap) is located on the anterior surface of the joint formed by the distal tibia and the tarsometatarsus. The latter in turn articulates posteriorly with the first metatarsal toe having a basal unit and phalanx 1. The forward toes are digits II, III, and IV. The medial toe is digit II, the middle toe is digit III, and the outer toe of each foot is digit IV. Each toe is formed by phalanx bones. Digit IV has phalanx 1, 2, 3, and 4; digit III has 1, 2, and 3; and digit II has 1 and 2. The toenails are distal phalanges (Chamberlain 1943). (See Fig. 1.6.)

The ribs articulate with the vertebrae and/or with the sternum. In the pigeon two pairs of ribs articulate dorsally with the posterior cervical vertebrae only. Four of five pairs of thoracic vertebral ribs join sternal ribs that articulate with the sternum. The fifth or posterior vertebral rib joins a floating sternal rib.

The sternum is the largest bone and it supports and protects the internal organs. It articulates anteriorly with the distal end of the coracoid. The wishbone or clavicle forms an anterior chest barrier as it joins the coracoids and scapula dorsally. (See Fig. 1.7.)

Myology. Muscles are highly specialized tissues characterized as striated body fibers, nonstriated or smooth fibers, and cross-striated cardiac fibers. Dark body fibers with a rich blood supply dominate the skeletal muscles of the pigeon. Smooth muscles are largely found in lungs and blood vessels. Specialized cross-striated muscle fibers are found only in the heart.

Bruises and hemorrhage are common findings in traumatized muscle tissues. Congestion is evident following toxic conditions, and hyaline degeneration of breast muscles may be noted in suffocation.

Digestive System. The upper digestive tract of the pigeon includes the mouth and its cavity, the tongue, the pharynx and tonsilar tissue in the soft palate, and the esophagus and bilobed crop with its milk glands.

The esophagus connects with the glandular stomach or proventriculus. According to Sturkie (1954), the stomach glands of birds have only one cell type, which secretes both acid and pepsinogen. The proventriculus empties into the muscular gizzard or ventriculus. The keratinized cuticular lining of the gizzard is a secreted layer of epithelial rods designed to withstand grinding action (Eglitis and

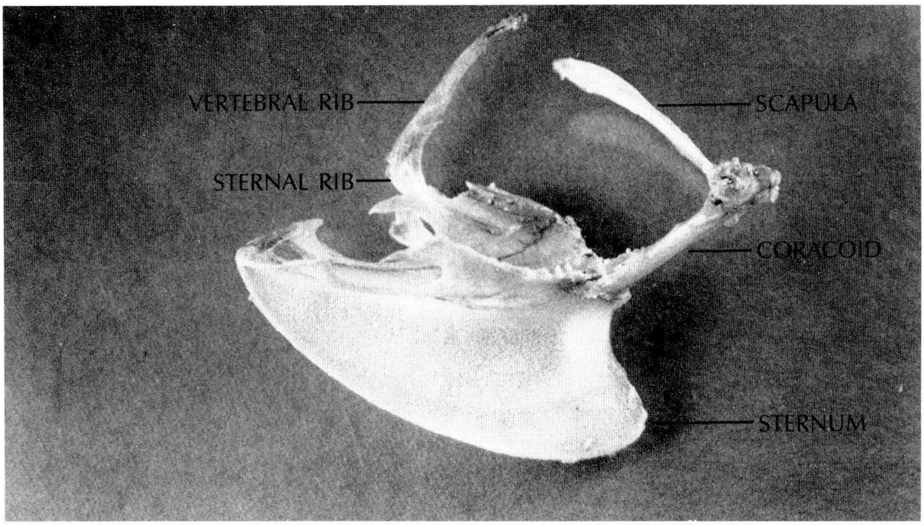

Fig. 1.7. Keel and attached bones.

Knouff 1962). Food passes from this organ to the small intestine, entering the loop of the duodenum, which holds the pancreas. Both the intestine and pancreas secrete digestive enzymes essential for the digestion of carbohydrates, proteins, and fats. The duodenum is adjacent to the liver, which provides bile for fat digestion. A bile duct and two or more pancreatic ducts enter the medial aspect of the duodenum. The small intestine is continued by the jejunum and ileum and empties into the large intestine at the junction with the rudimentary ceca. An arbitrary distinction is made between the jejunum and ileum at the degenerate stem of the former absorbed yolk sac of the hatched egg. The two ceca are tiny blind pouches that have lost their function in the process of evolution. The cloaca is divided into the urodaeum and coprodaeum. Ureters from the kidneys enter the dorsal urodaeum to deposit uric acid as the white cap on the droppings. This area also receives the egg from the distal end of the oviduct. (See Figs. 1.8 and 1.9.)

Respiratory System. The respiratory system pertains to those organs that are essential in breathing, including the rib cage. Outside air enters the nostrils and/or mouth and passes across the floor of the pharynx to the larynx and the entrance to the trachea. The trachea divides within the thoracic cavity to form the right and left bronchi. Just anterior to the point of divergence, vertical elastic mem-

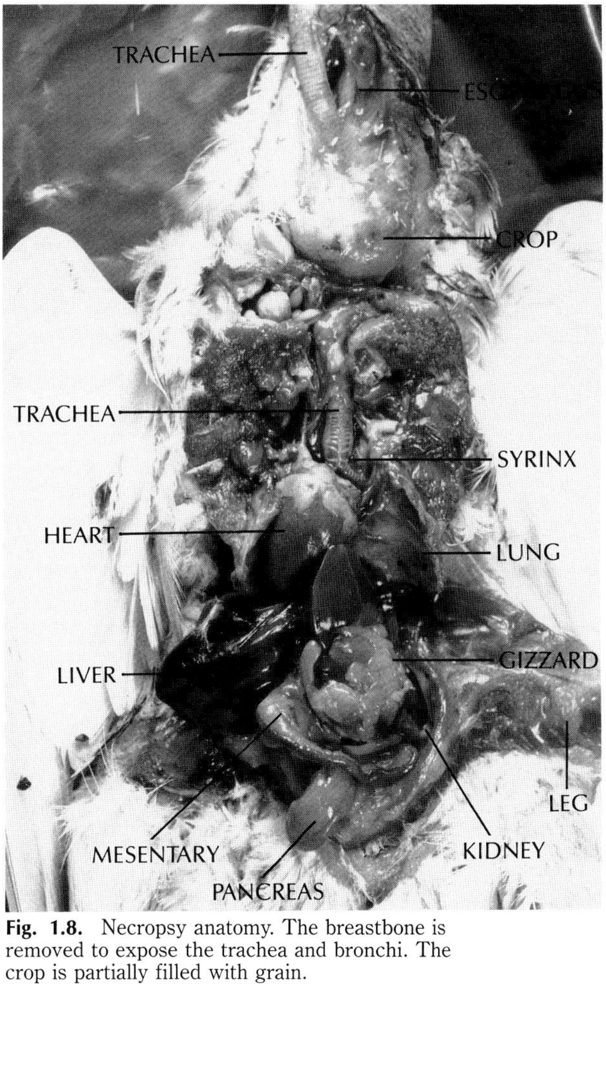

Fig. 1.8. Necropsy anatomy. The breastbone is removed to expose the trachea and bronchi. The crop is partially filled with grain.

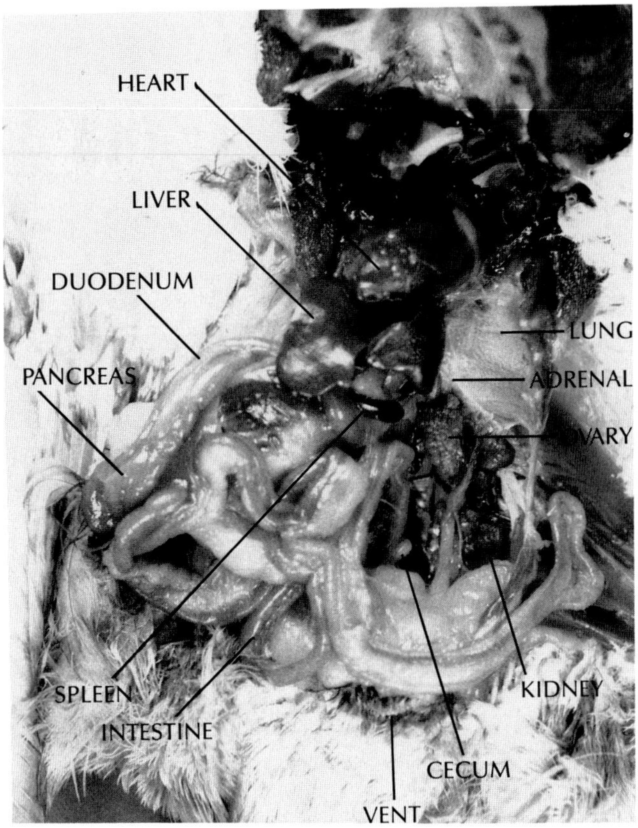

Fig. 1.9. Necropsy anatomy. The pancreas lies on the left side in the folds of the duodenum. The kidneys are exposed on the right.

branes replace cartilage rings and form either side of the tracheal wall. This is called the *syrinx* (voice box).

Each bronchus provides primary, secondary, and tertiary branches, which supply air to a lung. Primary and secondary branches are directed to the thin-walled paired anterior and posterior thoracic air sacs. The paired anterior thoracic sacs join the paired axillary sacs, which lie beneath the shoulders and wing axes. The axillary sacs in turn supply air to the wing bones. The anterior thoracic sacs also funnel air to the interclavicular sac, which lies anterior to the heart and clavicle. The latter sac connects with the paired cervical sacs feeding air to the neck and head. The anterior thoracic also feeds air to the paired abdominal air sacs. Air then proceeds to the cavities of the long bones of the legs. The lungs are applied to the wall of the rib cage and end at the last rib. There is no diaphragm. (See Fig 1.10.)

This system, which is only found in birds,

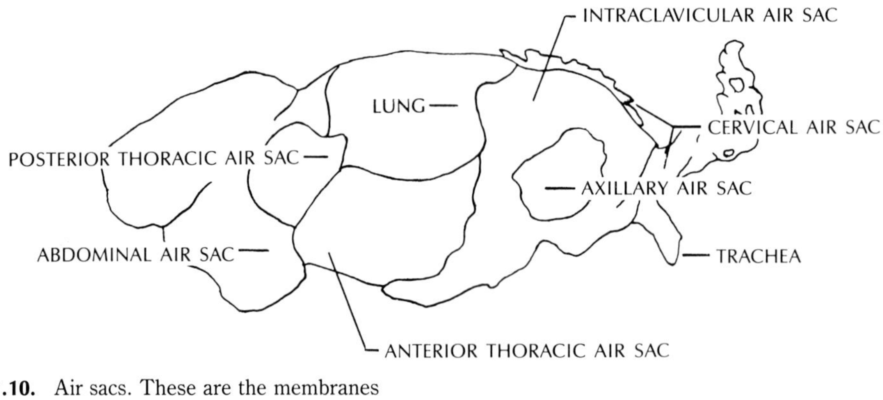

Fig. 1.10. Air sacs. These are the membranes that control air flow.

efficiently utilizes air. The air is heated once as it enters the body. Oxygen and carbon dioxide exchange occurs upon inhalation and continues until the air is exhaled. Obviously not all air is changed in each respiratory excursion. The system has one serious disadvantage. Disease organisms such as paramyxovirus quickly and easily pass throughout the body. In other animals air-oxygen exchange is limited to the lungs.

Excretory System. The excretory system is formed by two kidneys and their respective ureters. The kidneys are divided into three lobes: anterior, middle, and posterior. Many families of pigeons are tissue-deficient in one or more lobes. This is a defect that must not be carried on when mating birds. The respiratory system also serves in an excretory capacity in that it removes the waste products carbon dioxide and water.

Genital System. The genital system includes the sex organs and accessory structures. In the female, the left ovary is attached dorsally to the anterior lobe of the left kidney. The ovary, with its ova or follicles, is bound closely to the left adrenal gland. The *infundibulum* (funnel) of the oviduct arises posterior to the ovary and is continuous with the *magnum* (albumen-secreting part) and the narrow isthmus followed by the *uterus* (shell gland) and vagina.

The right ovary and oviduct remain undeveloped. Occasional malformed congenital abnormalities are observed in this area. The left ovary and oviduct may also experience infections and cysts or tumors.

The male reproductive tract includes paired testes, located anterior to the kidneys. Respective deferent ducts lead to the ejaculatory ducts and the cloaca. Sperm arise in seminiferous tubules of the testis and proceed to the rete testis and from there to the efferent duct and epididymus, which joins the deferent duct of each testis.

Circulatory System. The circulatory, or blood-vascular, system consists of a four-chambered heart and arteries, capillaries, and veins. The latter blood vessels pervade the entire body. The heart itself is enclosed by a pericardial sac.

Blood leaving the left ventricle of the heart enters the aorta and other arterial branches. It is fed to the anterior and posterior extremities and all organs of the body. Mesenteric portal veins of the intestine return blood to the liver. The posterior vena cava collects blood from the organs and extremities and returns it to the heart. The right side of the heart sends blood to the lungs by the pulmonary arteries. This oxygenated blood is returned to the heart by the pulmonary veins, which completes the blood pathway.

A lymphatic system is also present in birds. Lymph vessels closely follow veins and are found in all parts of the body. Lymph drainage follows the thoracic ducts on either side of the aorta and empties into the right and left anterior venae cavae (Lucas and Stettenheim 1965).

Lymphatic glands include paired thymus glands in the upper neck and the bursa Fabricius, which is closely applied to the dorsal aspect of the cloaca. The spleen, located above the gizzard and anterior to the adrenal, is also related. There are no lymph glands.

Nervous System. The central nervous system of pigeons includes the brain and spinal cord. The brain is composed of two anterior cerebral hemispheres, a *midbrain* (cerebellum) and a *medulla.* The latter is continued by the spinal cord, which is confined within the neural canal of the vertebral column. Birds have 12 cranial nerves but no cauda equina (Lucas and Stettenheim 1965). The cord ends in the last free tail vertebra.

The nervous system is further characterized by peripheral and sympathetic components. Peripheral nerves provide enervation to the extremities; Huber (1936) has listed 39 pairs of spinal nerves in pigeons. These paired spinal nerves proceed from either side of the neural canal to specific organs and extremities. Huber also described the relationship of the sympathetic trunk to the cranial and spinal nerves.

The sympathetic system is divided into sympathetic and parasympathetic fibers. These fibers regulate such functions as pupil activity. The sympathetic fibers dilate the pupil of the eye whereas parasympathetic fibers constrict the pupil.

Specific nerves should be mentioned. The *vagus nerves* (paired tenth cranial nerves) leave the brain to proceed beside the carotid arteries of the neck toward the thoracic cavity. Vagus nerves provide parasympathetic branches for the heart and lungs and other organs of the body. Interruption of function, for instance, may result in exaggerated breathing in birds as other mechanisms attempt to regulate respiration. The sciatic nerve follows the posterior medial aspect of the femur and lies in close contact with the femoral artery. It enervates the leg muscles. Trauma to these nerves usually alter leg movement and action. Faulty flight may likewise be attributed to wing brachial nerve injury.

Sensory System. Associated with the nervous system are the organs of special sense: the olfactory lobes of the brain, eyes, ears, semicircular canals, and organs of taste and touch.

ORGANS OF SMELL. The olfactory organs of smell are the paired external nares; the internal nares, which opens into the mouth; the turbinates; and the epithelial sensory receptors. Olfactory nerves conduct receptor impulses to the olfactory bulbs. Walter (1943) reported poor perception in pigeons. Earlier, Strong (1911) indicated that pigeons detected the odor of bergamot oil. More recent tests involving homing have been conducted to ascertain the ability of pigeons to respond to different odors.

EYE. The eye of the pigeon differs from the mammalian eye in that it has a pecten. The black-pigmented membranous pecten, which arises in the retina, is a folded 3- to 4-mm, almost square, flaglike structure in the posterior chamber of the eye. It originates on a white scleral ridge that overlies the attachment of the optic nerve. The pecten fans forward in the vitreous humor near the floor of the eyeball. (See Fig. 1.11.)

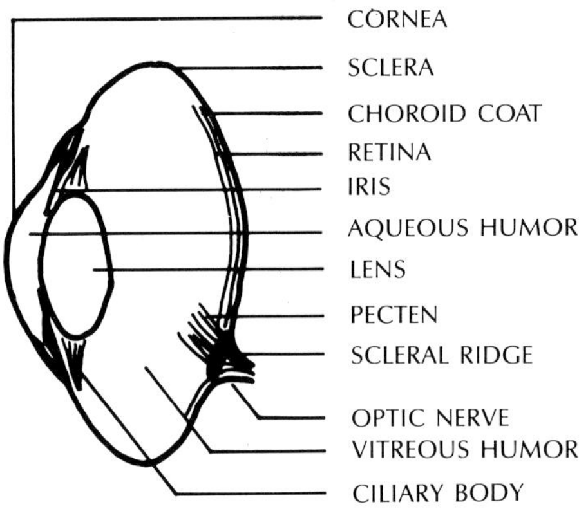

Fig. 1.11. Eye structure. The pecten floats like a flag in the vitreous humor.

The eyes have remarkable visual acuity and surpass human ability to see distant objects. The vision components include the eyeballs, optic nerves, upper and lower eyelids, conjunctiva, tear glands, ocular muscles, and a medial nictitating membrane for each eye, which is the third eyelid. The latter is usually concealed.

The lens of the eye accommodates to enable the pigeon to focus on subjects of varying distance. Walls (1942) reported the lens range to be 8–12 diopters. The eyeball itself is flattened to provide a wider field of vision. This is enhanced by lateral placement. When both eyes are used, the pigeon thus has vision of about 300 degrees, unlike human vision. Walls, however, indicated the binocular field of vision to be 24 degrees. This is when both eyes are focused on the object.

Color vision is a factor of light wavelength striking the cones of the retina. Walls (1942) has attested to the difficulty in determining the extent of color perception by birds. It is necessary to train pigeons before they can distinguish colors. Work in naval air sea rescue at Hawaii produced pigeons that were light-sensitive to yellow and orange distress colors. There are orange, red, and yellow-green oil droplets in the retina of pigeons (Wald and Zussman 1937). The presence of an ample quantity of red droplets in their retinas results in poor blue-violet recognition but good red perception.

EAR. The external opening to the ear canal is covered by feathers and located behind and below the eyes. The ear consists of an inner ear, a middle ear, and an external ear canal. The canal is directed downward beneath the braincase. It terminates medially at an oblique eardrum, a *tympanic membrane* that separates the outer and middle ear. The inner surface of the eardrum bears a small, elongated bone, the *columella,* comparable to the malleus of the human ear, which transmits the vibrations of the eardrum across the tympanic cavity to the stapes. The base of this bone is adapted to fit the membranous *fenstra vestibuli,* an oval opening to the central cavity or vestibule of the inner ear.

In pigeons the *lagena,* located in the inner ear, corresponds to the cochlea of mammals. It is a small, short, slightly curved ventral canal in which the sensory *organ of Corti* is found but is not well developed (Sturkie 1954). The human 8th cranial auditory nerve separates and conducts impulses to the brain from the vestibular and auditory receptors (Greep and Weiss 1973). The *lagena* nerve carries sound impulses from the hair cells of the basement membrane within the *membranous labyrinth* serving the lagena.

The membranous labyrinth is lodged within the *bony labyrinth* that houses organs of hearing and balance. The wall of this bony labyrinth is bathed with perilymph, and the space within the lining membrane is flooded with endolymph. Excess fluid drains to the spinal canal (Greep and Weiss 1973). Another duct, the *Eustachian tube,* provides drainage from the middle ear to the oral cavity and also equalizes air pressure.

The term *vestibular labyrinth* applies to the organs of equilibrium: the *semicircular canals*, the *utricule*, and the *saccule* (Greep and Weiss 1973). The sense of balance is mediated by the *Purkinje cells* of the cerebellum. These cells receive nerve impulses from the sensory hair cells of the *crista* in the ampulla of the three fluid-filled semicircular canals and the utricule and saccule of both balance centers. These are a part of the bony labyrinth located adjacent to each internal ear and beneath the brain. The superior vertical semicircular canal is positioned approximately at right angles to the posterior vertical canal and the lateral horizontal canal lies at right angles to both vertical canals.

ORGANS OF TASTE. Seventy percent of the taste buds of the pigeon are located on the surface of the tongue, but buds may be found on either side and on the ventral surface (Sturkie 1954). Moore and Elliot (1946) found in the pigeon tongue an average of 37 taste buds with a variation from 27 to 59. This number is much lower than in mammals, according to Duncan (1960).

ORGANS OF TOUCH. The sensory end organs, Vater-Pacini corpuscles, are vibration receptors (Schildmacher 1931) that are numerous in the dermal layer of the beak and near the follicles of contour feathers (Winkelmann and Myers 1961). Sensitivity is particularly evident when birds are molting.

Endocrine System. The endocrine system deals with secretory glands that have physiological regulatory functions. They produce chemical hormones that produce their effect in other organs. The glands are the *hypophysis* (pituitary) at the base of the brain, the *epiphysis* (pineal body) arising from the 3rd ventricle of the brain, the paired thyroids and parathyroids along the upper neck region, the *suprarenals* (paired adrenals) along the lumbar spine, the pancreatic alpha and beta islet cells in the loop of the duodenum, and the gonads anterior to the kidneys.

References and Selected Readings

Chamberlain, F. W. 1943. Atlas of Avian Anatomy. Lansing, Mich.: Hallenbeck.

Duncan, C. J. 1960. Ann Appl Biol 48:409.

Eglitis, I., and R. A. Knouff. 1962. Am J Anat 111:49.

Greep, R. O., and L. Weiss. 1973. *Histology.* 3rd ed. New York: McGraw-Hill.

Huber, G. C. 1936. J Comp Neurol 65:43.

Levi, W. M. 1941. *The Pigeon.* Columbia, S.C.: Bryan.

Lucas, A. M., and P. R. Stettenheim. 1965. In *Diseases of Poultry,* 5th ed., ed. H. E. Biester et al. Ames: Iowa State University Press.

Moore, C. A., and R. Elliot. 1946. J Comp Neurol 84:119.

Schildmacher, H. 1931. J Ornith 79:374.

Strong, R. M. 1911. J Morphol 22:619.

Sturkie, P. 1954. Avian Physiology. Menasha, Wis.: George Banta.

Wald, G., and H. Zussman. 1937. J Biol Chem 122:445.

Walls, G. L. 1942. The Vertebrate Eye. Bloomfield Hills, Mich.: Cranbrook Institute of Science.

Walter, W. G. 1943. Arch Nurland Physiol, Homme et Anim 27.

Winkelmann, R. K., and T. T. Myers. 1961. J Comp Neurol 177:27.

Physiology

Physiology is the science that deals with body functions, such as body temperature and heart rate. Deviations from normal levels provide a basis for evaluating the well-being of an individual. A brief discussion of some of these physiological norms for pigeons follows.

Blood cells are erythrocytes, or red blood cells (RBCs), leucocytes, or white blood cells (WBCs), and thrombocytes, or platelets. Pigeon RBCs are flattened, oval-shaped, biconcave discs that bear a central, irregular, oblong, cigar-shaped nucleus. In chickens they have a life span of 1–3 mo (Hevesy and Ottesen 1945). Groebbels (1932) reported pigeon RBCs to be 12.7/μm long by 7.5/μm wide and 3.7/μm thick. Adult pigeon RBCs number in millions/mm^3, 4.0 for cocks and 3.07 for hens (Groebbels 1932) or 3.23 for cocks and 3.09 for hens (Riddle and Braucher 1934). Natt's diluent (Natt and Herrick 1952) is preferred for counting both red and white cells in a Spencer bright line counting chamber or hemacytometer.

WBCs are divided into *granulocytes* (cells containing granules) and *agranulocytes* (cells containing no granules). Granulocytes are called heterophils, eosinophils, and basophils. Agranulocytes are large and small lymphocytes and monocytes. Thrombocytes, the smallest blood cells, are identified by a rounded nucleus and by one or more tiny, rounded, red granules near one end of the platelet. Routine evaluation of blood smears often reveal a visible lack of RBCs or a lack of hemoglobin within the cells. When infections are present, numerous heterophils with their red rod granules dominate the leucocyte picture. Basophils with round blue granules and eosinophils with tiny round red granules are seldom in large numbers. Shaw (1933) made pigeon leucocyte counts and reported 13,000 WBCs in the morning and 18,500 in the afternoon. Differential percentages also varied between morning and afternoon: heterophils 23% A.M. and 42.8% P.M.; eosinophils 2.2% A.M. and 1.9% P.M.; basophils 2.6% A.M. and 2.4% P.M.; monocytes 6.6% A.M. and 5.1% P.M.; and lymphocytes 65.6% A.M. and 47.8% P.M.

When flock problems arise, blood protein levels are often an indication of nutrition and health. Pigeon serum and plasma proteins are reported in g/100 ml by Mandel et al. (1947) as albumin 1.38, globulin 0.95, and total protein 2.30, and by McDonald and Riddle (1945) as 2.75–3.62 plasma protein (blood fluid before clotting).

The average heart rate of pigeons at rest is reported by Buchanan (1909) as 192 in a 240-g bird. Woodbury and Hamilton (1937) reported 221 in a 237-g pigeon and Stubel (1910) counted 244 in an adult bird.

The respiration rate varies depending on excitement and activity. Groebbels (1932) quoted reports of a respiration rate of 25–30 times a minute, but the sex was not given.

A special rectal thermometer with a temperature reading up to 115° F is required for pigeons. Groebbels (1932) reported the average rectal temperature to be 105.08–108.4° F in adult pigeons. Usually the temperature is near 105° F.

Body weight naturally varies with the pigeon breed and the health and nutrition of the bird, but Groebbels (1932) has recorded the average body weight to be 297 g and the heart weight 13.8 g/kg body weight. Brody (1945) reported the body weight as 0.278 kg.

References

Brody, S. 1945. *Bioenergetics and Growth.* New York: Reinhold.

Buchanan, F. 1909. J Physiol 38:62.

Groebbels, F. 1932. *Der Vogel Erster Band: Atmungswelt und Nahrungswelt.* Berlin: Verlag von Gebruder Borntraeger.

Hevesy, G., and J. Ottesen. 1945. Nature (London) 156:534.

Mandel, P., J. Clavert, and L. Mandel. 1947. Soc Biol (Paris) 141:678.

McDonald, M. R., and O. Riddle. 1945. J Biol Chem 159:445.

Natt, M. P., and C. H. Herrick. 1952. Poult Sci 31:735–37.

Riddle, O., and P. F. Braucher. 1934. Am J Physiol 108:554.

Shaw, F. R. 1933. J Pathol and Bacteriol 37:411.

Stubel, H. S. 1910. Arch ges Physiol (Pflugers) 135:249.

Woodbury, R. A., and W. F. Hamilton. 1937. J Physiol 119:663.

2

Feathers and Feather Problems

In pigeons, feathers assume a very vital part of the anatomy and are thus considered separately. Pigeons are covered with feathers over the entire body except for feet, legs, beak, cere, and eyes. Specific breeds and varieties of show birds are the exception. They may also have feathers over legs and feet. The plumage is interrupted and separated into *pterylae* (tracts), which are separated by featherless *apteria* (spaces). Bird feathers are modified epidermal structures that may have evolved from reptilian or dinosaur scales. Emerging feathers arise from dermal papilla in the skin epithelium. They provide protection from injury, heat, cold, and rain and enable flight. There are several types: flight quills, contour feathers, down, fluff, filoplumes, and pinfeathers.

Feather Types. Flight quill feathers, located on the posterior aspect of the wings and tail, are known respectively as the *remiges* and *rectrices.* Usually there are 10 primary flights, but the number varies between 9 and 11. They are located on the outer portion of the wing and are counted from the center axis or joint of the wing outward. The 10 or 11 secondary wing flights extend from the axis of the wing inward toward the body. In addition there are usually six paired tail flight feathers, which are also quill feathers. The number may vary from 10 to 16 in the White Carneau but Oriental Rollers, Danzig Highfliers, and Fantails may have more (Levi 1941).

The *calamus* (quill) extends only to the web of the *vexillum* (flat vane). At either end of the quill is a pin-sized circular hole through which the axial artery and vein nourish the growing feather. These openings are known as the *inferior* and *superior umbilici.* As the feather matures, the nutritive pulp within the quill degenerates and is resorbed, leaving a comparatively hollow calamus free of a blood supply. This contributes to strength and lightness and prevents blood loss if feathers become damaged.

The quill is continued by the *rachis* (central shaft), which supports the vane. Barbs or rays extend from either side of the shaft at about a 45-degree angle, herringbone style, and in turn barbules project forward and backward at about a 45-degree angle from either side of the barb. The barbules are interconnected by tiny hooklets or barbicels projecting downward from the barbules. (See Fig. 2.1.)

Streamling *plumae* (contour feathers) are short, smaller feathers that are directed posteriorly and cover most of the body and wings, providing most of the coloration.

Plumulae (down feathers), the soft, hairy, infant feathers of the nestling, are lost as the

Fig. 2.1. Feather barbs and barbules. The hooks on the barbules hold the feather together.

bird grows older. Their barbs are devoid of hooked barbules. These are replaced by fluff feathers, which form the warm, fluffy, tufted undercoat of the adult. They are found in several locations but chiefly on the sides of the bird.

Filoplumes are hairlike feathers about 1-2 in. long that have a very fine shaft and are often tufted at the free end. They are more prominent in the tail area, and the sensory nerve endings at their base may aid in controlling the posture of the contour feathers for flight and insulation. *Pinfeathers* are short, stubby, enshrouded, furled, immature, developing feathers that replace molted feathers. These new feathers are well developed in nestlings by 6–10 wk of age.

Molting. Pigeons replace all their feathers at least once a year by molting. The molted feather is forced out by proliferation of the epidermal collar. If a feather is pulled out it may be replaced immediately or at the next molt. Broken or cut feathers are not replaced until the next molt unless the quill is pulled out. Usually adult birds shed their feathers in late summer or fall, but heavy molts may be initiated at any time.

Flight feathers are molted independently from body feathers. The primaries are molted singly and in order, with the number 1 primary being molted first at 6–7 wk of age. The secondaries are molted indiscriminately over a period 3–8 mo of age. The last primary and the last pair of tail flights are molted about the same time, 6–7 mo of age. The primaries are usually lost over a period of approximately 6 mo, which means that in racing birds, primary flight feather replacement takes about 2½ wk per feather. The tail flights are molted in pairs, one from either side of the tail, often on the same day and over a period of 3 mo. This occurs 3½–6½ mo of age(Levi 1941). The process may be interrupted at any time to be continued at a later date with molting of unmolted feathers. The body molt persists year-round with the normal gradual replacement of small contour and fluff feathers.

Factors Controlling Molting. The mechanism that controls molting is not fully understood. Several factors are involved. Molting can be induced in 7–8 days in chickens by the administration of thyroxine or desiccated thyroid (Van der Meulen 1939). According to Sturkie (1965), it is the feather papilla that is stimulated. The removal of the thyroids, which are separate from the parathyroids in pigeons, is described by Marvin and Smith (1943). Surgical removal of the thyroid reduces the rate of feather growth and causes structural changes (Blivaiss 1947). Such male and female feathers become fringed and elongated and lack barbules (Chu 1940). The size and shape of feathers is also influenced by gonadal hormones (Sturkie 1965). Prolactin from the anterior pituitary may also initiate molt and may be the immediate stimulation of new feather growth following a disruption of ovarian activity (Juhn and Harris 1958). Progesterone from the corpora lutea of the ovary may inhibit or retard ovulation in the pigeon (Dunham and Riddle 1962) and will induce molting (Tanabe et al. 1957).

Stress can trigger partial or complete molts. Stressful environmental factors, such as lack of feed or water, rapid and prolonged rises in temperature, injury, and disease, can initiate or enhance molting. Cold, however, deters feather loss. Removal of feed and water for 1 day or drastically reducing the amount of feed for a period is usually sufficient stress to initiate a molt. Birds should be returned to a full ration immediately after the molt starts to prevent a loss of body weight. Changes in the length of daylight and artificial light also appear to have an effect. Adult mated birds drop their first primaries in March in the northern hemisphere. March 21 is the time when days and nights are of equal length. As daylight increases in March it appears to affect the pituitary gland, which stimulates the onset of molting.

Feather Problems. Young, growing flight feathers are very tender, and because of capillary vascularity of the unfurling pinfeathers, they are easily injured. A blood clot resulting from injury may form within the quill of a growing primary feather. As the feather grows and *keratinizes* (becomes harder), its blood supply is interrupted, thus preventing the further resorption of the clot. This blackened, blood-filled quill is termed a *blood quill.* If the feather is plucked, damage occurs to the dermal papilla and the epidermal collar, resulting in bleeding into the empty follicle, but a new normal feather will emerge.

There is also some indication that capillary fragility within the quill results in quill hemorrhage and clot formation. This appears to be caused by inadequate nutrition of rapidly growing feathers. Supplementation with sulfur-bearing amino acids and vitamin K may be indicated.

The undersurface of the rachis has a reinforcing concave V groove along most of its length, which ends at the junction with the calamus. Apparently nutritional deficiencies and/or genetic defects alter the structure of the quill and rachis in feathers of some birds, and cracks develop close to the body in weakened feathers. These *split* or *cracked quills* result from trauma.

Numerous pinhead holes are sometimes observed in feather vanes. These appear to be caused by large body-louse infestations. Mites suck blood and seldom alter feathers.

Worn flight feathers are often observed in racers; this results from an inadequate supply of *bloom* (a fine, white keratin powder). In healthy birds the skin sheds bloom, which waterproofs and lubricates the feathers, improving the efficiency of flight. Nutritional and infectious disease factors alter the general health and condition of the skin and feathers and thus contribute to a lack of bloom.

A waterproof, lipoidal, sebaceous secretion is also provided by the *uropygial gland* (preen or oil gland) located dorsally at the base of the tail in some breeds. The White Carneau and the Rumpless possess no uropygial gland (Levi 1941).

Abnormal feather structure or color may be due to nutritional problems involving arginine, cystine, lysine, methionine, copper, iron, sulfur, and zinc. Diet also appears to affect the duration of the molt as well. In addition, when young growing squeakers are deprived of parent crop milk for a couple of days before they have learned to eat and drink on their own, nutritional signs may be evident weeks later in weakened defective flight feathers. Grau et al. (1987) showed lysine deficiency limited growth but not melanin deposition in feathers.

Feather pulling and self-inflicted cannibalism can be the result of fungus infections of the skin (Tudor 1983). Proper loft moisture control and individual bird treatment is usually sufficient to overcome the problem. Depluming mites are seldom involved. (See Fig. 2.2.)

Quill mites may be observed as a brown powder inside the feather quill. These seldom cause a problem, but individual treatment is indicated.

Feather mites are most often observed near the vent area. Tiny, moving black specks serve to establish their presence.

Fantails and Homers have been observed to carry a genetic Mendelian recessive porcupine feather defect (Cole and Hawkins 1930). The feather web development is partially arrested and the feather fails to spread out as it passes the sheath. It remains rolled up.

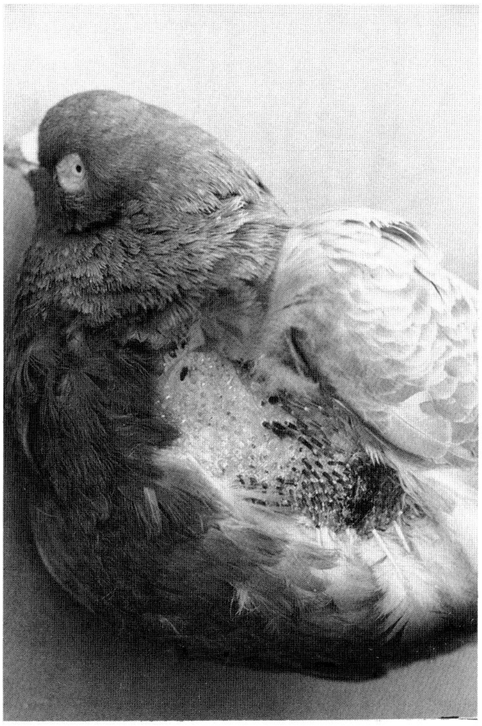

Fig. 2.2. Fungus infection of the skin and feathers, *Mucor circinelloides.* Wet buildings encourage fungi.

References

Blivaiss, B. B. 1947. Physiol Zool 104:267.

Chu, J. P. 1940. J Genet 39:493.

Cole, L. J., and L. E. Hawkins. 1930. J Hered 21:51–60.

Dunham, H. H., and O. Riddle. 1962. Physiol Zool 15:383.

Grau, C. R., et al. 1987. Poult Sci 66(Suppl):106.

Juhn, M., and P. C. Harris. 1958. Proc Soc Exp Biol Med 98:669.

Levi, W. M. 1941. *The Pigeon.* Columbia, S.C.: R. L. Bryan.

Marvin, H. N., and G. Smith. 1943. Endocrinol 32:87.

Sturkie, P. 1965. *Avian Physiology.* 2d ed. Ithaca, N.Y.: Cornell University Press.

Tanabe, Y., K. Himeno, and H. Nozaki. 1957. Endocrinol 61:661.

Tudor, D. C. 1983. Vet Med Small Anim Clin 78:249.

Van der Meulen, J. B. 1939. Proc VII World Poult Congr, p. 109.

II

INFECTIOUS DISEASES

Infectious diseases are due to the growth or action of microorganisms or parasites that can spread from one individual to another. In pigeons these diseases include viral, bacterial, mycoplasmal, and mycotic infections and internal and external parasites.

3

Viral Diseases

Virus organisms are tiny, filterable, submicroscopic disease-producing agents composed of RNA or DNA that can only reproduce within living cells. Twelve viral diseases have been observed in pigeons: pigeon pox, paramyxovirus infection, Newcastle disease, herpesvirus infection, arbovirus infection, avian influenza, avian adenovirus infection, reovirus and rotavirus infections, rubella, rabies, and infectious bronchitis.

Pigeon Pox

Definition and Synonyms. Pigeon pox is an acute or subacute, mosquito-borne, slow-spreading, epitheliotropic virus disease of birds that is characterized by progressive wartlike eruptions forming scabs on the exposed surfaces of the body and thickened, yellowish, diphtheritic patches on mucous membranes. Pigeon pox is not a public health problem because it does not affect humans.

Pigeon pox is otherwise known as bird pox, contagious epithelioma, sorehead, and avian diphtheria. Mucous membrane infection is called wet pox.

Cause and Classification. The virus is classified in Group 2 in the family Poxviridae and in the genus *Avipoxvirus* (Matthews 1979). The genus includes fowl, turkey, pigeon, canary, quail, starling, junco, and sparrow pox as separate types of pox in birds (Fenner 1979). The avipox genus is separate and distinct from the five other pox genera that affect humans, animals, and fish (Fenner et al. 1974). It is possible that all pox strains had a common origin. Serological cross-relationships do exist between the avian poxviruses but not, for instance, between avian and sheep or goat pox (Uppal and Nilakantan 1970).

Nature. Pigeon pox as a disease is caused by a virus similar to fowl pox (Skalinskii et al. 1965), which is a double-stranded adenine-thymine-type DNA virus (Randall et al. 1962). Fenner and Burnet (1957) described the avipox group viruses as large, brick-shaped complexes with measurements 250–330 nm long and 200–250 nm wide. Randall et al. (1964) records fowl poxvirus as 258 nm by 354 nm. Hyde et al. (1967) and Gafford and Randall (1967) describe it as the largest viral nucleic acid molecule yet found. The inner, biconcave, genome-containing core, or nucleoid, is enclosed within an outer membrane envelope formed from the infected cell membrane as it emerges as a budding particle from the cytoplasm of the cell (Arhelger and Randall 1964). (See Figs. 3.1 and 3.2.)

Infectivity, Hosts, and Relationships. Avian poxvirus is prevalent worldwide in birds only, with one exception. Fowl pox has been recovered from a rhinoceros (Mayr and Mahnel 1970). Pox infection has been found in about 60 species of birds (Tripathy and Cunningham 1984). Relationships between each of the avian pox agents have been noted by numerous workers. Pigeon poxvirus, for example, has been reported to affect other avian species and conversely other types of poxvirus may affect pigeons. Gelenczei and Lasher (1968) and Mayr (1963) summarized these relationships in Table 3.1 when individual virus strains were applied on different hosts.

Table 3.1 and ensuing reports are somewhat

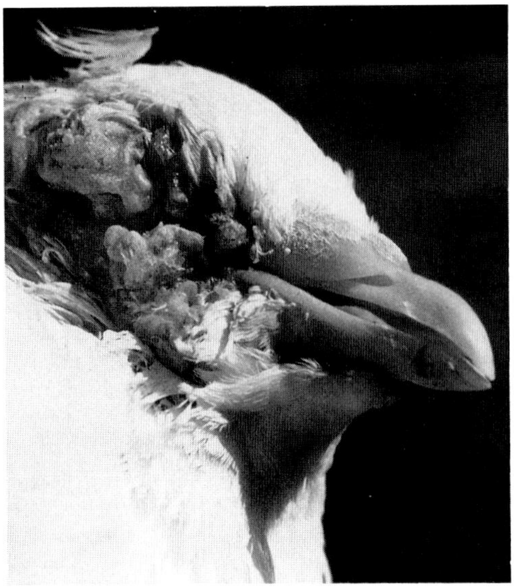

Fig. 3.1. Pigeon pox in the eye, transmitted by mosquitoes.

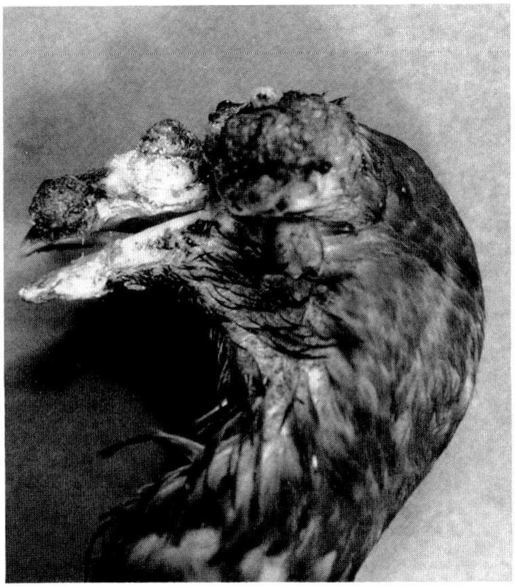

Fig. 3.2. Pigeon pox in the eye and beak. The eye must be kept open to prevent eye destruction.

confusing but provide a better understanding of the nature of the disease problem. Basically pigeon pox was found to be quite infectious for pigeons and less pathogenic for chickens and turkeys, producing only cutaneous lesions and no generalized infection. It was not infectious for ducks or canaries. It was shown that the type of poxvirus, the amount and method of delivery, and the resistance of the bird largely determined the extent of infection.

Prior to these findings, Beaudette and Hudson (1941) inoculated with turkey pox 10 chickens previously vaccinated with pigeon pox. One bird presented a definite reaction, and 4 were questionable. This means that the pigeon pox vaccination was really not effective against turkey pox. Of 10 chickens reinoculated with canary pox following pigeon pox, 1 had a questionable take. From this it is clear that pigeon pox largely protects against canary pox reinoculation. Of 11 chickens reinoculated with fowl pox after pigeon pox, 3 had definite takes; therefore pigeon pox did not protect against fowl pox in the dose given. (It has since been learned that the immunity in chickens to pigeon poxvirus is proportional to the number of feather follicles inoculated.) Of 10 chickens reinoculated with pigeon pox, no reaction developed; thus pigeon pox protects against a pigeon pox challenge. This indicates that pigeon pox produces a variable immune response in chickens to a challenge with pigeon, turkey, canary, and fowl pox. It suggests a fairly close relationship between strains of pigeon and canary poxviruses.

These relationships can be further studied in

Table 3.1. Host relationships

	Hosts/Virus Strain									
	Pigeon		Chicken		Turkey		Canary		Duck	
Host	C	IV	C	IV	C	IV	C	IV	C	IV
Pigeon	+G	+G	+	0	+	0	0	0	0	0
Fowl[a]	+	0	+G	+G	+G	+G	0	0	0	0
Turkey	+G	+G	+	+G	+	+G	0	0	+	+G
Canary	+G	+G	+	0	+	0	+D	+D	+	0

Note: C = cutaneous viral infection, IV = intravenous viral infection, + = local pox lesion, 0 = no signs, +G = local pox lesion and generalized pox, +D = local pox lesion and death.

[a]Beaudette (1941).

the reports of other workers although their work does not always agree. Doyle and Minett (1927) were able to adapt a strain of fowl pox to the pigeon by frequent serial passages. Brandly and Bushnell (1932) could not infect pigeons with fowl pox but did infect fowl with pigeon pox. Tietz (1932) was unable to infect canaries, blackbirds, siskins, and ricebirds with pigeon pox, but 1 of 2 linnets showed several pinhead-sized nodules; 1 of 2 sparrows showed three swollen follicles and 1 starling showed several follicle swellings. Canaries, pigeons, and fowl were reported infected with an apparent canary strain by Eberbeck and Kayser (1932). They also cited the report of Stadie (1931) concerning an epizootic of pox in pigeons that also affected canaries and bullfinches. A canary poxvirus inoculated into greenfinches, starlings, and a bullfinch resulted in transient swellings in the starlings and 1 greenfinch. The same virus was inoculated into pigeons. One lot showed a limited reaction, but when challenged with pigeon virus, no immunity was evident. Burnet (1933), using a Kikuth strain of canary virus, could not infect pigeons, 1-day-old chicks, fowl, or budgerigars intramuscularly, but sparrows were susceptible and died. He also could not infect canaries with pigeon virus. Brunett (1934) found a severe reaction in turkeys from a strain of pigeon pox, but it produced no immunity to either turkey or fowl pox. A strain of turkey pox from a natural case in turkeys was able to infect turkeys, chickens, and pigeons. Irons (1934) reported infection of canaries, snow buntings, English and ground sparrows, pigeons, and chickens with a G strain of pox from a wild pigeon. No other pigeon strains studied would infect English sparrows. He also noted that all the strains of pigeon pox were infectious for chickens. He was unable to infect pigeons with turkey virus or to infect crows, blackbirds, ducks, guinea fowl, monkey-faced owls, screech owls, and starlings with pigeon virus. One strain of fowl pox was transmitted to pigeons, with gradually increasing virulence. Two other fowl pox strains were not. Reis and Nobrega (1937) reported that strains of passerine pox from the wild canary were not infective for pigeons or fowl, but another wild canary virus caused specific inclusions in chicks, canaries, and pigeons. Eight pigeons were infected on the breast with this virus. A month later, 4 were immune to challenge with pigeon pox and 4 were immune to the canary virus. Brandly and Dunlap (1938) reported that pigeon poxvirus will infect turkeys but produced no measurable immunity to fowl pox. Grosso and Prieto (1939) reported a mild transitory lesion in pigeons and chicks with a canary virus. Beaudette (1941) reported that fowl pox can be adapted to the pigeon by frequent serial passage and if grown in this species for a prolonged period will eventually not produce a generalized disease in the chicken. Such a virus modified by passage in pigeons cannot be differentiated from pox that caused a natural outbreak in pigeons. Pigeon pox has a much greater degree of infectivity for pigeons than for chickens. He further indicated that pigeon pox solidly immunizes the pigeon against pigeon poxvirus but does not produce solid immunity against fowl pox in chickens. Also, pigeon pox is less likely to immunize turkeys than fowl pox. Jansen (1942) infected a jackdaw, pigeons, fowl, and canaries with canary poxvirus. No immunity to the jackdaw strain of canary pox resulted following vaccination with fowl pox or pigeon pox, but immunity was present following canary pox vaccination. McGaughey and Burnet (1945) studied an apparent canary poxvirus that was fatal in sparrows and canaries but produced only localized lesions in fowl and pigeons. Kossack and Hanson (1954) reported a natural outbreak of pigeon pox in mourning doves that was transmitted to ringdoves by cohabitation. Jactot et al. (1956) infected pigeons, sparrows, and fowl with canary pox, but no immunity developed in pigeons to pigeon pox. He also found that pigeon pox immune pigeons were susceptible to canary pox. Kato et al. (1965) infected chickens with sparrow pox. Richter (1969) passed pigeon pox in ducks by the intrafollicular route and also produced immunity in pigeons. El-Dahaby et al. (1971) tried unsuccessfully to infect wild doves and sparrows via the wing web with pigeon pox; the feather follicle route, however, was effective in both species although sparrows had only a mild infection. Olah and Palatka (1971) vaccinated pheasants with pigeon pox by the feather follicle method, but only half of the birds were immune to fowl pox.

Distribution

GEOGRAPHY. Pigeon pox is commonly found worldwide. It is more often found near water or stagnant reservoirs where mosquitoes breed.

SEASON. Pox may appear at any time but is more common in the fall in nontropical climates when infected mosquitoes are prevalent.

AGE, SEX AND BREED. All ages, sexes, and breeds contract the virus. The disease is usually found in older pigeons, but squabs may be infected and mortality may reach 50% overall.

Transmission. There is no evidence of a dormant period or virus replication within mosquitoes (DaMassa 1966). *Aedes* and *Culex* mosquitoes may mechanically transmit the virus in the proboscis and remain infected for 16–19 days according to Kligler and Ashner (1929). Blanc and Caminopetros (1930) showed that *C. pipiens* could transmit for 58 days, transferring infection from pigeon to pigeon for 38 days, and *A. vexans* was able to transmit the virus for at least 27 days. It is also possible for mites and biting flies to spread the infection to a bird. Mites may also spread poxvirus on an already infected bird, which results in multiple lesions. Once introduced into a closed, dry, dusty pen, poxvirus may be inhaled and thus establish throat or respiratory infection. Direct bird-to-bird transmission may occur if aggressive picking occurs. Feed, feed hoppers, and bathwater or waterpans can become contaminated with pox scabs or curdy exudate from the mouth. Each of these modes of transmission establishes infection only if a scratch or break in the skin or mucous membrane permits introduction of the virus. Poxvirus grows only in wounds or traumatized epithelial tissue, not on intact skin or mucous membranes.

Doyle and Minett (1927) and Doyle (1930) have indicated that pigeon poxvirus may be found in internal organs and may persist for a considerable length of time in these organs in recovered pigeons. These findings have not been repeated. If the infection is extensive on a pigeon, as indicated by the number and size of lesions, virus may be expected in the bloodstream, with secondary lesions developing wherever a break in the skin occurs. This may be at the site of feather loss or bites of louse flies or other parasites. Whenever skin or mucous membrane infection is gone, the chance of residual internal organ infection is remote. Recovered carriers have not been observed.

Natural transmission of the disease may result from the introduction of infected but not visibly affected birds returning from races or shows. There is no true continuous carrier state. Birds in the incubation stage of the disease may carry the virus but not show it. Birds in late stages of the disease may shed skin flakes and scabs from pox lesions. These scabs may continue to retain live virus for months in the dry state in pigeon baskets, transport vehicles, and lofts.

Epidemiology

DISTRIBUTION IN BODY. The virus is usually confined to the skin, but mucous membranes of the mouth, throat, crop, eye, and trachea may be involved. If birds receive too much virus and are severely affected, virus may be present in the bloodstream. This may occur as a result of a natural infection or from vaccination.

ENTRY AND EXIT. Poxvirus enters through any break in the skin or mucous membrane. This may be a mosquito bite puncture or a scratch. The virus leaves in dried scabs that fall from the skin or in the curdy exudate of the mouth or eye.

PATHOGENICITY. The degree of virulence largely depends on the strain and the titer of the challenge virus.

Incubation. Lesions take about 5–7 days to develop.

Course. Scabs usually fall off in 3–4 weeks.

Morbidity. The extent of the disease within a loft depends on the mosquito population and/or the dryness and dustiness of a loft.

Mortality. Skin pox seldom causes appreciable losses, but tracheal pox usually kills the bird because the tracheal plug excludes air.

Signs and Necropsy. Pox infection causes changes in the skin. The characteristic lesions on the upper surface of the skin, feet and legs, and base of the beak start as small, whitish, blisterlike foci that rapidly increase in size within a week and become yellowish. These form warty growths, which join and form large, rough, brownish scabs that usually last 3–4 weeks. If the dried scab is removed, the undersurface may be oozing, moist, and bleeding. If the scab falls off, a smooth scar may be present.

Frequently mucous membranes of the mouth, throat, crop, eye, or trachea become involved. Diphtheroid, raised, yellowish gray, soft, curdy masses of adherent dead tissue appear on the mucous membranes and block the windpipe. The corners of the mouth are often affected first. When lesions are removed, bleeding tissue is exposed. In the eye, a cheesy, yellow, swollen mass replaces the eyeball. Infection at the edges of the eyelids often serves to initiate the process.

The disease often appears as a combination skin and mucous membrane infection. Secondary signs include loss of appetite and weight; sticky, viscid, cheesy material in the eyes; and stained hackle feathers. No internal signs other than those of the throat, sinuses, trachea, crop, and/or esophagus may be found.

Micropathology. In stained tissue sections of thickened skin or mucous membrane lesions, eosinophilic cytoplasmic inclusions may be observed by light microscopy. Brandly (1941) studied the growth of pigeon pox on the chorioallantoic membrane (CAM) of chicken eggs and noted that the virus produced infection and thus cellular changes in all three germ layers. Most of the proliferation occurs in the ectodermal layer.

Diagnosis. A diagnosis of pox is made by the visual recognition of skin or throat lesions, by the recovery of the virus, and by the identification of Bollinger bodies.

The labeling of avian poxvirus is not always accurate when based on the host of origin. Any poxvirus recovered from a pigeon is not necessarily pigeon pox. The virus must be tested on birds other than pigeons, and its true host or identity recognized by specific antibodies and by the degree of immunity produced in each of the hosts tested. In addition, other tests may be essential in establishing viral identity: immunofluorescence, complement fixation, hemagglutination-inhibition (Mathew 1967), immunoperoxidase, hemagglutination (Garg et al. 1967; Tripathy et al. 1970), and immunodiffusion (Tsubahara and Kato 1959; Woernle 1966). The latter test may be used to distinguish fowl poxviruses and pigeon poxviruses or antibody from those of other avian viruses (Woernle 1966).

Differential Diagnosis. Only advanced trichomonad lesions resemble pigeon pox.

Biological Properties

TISSUE AFFINITY. Poxvirus grows on epithelial tissues only. This includes skin and mucous membranes of the bird. Feather follicle cells have a special affinity for the virus. In the embryo the ectodermal layer is the tissue of choice.

INCLUSION BODIES. Intracytoplasmic inclusions called Bollinger bodies are 50% extractable lipid and weigh on the average 6.1×10^7 mg (Tripathy and Cunningham 1984). Inclusions appear 72 hr after chicken skin infection (Arhelger et al. 1962) and 96 hr after embryo CAM infection (Arhelger and Randall 1964).

STAINING. Hematoxylin-eosin stains inclusion bodies in tissue sections quite well.

STABILITY. The virus withstands drying but is destroyed by most standard disinfectants. Since the virus contains about 30% fat (Arhelger and Randall 1964), any good hot detergent will inactivate most virus particles in a loft. Kim et al. (1968) report that all virus is destroyed in 192 hr at 25°C or 96 hr at 37°C and in 30 min with 5% phenol or 3% formalin at 20°C.

CULTURE. The virus may be grown on the CAM of 10-day-old chicken embryos maintained at incubation temperature. Localized, opaque zone thickenings develop at the point of inoculation; they are 1–3 mm in diameter. More virus in the inoculum increases the membrane thickening. Scattered pinhead to millet-sized, globular lesions may develop at various points on the membrane with generalized infection. The greatest concentration of virus appears at least 5 days following inoculation.

Pigeon pox may be grown in tissue culture on chicken embryo fibroblast cell cultures (Mayr 1963) and duck embryo fibroblasts (Gelenczei and Lasher 1968). Pigeon pox produces plaques 1–3 mm in diameter with a lysis not produced by other viruses (Tripathy and Cunningham 1984).

Prevention and Vaccination. Chicken embryo, egg-propagated, or tissue culture pigeon pox vaccine specifically for pigeons is the only effective vaccine (Woodward and Tudor 1973). The use of so-called pigeon pox vaccine designated for use on chickens will not immunize pigeons.

The reconstituted pigeon pox vaccine should be applied upward with a small stiff brush in no more than four or five defeathered follicles on the leg or breast. Pigeons may be vaccinated in the wing web with double needles, but most fanciers avoid this method because it may cause wing swelling. Follicle or inoculation site swelling will occur in 5–7 days. A *take* (a local reaction indicating a successful vaccination) is present when thickening, reddening, swelling, and often scab formation develop at the site of the virus application. Scabs may form with certain pigeon pox vaccines and last 3–4 wk. Birds not demonstrating a take should be revaccinated. The lack of a reaction means that the bird was immune, the vaccine was improperly applied, or the vaccine was not potent. A vaccine may be unsatisfactory if it is used beyond the date of expiration. It should not be saved to be used later after it has been mixed.

Immunity develops as the reaction develops, and antibodies reach a workable level 10–14 days after vaccination. Swelling begins to recede about the third week as immunity reaches a maximum.

All healthy birds 4–6 wk of age and older should be vaccinated. There is no upper age limit. Treatments for worms or other diseases should not be undertaken during the course of the vaccination reaction. The entire flock should be vaccinated at the same time, or vaccinated birds should be segregated from nonvaccinated until after immunity has developed and the swelling has gone down. Segregation implies mosquito-screened pens 50 ft apart. Vaccine should be applied in the spring or early summer or before the disease becomes prevalent in mosquitoes. In tropical climates vaccination may be done at any time, since the disease may be present in mosquitoes throughout the year.

Emergency vaccination of newly infected flocks may be done to reduce the spread of the disease in a loft. All noninfected birds should be segregated and vaccinated at the same time. Visibly infected birds should not be vaccinated.

Vaccination should be conducted each year on all birds before the racing or show season, and birds should not be entered in shows or races until pox swellings have disappeared. Newly purchased birds and those returning from shows or races should be isolated for 2 wk until their health has been established.

Certain precautions should be taken. The directions with the vaccine should be followed. If other infections are present, vaccination should be delayed. The vaccine is applied to only one area. Clothing must be changed if vaccine is spilled on it. Litter spills should be cleaned up and all unused vaccine and containers properly disposed of.

If more than the recommended follicles are inoculated, secondary lesions may develop at other wound locations on the body. Fever may be expected about 5–7 days after vaccination. This may reduce feed consumption. Vaccinating during the coolest portion of the day reduces postvaccination stress.

The potency of a vaccine depends on the number of living poxvirus particles. Beaudette (1941) specified 80 mg dried CAM/4cm^3 diluent for 100 birds. Winterfield and Hitchner (1965) have since reported 10^{-5} embryo infective doses/ml as the prerequisite standard for pox vaccine. This is called titer or potency. The maintenance of such a titer depends on adequate initial drying under vacuum and proper refrigeration during storage and transit.

Vaccination problems arise when a vaccine is improperly handled or stored. Vaccines must be kept under constant refrigeration. Any break in the seal of the vial will also reduce the potency of the vaccine. Another problem may arise from free-flying birds, which may introduce another virus type. Jactot et al. (1956) have shown that pigeons immunized with pigeon pox were not immune to canary pox. It is thus possible for properly immunized pigeons to become infected with a poxvirus from free-flying birds. In the prevention of disease it is essential to exclude free-flying birds from lofts and exercise yards.

Control. If pox infection is present in a loft, segregate visibly infected birds. Vaccinate only noninfected birds. All birds from an infected loft and recently vaccinated birds should kept at home for 4 wk. The daily bath routine should be avoided until the vaccination scabs are gone. Waterpans may be disinfected with sodium hypochlorite at 2 tablespoons per pint. Pans should then be rinsed and filled with fresh water. Disinfectant should not be used in the drinking water.

Seriously infected birds should be destroyed or isolated, and noninfected birds fed and watered first. Persons vaccinating, feeding, or caring for noninfected birds should wear clean, washed clothes. Feed should be covered and stored away from infected birds. Lofts should be clean and nondusty, with good ventilation.

Treatment. Treatment is seldom effective.

Eye infection may be relieved by keeping the eye open and the eye bathed daily with saltwater (1 level teaspoonful of salt per pint of water), after which several drops of mineral oil is applied under the eyelids. If the eyelids remain closed and pasted together, the eye is usually lost. Daily applications of tincture of merthiolate at 1/1000 may be helpful on early skin lesions but not in the eyes. No other treatments are reported effective.

References

Arhelger, R. B., and C. C. Randall. 1964. Virology 22:59–66.

Arhelger, R. B., R. W. Darlington, L. G. Gafford and C. C. Randall. 1962. Lab Invest 11:814–25.

Beaudette, F. R. 1941. Proc US Livestock Sanitary Assoc 127–41.

Beaudette, F. R., and C. B. Hudson. 1941. Poult Sci 20:79–82.

Blanc, G., and J. Caminopetros. 1930. C R Acad Sci (III) 190:954.

Brandly, C. A. 1941. Ill Agric Exp Sta Bull 478.

Brandly, C. A., and L. B. Bushnell. 1932. J Am Vet Med Assoc 80:782.

Brandly, C. A., and G. L. Dunlap. 1938. Poult Sci 17:511–15.

Brunett, E. L. 1934. Rep N Y State Vet Coll 1932–33, p. 69.

Burnet, F. M. 1933. J Pathol Bacteriol 37:107–22.

DaMassa, A. J. 1966. Avian Dis 10:57–66.

Doyle, T. M. 1930. Rep 11th Int Vet Congr 3:675.

Doyle, T. M., and F. C. Minett. 1927. J Comp Pathol Ther 40:247.

Eberbeck, E., and W. Kayser. 1932. Arch Wiss Prakt Tierheilkd 65:307–10.

El-Dahaby, H., A. H. El Sabbagh, and M. I. Nassar. 1971. J Egypt Vet Med Assoc 30:97–104.

Fenner, F. 1979. Intervirology 11:137–57.

Fenner, F., and F. M. Burnet. 1957. Virology 4:305–14.

Fenner, F., et al. 1974. Intervirology 3:193–98.

Gafford, L. G., and C. C. Randall. 1967. J Mol Biol 26:303–10.

Garg, S. K., M. S. Sethi, and S. K. Negi. 1967. In-

dian J Microbiol 73:101-2.
Gelenczei, E. F., and H. N. Lasher. 1968. Avian Dis 12:142-50.
Grosso, A. M., and C. Prieto. 1939. Univ B Aires Inst Enferm Infecc 1:4.
Hyde, J. M., L. G. Gafford, and C. C. Randall. 1967. Virology 33:112.
Irons, V. 1934. Am J Hyg 20:329-51.
Jactot, H., A. Vallie, and L. Reinie. 1956. Ann Inst Pasteur 90:28.
Jansen, J. 1942. Tijdschr Diergeneeskd 69:128-31.
Kato, K., T. Horiuchi, and H. Tsubahara. 1965. Nat Inst Anim Health Q Tokyo 5:130-37.
Kim, S. J., S. Namgoong, and C. K. Lee. 1968. Res Rep Rural Dev Korea 11:69-73.
Kligler, I. J., and M. Ashner. 1929. Br J Exp Pathol 10:347.
Kossack, C. W., and H. C. Hanson. 1954. J Am Vet Med Assoc 124:199-200.
Mathew, T. 1967. Indian J Pathol Bacteriol 10:1-8.
Matthews, R. E. F. 1979. Intervirology 12:160-64.
Mayr, A. 1963. Berl Munch Tierarztl Wochenschr 76:316-24.
Mayr, A., and H. Mahnel. 1970. Arch Gesamte Virusforsch 31:51-60.
McGaughey, C. H., and F. M. Burnet. 1945. J Comp Pathol Ther 55:201-5.
Olah, P., and Z. Palatka. 1971. Magyar Allatorv Lapja 26:152.
Randall, C. C., L. G. Gafford, and R. W. Darlington. 1962. J Bacteriol 83:1037-41.
Randall, C. C., et al. 1964. J Bacteriol 87:939-44.
Reis, J., and P. Nobrega. 1937. Arch Inst Biol, Sao Paulo 8:211-14.
Richter, J. H. M. 1969. Tijdschr Diergeneeskd 94:813-18.
Skalinskii, E. I., L. S. Ageeva, and V. E. Teymlyakov. 1965. Veterinariya (Moscow) 42:20-24.
Stadie. 1931. Dtsch Weidwerk Ausq A 21:572.
Tietz, G. 1932. Arch Wiss Prakt Tierheilkd 65:244-55.
Tripathy, D. N., L. E. Hanson, and W. L. Myers. 1970. Avian Dis 14:29-38.
Tripathy, D. N., and C. H. Cunningham. 1984. In *Diseases of Poultry,* 8th ed., ed. M. S. Hofstad, H. J. Barnes, et al. Ames: Iowa State University Press.
Tsubahara, H., and K. Kato. 1959. Proc Jpn J Vet Sci 21:6.
Uppal, P. K. and P. R. Nilakantan. 1970. J Hyg 68:349-58.
Winterfield, R. W., and S. B. Hitchner. 1965. Avian Dis 9:237-41.
Woernle, H. 1966. Veterinarian 4:17-28.
Woodward, H., and D. Tudor. 1973. Poult Sci 11:4.

Paramyxovirus Infection

Definition and Synonyms. Paramyxovirus infection is an acute, infectious, febrile virus disease of pigeons, characterized by depression, loose droppings, nervous tremors, paralysis, twisted necks, and mortality. Paramyxo, parainfluenza, and PMV-1 are synonyms.

Cause and Classification. Paramyxovirus infection is caused by a virus in the Paramyxoviridae family, which includes three genera: *Paramyxovirus* (PMV), *Morbillivirus* (M), and *Pneumovirus* (P) (Braude et al. 1981). Myxoviruses are named from the Greek *myxa* (referring to nasal secretion or slime) because of the affinity of these viruses for mucoproteins (Andrews et al. 1955). The pigeon viruses classified as Parainfluenza 1 (Alexander 1982) is a Newcastle strain of virus (Meulemans 1984; Peterson 1985).

The following viral classification gives the relationships with some of the other members of the genus *Paramyxovirus* (there are nine parainfluenza groups in this genus).

1. Parainfluenza 1. Only groups 1 and 7 pertain to birds. This group includes four viruses:
   Newcastle–fowl, pigeons, birds, humans.
   PMV 1–pigeons, chickens (Alexander 1982) Sendai (Japanese hemagglutinating)–humans, swine, mice (Lyght 1966)
   HA2 (American virus) children (Lyght 1966; Braude et al. 1981)
2. Parainfluenza 7
   Tenn/USA/4/75–dove (Alexander et al. 1981)
   Otaru/Japan/76–rock pigeon (Alexander et al. 1984; Kida and Yanagawa 1979)

Nature

FORM. All paramyxoviruses consist of a single strand of RNA bounded by a protein arranged in a helical or spiral form. The nucleocapsid is covered by a lipoprotein outer envelope which forms an irregular, spherical virus particle. The ether-sensitive envelope is covered with short spikes or projections 80 Å long. Viral components are assembled at the host cell membrane and virions are released by budding (Andrews and Pereira 1972). Most characteristics and properties of NDV are shared by other members of the paramyxovirus genus.

REPRODUCTION. Paramyxoviruses are negative-strand viruses, in that replication requires transcription of the virion RNA by a virus polymerase before the RNA can act as a messenger for the virus protein (Alexander 1982).

CHARACTERISTICS. Parainfluenza (P) viruses have three distinct type-specific antigens: hemagglutinin (HA) and neuramidase (N) surface antigens and an internal nucleocapsid antigen (Davis et al. 1973). The viruses possess two surface glycoproteins, one with HA and N activity, the other with cell fusion and hemolysin activity. The diseases of the M genus are antigenically related but do not cross-react with the P viruses. Morbilliviruses lack N and hemagglutinate only red blood cells of old world monkeys (Braude et al. 1981). Pneumoviruses lack HA, N, and hemolysin. Newcastle disease virus (NDV) is immunologically distinct from other P viruses, but sera of convalescent human mumps cases show hemagglutination inhibition (HAI) against NDV (Braude et al. 1981). Differentiation of avian paramyxovirus isolates may be made by HAI (Alexander 1982) and a comparison of other properties including structural profiles of the polypeptides (Alexander and Collins 1981). There is no cross-reaction between nucleocapsid and envelope antigens of NDV and other P viruses (Braude et al. 1981). Unlike influenza, these viruses are antigenically stable and no genetic recombinations occur on a routine basis. Each virus shares antigens but there is no single antigen common to all P viruses (Davis et al. 1973).

Distribution

GEOGRAPHY AND HISTORY. Parainfluenza viruses were first recognized in 1957 as causes of human respiratory disease (Davis et al. 1973). Avian paramyxoviruses of serotype 1 (PMV-1) were known to infect racing pigeons and other members of the Columbidae family (Lancaster and Alexander 1975). Racing pigeon PMV-1 was first reported in Sudan and Egypt (BSAVA 1984; Alexander 1985). PMV-1 spread throughout Europe during 1981–1983 (Alexander and Parsons 1984). By 1982 it was reported in the province of Emilia in Italy by Biancifiori and Fioroni (1983) and in Belgium by Viaene et al. (1983). Vindevogel in Belgium (1982) observed the disease in two pigeons from Italy in 1981. In 1984 Prip in Denmark noted the disease, as well as Richter (1983) and Richter et al. (1983), who reported it in Germany. England had its first outbreak in May of 1983 (Alexander 1985). In April 1984 Israel experienced outbreaks following the importation of show pigeons from West Germany (Weisman et al. 1984), and vonBroos and Singer reported it in 1984 in Bavaria.

The disease first appeared in Brooklyn, N.Y., in April 1984 and spread to New Jersey and Connecticut. Later in 1984 and 1985 it was diagnosed in Delaware, Maryland, Pennsylvania, Ohio, Vermont, Maine, and Massachusetts. Canada experienced the disease in Ontario in June 1985 (Boulianne 1990). Western states have since encountered the disease. A total of 12 states were reported infected by Pearson et al. (1987), and seven of eight isolates studied from these states were indistinguishable from those of Europe and Great Britain. Ide in Canada (1990) noted an increase in virulence for chickens between viruses recovered in 1985 and 1989.

There were 192 outbreaks involving all 29 counties of Great Britain between June and December 1983 (Alexander 1985; Alexander and Parsons 1984; Alexander et al. 1984, a, b). The disease continued to spread and 810 cases were reported in 1984. Belgium had its first case in July 1983. Numerous cases followed during the summers of 1983–1984 (Meulemans 1984). In the United States many unvaccinated flocks developed infection during the racing season.

SEASON. The disease may appear throughout the year, but in England most of the cases occurred in August through October. In the United States in 1984, cases appeared in April through June and again in September and October.

AGE, SEX, AND BREED. The infection often affects young birds but all ages are susceptible. There is no sex or breed difference.

SPECIES. Aside from natural infection of pigeons and chickens Biancifiori and Fiorini (1983) reported the virus to be of low pathogenicity for quail. In 1989 the author also examined guineas with torticollis, from which the virus was isolated by the USDA Diagnostic Laboratory, Ames, Iowa. This appears to be the first report in guineas.

CARRIERS. Subclinical disease occurs.

Transmission. Direct and indirect contact serve to transmit infection. Direct contact with infected racing birds in crates during training flights and at shows permits exchange of infection. Indirect contact with contaminated feed or water, crates, equipment, and vehicles also permits transfer of virus. Alexander et al. (1984, a b c) infected pigeons intranasally and contact infected other pigeons. One contact pigeon excreted virus 31 days postinoculation. This bird first excreted virus on day 16 but showed no immune activity until day 27. Ten 3-wk-old chickens in contact with these infected pigeons excreted virus in their droppings and showed an HAI response but no clinical signs. By introducing the virus four times intramuscularly in 2-wk-old chickens, the virulence was increased for both pigeons and chickens when given intravenously. Nervous signs and high mortality were noted. Pearson et al. (1987) inoculated pigeons intravenously and intramuscularly with the development of typical signs, but only one intranasally inoculated pigeon showed clinical signs.

Epidemiology

DISTRIBUTION IN BODY. The virus spreads throughout the body. In pigeons with nervous signs, the virus may be recovered from the brain and spleen. Virus has been recovered from kidney, pancreas, and lungs by Biancifiori and Fioroni (1983).

ENTRY AND EXIT. Ingestion and/or inhalation serve as the means of entry. Feral pigeons at the Liverpool docks in England apparently contaminated rice bran with their droppings. The bran was fed to chickens, which contracted the infection (Alexander et al. 1984b; Cullen 1984).

Droppings and discharges carry the virus from the body.

PATHOGENICITY. The virus is highly infectious for pigeons but it spreads slowly. Alexander and Parsons (1984) in Weybridge, Surrey, tested 40 PMV-1 pigeon isolates and found that those with low intravenous pathogenicity for chickens gave a much stronger intravenous effect in adult pigeons. Virulence for chickens and pigeons was increased in one isolate by the repeated intravenous passage of the virus 3 to 4 times in chickens. When inoculated intranasally into pigeons, their isolate, 561/83, caused an antibody response and excretion of virus but no clinical signs. Contact chickens with infected pigeons were infected, but no signs were evident. Intravenous inoculation of pigeons and chickens with other isolates resulted in typical clinical signs. Pearson et al. (1987) observed that 6-wk-old chickens remained healthy following inoculation with 10 pigeon isolates by cloacal, intranasal, or caudal thoracic air sac routes but 1-day-old chicks reacted as with velogenic NDV when 4 of 6 isolates were inoculated intravenously. They also reported that pigeons receiving the LaSota strain of NDV did not develop signs of disease but 14 pigeons receiving pigeon PMV-1 isolates intramuscularly developed typical infection.

Incubation. The period can be as long as 4 wk (Eskelund 1985) but is usually 1–2 wk. Alexander (1985) also reported 1–2 wk but placed the upper limit at 6 wk or more. In work done by Alexander et al. (1984a), virus excretion occurred in 6–11 days following intranasal inoculation of pigeons and 11–16 days after contact exposure, but neither group showed clinical signs. Senne (1985) tested several PMV-1 isolates and observed that a high percentage of inoculated pigeons shed virus 4 days after inoculation but did not continue longer than 18 days.

Course. The duration of the infection depends on the virulence of the virus to which the birds are exposed, the age of the pigeons, and their prior exposure or resistance.

Morbidity. A few birds in a loft may never show signs of infection, but 20–80% of the loft may develop signs of the disease at one time. In Italy, Biancifiori and Fiorini (1983) reported illness in 80% of their experimentally infected pigeons.

Mortality. Up to 90% of loft birds with nervous signs may die, but culling by fanciers makes it difficult to get accurate figures. Peterson (1985) indicated that signs and mortality vary from 10% to 100%. Tangredi (1985) observed almost 100% loss in juvenile pigeons. Pearson et al. (1987) reported the mean death time for inoculated pigeons as 9.4 days with a range of 4–25 days.

Signs. The disease is somewhat similar to Newcastle disease. Birds show loose, watery, often greenish droppings, inappetence, depression, general loss of condition, reluctance to move, ruffled feathers, excessive thirst, and incoordination. Difficulty in flying, walking, or eating, trembling of wings and head, partial paralysis of wings and legs, and twisting of the neck, together with high mortality, is noted in some cases. Alexander et al. (1984a) observed that intranasally infected pigeons showed a general unwillingness to fly during week 2–3 postinoculation.

Necropsy. Necropsy signs in themselves are not diagnostic. Diffuse intestinal mucosal hemorrhages, cloacal, pancreatic, and proventriculus hemorrhages, air sac involvement, lung congestion, and spleen enlargement may be observed. Other secondary findings include loss of weight, enteritis, a greenish, off-color liver, and congestion of liver, spleen, and kidneys.

Diagnosis. In establishing a diagnosis the virus must first be isolated in chicken embryos or by cell culture methods or be identified in tissues of infected birds by the immunofluorescent antibody technique.

Seriological methods, in addition to other procedures, must be employed in the identification of the virus. Alexander et al. (1984a) differentiated PMV-1 from Newcastle by the use of chicken antisera and mouse monoclonal antibodies. This involved the injection of separate mouse colonies with PMV-1 and with Ulster 2C NDV for the production of specific antibodies to the viruses. The mouse ascitic fluid antibodies are then cross-tested with samples of NDV and paramyxoviruses to determine the degree of neutralization. The pigeon viruses formed a distinct subgroup of PMV-1 because of "their ability to cause binding of 3 of 9 mouse monoclonal antibodies to pigeon virus infected Madin-Darby bovine kidney cells. The pigeon viruses had a monoclonal antibody binding pattern which showed a variation of only one epitope (HN-1) from a group of isolates comprising prepandemic and postpandemic viscerotropic velogenic Newcastle disease virus." (Russell and Alexander 1983). Kida and Yanagawa (1979) used the double immunodiffusion test to classify the agent on the basis of the antigenic specificity of M protein antigens.

Hemagglutination characterizes the paramyxoviruses. PMV-1 does not, however, produce a high level of agglutination but it helps to differentiate it from herpesvirus, which does not agglutinate red blood cells (RBC). Agglutination is caused by the attachment of hemagglutinating spikes of the virion's envelope to receptors on the membrane of the RBC. The use of specific antiserum from pigeons previously infected with PMV-1 can neutralize the virus and inhibit agglutination. Thus the HAI test is also used in helping to establish a diagnosis.

Differential Diagnosis. Herpes, NDV, and salmonellosis must be considered in any diagnosis. If RBC agglutination occurs, a distinction must be made between NDV and pigeon PMV-1. Alexander et al. (1984a) used mouse monoclonal antibodies to make a differentiation. Russell and Alexander (1983) clarified the differences between LaSota NDV and pigeon virus. They learned that LaSota virus had 7 specific antigen components and the pigeon virus had 3. Only 2 of the components of pigeon PMV-1 were shared with those of LaSota.

Biological Properties

STRAINS. Meulemans (1984) in Belgium classified 20 isolates obtained in August through November 1983 as 2 lentogenic, 17 mesogenic, and 1 velogenic. The intracranial pathology index (ICPI) of mesogenic strains and the velogenic strains showed little variation (1.00–1.50), whereas the intravenous index varied from 0 to 1.67. These indexes simply provide an indication of the harm produced by the inoculation of each virus isolate. In the ICPI, 1 means the bird is sick and 2 means dead.

INCLUSION BODIES. No inclusions are reported.

STABILITY. The virus is acid and ether-sensitive. The usual disinfectants are effective in destroying the virus. It is also sensitive to heat.

CULTURE MEDIA. Ten-day-old chicken embryos may be used for virus isolation. The median death time as reported by Peterson (1985) was over 100 hr. Gelb et al. (1987) reported 88 to 109 hr. Day-old chicks inoculated intracerebrally showed sudden death or nervous signs (Alexander and Parsons 1984).

Prevention and Control. The very nature of showing and racing pigeons entails the mixing of birds from many lofts, which is ideal for the transmission of disease. The problem is compounded by the fact that the disease can have a long incubation period. This means that apparently healthy birds incubating the infection can spread the virus without the owner's knowledge.

Exclusion of wild birds and feral pigeons from the loft is necessary to prevent introduction of virus. Newly purchased pigeons and those returning from a race or show must be held in isolation until they are shown to be healthy. Prompt disposal of infected birds seldom controls the spread of disease. Immediate vaccination with a killed oil-adjuvant pigeon PMV-1 vaccine may reduce losses if given soon enough.

In the United States the use of Type 1 killed avian paramyxovirus (Maine Biological Laboratory) is effective in preventing and controlling the disease. The sterile product contains no live virus particles and cannot infect pigeons or cause other diseases. The oil-emulsion vaccine acts by slowly releasing the killed virus.

The vaccine should be held under refrigeration, below 45°F but not frozen. It should be slowly warmed to room temperature (70–80°F) and shaken prior to use. Warming makes it flow easier and reduces cold shock to the bird. The use of a sterile 18-gauge needle in the vaccine bottle speeds the filling of the syringes or automatic injecting device. The syringes, 20-gauge needles, and injecting devices must be sterilized before use. They must be washed clean, rinsed, and boiled for 20–30 min prior to use. Steam sterilization in a pressure cooker for 15–30 min at 15 lb pressure is preferred. The vaccine bottle cap should be wiped clean with 70% rubbing alcohol before insertion of the 18-gauge needle and syringe filling. If the needles are wiped clean with an alcohol-impregnated cotton swab between birds and the syringes are kept in a clean tray, the syringes may be refilled about three times and reused without being resterilized. The sterile needle shaft must not be handled with fingers. Finger contact with the needle hub with the ferrule in place is permitted during syringe refilling. The syringe without the needle is connected to the 18-gauge needle in the vaccine bottle for filling.

Disease-free birds may be successfully vaccinated to prevent the disease. Paramyxovirus-infected birds are seldom aided by the vaccine. All young birds 4 wk of age or older should be vaccinated prior to the racing or show season. Some young birds are not satisfactorily immunized by only one vaccination. All birds, young and old, should be revaccinated yearly prior to the breeding season. Flocks in the throes of an outbreak may be vaccinated, but only those birds with no evidence of infection should be injected.

The dose of vaccine is 0.5 ml (1/2 ml) injected subcutaneously in the lower neck region. Small breeds may receive 0.25 ml (1/4 ml).

Vaccination is conducted by the vaccinator and an assistant. The assistant seated opposite the right-handed vaccinator holds the pigeon in the right hand and grasps its head between the thumb and two fingers of the left hand. The vaccinator grasps the skin on the top of the neck with fingers of the left hand, lifting the skin. The syringe in the right hand of the vaccinator delivers the vaccine posteriorly under the skin behind the head in the top central portion of the neck. Vaccine placed deeply on either side of the neck adjacent to the vagus nerve and blood vessels may cause death. Adverse reactions seldom occur; however, it is advisable to practice on less valuable birds and give the birds a day to rest following vaccination.

If the vaccine is accidentally injected into the hand or finger, a physician should be consulted. The manufacturer's directions should be followed closely.

In England the Galaxo (1984) killed pigeon vaccine has been used with success. Viaene et al. (1984) in Belgium and Lumeij and Stam (1985) in the Netherlands reported reliable protection with a similar product. Weisman et al. (1984) in Israel used inactivated oil-based Newcastle and intraocular live virus LaSota or B1 strains with success. The live LaSota and B1 strains were not effective in curtailing the infection in New York when the disease started. Pearson et al. (1987) were unable to recover virus after chickens were inoculated with pigeon PMV-1 by the cloacal, intramuscular, or intranasal routes. Gelb et al. (1987) found complete cross-protection in challenge studies in chickens using the B1 NDV and pigeon PMV-1, but PMV-1 and NDV were readily distinguishable using a NDV monoclonal antibody, 1D12.

Treatment. No specific treatment is effective. Birds should easily reach food and water containers. Birds should be kept comfortable, conserving their energy by letting them rest. It is not necessary to depopulate infected birds. Obviously paralyzed birds may be culled. A percentage of birds will recover if the disease is permitted to run its course.

References and Selected Readings

Alexander, D. J. 1982. World Poult Sci J 38:97.

______. 1984. Personal communication.

______. 1985. Paramyxovirus Symposium, March 3, Davis, Calif.

Alexander, D. J., and M. S. Collins. 1981. Arch Virol 67:309–23.

Alexander, D. J., and G. Parsons. 1984. Vet Rec 114:466–69.

Alexander, D. J., V. S. Hinshaw, and M. S. Collins. 1981. Arch Virol 68:265.

Alexander, D. J., G. Parsons, and R. Marshall. 1984b. Vet Rec 115:601–2.

Alexander, D. J., P. H. Russell, and M. S. Collins. 1984a. Vet Rec 114:44–46.

Andrews, C. H., and H. G. Pereira. 1972. *Viruses of Vertebrates.* 3d ed. Baltimore: Williams & Wilkins.

Andrews, C. H., F. B. Bang, and F. M. Burnet. 1955. Virology 1:176.

Bankowski, R. A., R. Corstvet, and G. T. Clarke. 1960. Science 132:292.

Biancifiori, F., and A. Fioroni. 1983. Comp Immunol Microbiol Infect Dis 6:247–52.

Boulianne, M. 1990. Proc 62d Northeast Conference on Avian Disease, Guelph, Ontario, p. 7.

Braude, A. I., C. E. Davis, and J. Fierer. 1981. *Medical Microbiology and Infectious Diseases.* Vol. 2. Philadelphia: Saunders, p. 1935.

BSAVA. 1984. Vet Rec 114:232.

Cullen, G. 1984. Personal communication.

Davis, B. D., et al. 1973. *Microbiology* 2d ed. Hagerstown, Md.: Harper & Rowe, pp. 1331–60.

Eskelund, K. 1985. Personal communication.

Gelb, J., P. A. Fries, and F. S. Peterson. 1987. Avian Dis 31:601.

Galaxo. 1984. Pigeons and Paramyxo. Uxbridge, Middlesex, UK: Galaxo Animal Health Laboratory.

Ide, P. 1990. Proc. 62d Northeast Conference on Avian Disease, Guelph, Ontario, p. 10.

Kida, H., and R. Yanagawa. 1979. Zentralbl Bakteriol [Orig A] 245:421–28.

Lancaster, J., and D. J. Alexander. 1975. Can Dep Agric Monogr 11, p. 26.

Lumeij, J. T., and J. W. E. Stam. 1985. Vet 7:60–65.

Lyght, C. E. 1966. *The Merck Manual.* 11th ed. Rahway, N.J.: Merck, p. 763.

Meulemans, G. 1984. Personal communication.

Nerome, K., et al. 1978. J Gen Virol 38:293.

Pearson, J. E., et al. 1987. Avian Dis 31:105–11.

Peterson, I. L. 1985. Northeast Conference Avian Disease, Columbia, Md.

Prip, M. 1984. Personal communication.

Richter, V. R. 1983. Dtsch Veterinaermed Ges, pp. 86–95.

Richter, V. R., J. Kosters, and K. Kramer. 1983. Prakt Tieraerztl 64:915.

Rosenberger, A. K., W. Krauss, and R. D. Slemons. 1974. Avian Dis 18:610–13.

Russell, P. H., and R. D. Alexander. 1983. Arch Virol 75:243.
Sandhu, T., and V. Hinshaw. 1981. Proc 1st International Symposium on Avian Influenza, Beltsville, Md., pp. 93-99.
Senne, D. A. 1985. Paramyxovirus Symposium, Davis, Calif., Mar. 3.
Shortridge, K. F., and D. J. Alexander. 1978. Res Vet Sci 25:128.
Shortridge, K. F., D. J. Alexander, and M. S. Collins. 1980. J Gen Virol 49:255.
Tangredi, B. P. 1985. Avian Dis 29:1252.
Viaene, N., et al. 1983. Vlaams Diergeneeskd Tijdschr 52:278.
Viaene, N., et al. 1984. Vlaams Diergeneeskd Tijdschr 53:45-52.
Vindevogel, H. 1982. Ann Med Vet 126:5.
von Broos, H. W., and H. Singer. 1984. Tierarztl Umsch 7:557.
Weisman, Y., et al. 1984. Vet Rec 115:605.

Newcastle Disease

Definition and Synonyms. Newcastle is an acute highly infectious transmissible respiratory paramyxovirus disease of chickens and other birds, but seldom a disease of pigeons. When exposed to other birds with virulent, hot Newcastle disease virus (NDV), pigeons may contract the disease and exhibit hemorrhages of the intestine, paralysis, and death. Other birds may present signs of coughing or respiratory distress, followed by nervous signs in some cases.

The disease is also known as avian pneumoencephalitis, Doyle's disease, pseudo-fowl pest, ranikhet, and geflugelpest (Beaudette 1943).

Cause and Classification. Newcastle disease is caused by a virus in the Paramyxoviridae family, which includes three genera. NDV is classified under the genus *Paramyxovirus* (Braude et al. 1981). For further information relating to classification, see the section on paramyxovirus.

Nature

FORM. The Newcastle virion, which is about 180 nm in diameter, is composed of a single strand of RNA enclosed within a double-stranded, left-handed-spiral nucleocapsid forming a hollow tube 180 Å in diameter. The envelope covering the nucleocapsid bears spikes projecting from it 80 Å in length (Lancaster and Alexander 1975).

REPRODUCTION. NDV virions adsorb to the host cell, and the envelope fuses with the cell membrane, permitting the entrance of the nucleocapsid into the cell (Hanson 1978).NDV is a negative-stranded virus. Replication requires transcription of the virus RNA by the viral polymerase before the RNA can serve as a messenger for the viral protein (Alexander 1982). Replication of the virus in the cell cytoplasm reaches a maximum in 5-8 hr (Hanson 1978).

CHARACTERISTICS. The envelope contains the components that initiate hemagglutinin (HA) and virus neutralizing antibodies in the host. The HA are associated with the spikes (Rott 1964), and neuramidase activity can not be separated from the HA (Scheid and Choppin 1973). Two envelope glycoproteins and seven polypeptides have been reported by Alexander and Collins (1981).For the virus to be infectious, the surface glycoproteins must separate (Nagai et al. 1979).

Distribution

GEOGRAPHY AND HISTORY. Newcastle disease was first reported in 1926 in poultry in Batavia, Java, by Kraneveld (1926), in Newcastle, England, by Doyle (1927), and in Korea by Konno et al. (1929).It has since been found throughout the world. It was identified in the United States in 1944 (Brandly et al. 1944).A subsequent report has since indicated its presence here in 1938 (Beaudette and Hudson 1956). Naturally occurring pigeon infections have been reported in the Dutch East Indies (Picard 1928), India (Iyer 1939), South Africa (Kaschula 1952), Germany (Ulbrich and Sodan 1965; Hillbrich 1972), Egypt (Ahmed and Reda 1967), Japan (Kogo et al. 1969), England (Stewart 1971; Wilson 1986), and Belgium (Vindevogel et al. 1972).

INCIDENCE. Pigeons are relatively resistant to natural infection.This fact is supported by the few reports of spontaneous outbreaks in pigeons. New Jersey laboratories have not clinically diagnosed or recovered NDV from pigeons, and this includes the peak period of the poultry disease in New Jersey from 1945 to 1955. Until Vindevogel et al. (1982) reported the recovery of a lentogenic strain from a pigeon, only velogenic strains were isolated (Richter and Goosens 1971; Stewart 1971; Vindevogel et al. 1972; Hillbrich 1972; Utterbach and Schwartz 1973).

SEASON. The disease may occur at any time of the year.

AGE, SEX, AND BREED. The disease can occur at any age. There appears to be no predisposition to either sex, and all breeds appear equally susceptible.

SPECIES. Chickens are most susceptible, but many domestic and wild birds contract the infection naturally. Pigeons, ducks, and geese are quite resistant. Lancaster and Alexander (1975) list many species naturally infected: American widgeon, bamboo partridge, barn owl, canary, Chinese spotted dove, cockatoo, cormorant, dwarf turtledove, European martin, European starling, francolin, guinea fowl, gannet, kingfisher, parrot, great horned owl, Gouldian finch, hornbill, sparrow, jackdaw, kestrel, king penguin, laughing dove, little owl, macaw, osprey, ostrich, parakeet, peafowl, pintail, quail, raven, ringed dove, ring-necked pheasant, shag, secretary bird, toucan, vulture, white-tailed eagle, and willow grouse.

Human conjunctivitis often occurs in laboratory workers. Poultry eviscerators and others in close contact with the virus may develop a generalized infection (Beaudette 1943, 1949, 1950, 1951; Brandly 1952).

CARRIERS. Infected birds shed the virus in their discharges. They may also serve as mechanical carriers. Research by Kaschula (1952) demonstrated that in pigeons the oral administration of virulent virus resulted in virus excretion in the feces on the 12th day but the birds did not sicken. This study proved that pigeons can be a carrier of the virus and not show evidence of infection. Sharman and Walker (1973) also suggested that pigeons could carry viscerotropic velogenic Newcastle disease (VVND) virus from farm to farm.

Chickens appear to be the primary sources of lentogenic and mesogenic types, and parrots, parakeets, mynahs, pheasants, and partridge are the principal source of VVND. Arthropods have not been incriminated as vectors.

Transmission. The virus is transmitted by direct and indirect contact with infected birds, contaminated clothing, and/or equipment. It is present in airborne dust or aerosol particles.

Epidemiology

DISTRIBUTION IN BODY. In the preclinical and early clinical stages of the disease, the virus may be found in all fluids and tissues of the body and in secretions and excretions.

ENTRY AND EXIT. The virus gains entrance into the pigeon by inhalation, ingestion, or by way of the eye membranes and leaves the body in the feces and respiratory discharges.

PATHOGENICITY. Beaudette (1943, 1949, 1950, 1951) summarized and reported attempts by several workers to infect pigeons. Others have shown that pigeons are susceptible to inoculation by various routes. (Picard 1928; Doyle 1927, 1935; Dobson 1939; Pomeroy and Fenstermacher 1948; Kaschula 1951, 1952; Popovic 1951; Placidi and Santucci 1953; Walker et al. 1954; Reuss 1961). Doyle (1927) first showed that pigeons were highly susceptible when injected intramuscularly, but he failed to produce the disease by natural means.

Ahmed and Reda(1967) observed that lentogenic strains administered to 4- to 6-wk-old pigeons by the intracerebral or intramuscular routes failed to provoke either outward clinical signs or seriological hemagglutination inhibition (HI) antibody response within the bloodstream. This is in agreement with Olah and Palatka (1963). Ahmed further stated that mesogenic strains in pigeons produced acute illness that often terminated fatally 4–12 days after inoculation by the intracerebral or intramuscular route and velogenic strains killed all pigeons.

Erickson et al. (1980) reported that racing homer pigeons were contact exposed and inoculated intraocularly with four strains of VVND and mortality resulted in 7 of 21 juvenile birds and 1 of 10 adult birds. Virus shedding was noted in the juveniles from day 3 to 21, whereas in the adults it occurred from day 3 to 7. Antibody immunity levels (HI titers) peaked in 1–2 wk in both age groups. It was interesting that pigeons required 10,000 times more VVND-Largo strain virus to kill them than did chickens, yet both were equally susceptible to infection. This points to the inherent resistance of pigeons to this disease. Even when exposed to velogenic strains, many pigeons survive and in such birds shedding of virus may be expected even before the onset of clinical disease.

Incubation Period. It takes approximately 3–4 days for the clinical disease to develop in chickens.

Course. The clinical stage of the disease in chickens will vary, but is usually 2–3 wk. In a flock, the course is longer because not all birds contract the infection at the same time. Ahmed and Reda (1967) administered 8 strains of NDV by intracerebral and intramuscular routes to 4- to 6-wk-old pigeons. Mesogenic and velogenic virulent strains often terminated fatally in 4–12 days postinjection. Lentogenic strains provoked no outward clinical signs or an HAI antibody response.

Morbidity. Mild lentogenic strains are unlikely to produce any evidence of clinical illness. Other more virulent strains may produce acute illness.

Mortality. The death loss depends on the strain of virus to which the birds are exposed. Mesogenic and virulent velogenic strains can be fatal.

Signs. Evidence of infection is seldom observed in pigeons following exposure to mild strains, but paralysis or death may occur with virulent strains.

Necropsy. When pigeons are exposed to VVND, according to Erickson et al. (1980), hemorrhages of the gizzard and intestine occur below the internal surface membrane or submucosa, but do not extend deeply to the mucosa. Kaschula (1951) reported that all strains of the virus that produce apparent disease in the pigeon resulted in progressive paralysis ending in death. Recent use of the LaSota strain in the New York area with the hope that it would control paramyxovirus disease often resulted in mild air sac involvement with few outward clinical signs.

Diagnosis. A dependable diagnosis can be established only by the isolation of the virus and demonstration that the recovered virus can produce the disease observed.

The virus may be recovered from tracheal exudate of infected birds or be isolated from a variety of tissues, such as spleen, brain, yolk, and bone marrow. Diagnosis also involves the seriological demonstration of NDV hemagglutinins. The virus neutralization (VN) test or HAI test may be conducted to show evidence of past exposure. Serum neutralization, fluorescent antibody, and complement fixation tests may be used to demonstrate NDV.

Differential Diagnosis. Paramyxovirus and herpesvirus must be excluded.

Biological Properties

STRAINS. It is generally accepted that the different strains of NDV afflicting birds are of three types: lentogenic, mesogenic, and velogenic. Russell and Alexander (1983) identified 8 distinct antigenic NC groups by their ability to induce binding of 9 mouse monoclonal antibodies against Ulster 2C NCDV. Strains are classified according to their virulence or ability to produce disease. The mildest strains are of the lentogenic type and include B1 and LaSota strains, both of which were isolated in the Poultry Diagnostic Laboratory at Rutgers University. Ulster is also included in this group (Beard and Hanson 1984). The B1 was first described by Hitchner and Johnson (1948). Roaken, a mesogenic-type New Jersey strain, was originally described by Beaudette and Black (1946). It was intermediate in virulence whereas VVND, described by Beach (1942), was the most severe and includes such strains as Boney, Herts, Fontana Ca-1083, Largo, P1307, and P5658. The latter strain was isolated from racing homer pigeons by the USDA Diagnostic Laboratory, Ames, Iowa.

HEMATOLOGY. Human, guinea pig, and mouse red blood cells (RBC) are agglutinated by all strains of NDV, but some strains do not agglutinate RBC of cattle, sheep, goats, and swine (Winslow et al. 1950). Partial agglutination occurs with RBC of birds, reptiles, and amphibia (Placidi and Santucci 1956). Some NDV strains agglutinate a variety of avian and mammalian RBC, but all strains of NDV adsorb to and agglutinate fowl RBC (Andrews and Pereira 1972).

The hemagglutinins appear to be in the surface projections and are mucoprotein receptors on the virion envelope. For hemolytic activity the receptors must be present (Davis et al. 1973). During hemagglutination NDV glycoprotein HA attach to the RBC or other cell surface receptors. This is followed by neuramidase destruction of the receptors and the release of NDV from the cell (Ackermann 1964).

TISSUE AFFINITY. The virus may be isolated from many parts of the body, but tracheal exudate, brain, and spleen tissue are the most reliable sources. The agent has an affinity for respiratory epithelial tissue, but it may also become localized in nerve tissue.

STAINING. Tissue culture monolayers may be stained with neutral red. Infected cells accept the stain more readily than do normal cells (Schloer and Hanson 1968).

STABILITY. NDV is quite resistant over several hours to a change in pH over a wide range of values, pH 2–10 (Moses et al. 1947). Viricidal chemicals, however, will destroy NDV with reasonable rapidity (Cunningham 1948) if not protected by proteinaceous organic matter that inactivates the disinfectant (Walker et al. 1953).

Dry cold preserves indefinitely. Incubator heat of 37°C will normally destroy the virus over a period of days (Hanson 1978). Ultraviolet sun rays will also destroy the virus (Levinson et al. 1944).

CULTURE MEDIA. Ten-day-old embryonated chicken eggs are useful in the isolation of NDV. The embryos should be free of other pathogens and antibodies to NDV. Tracheal exudate and/or spleen tissue taken from infected birds may be ground in broth and inoculated into the chorioallantoic sac of the embryo. Each egg should receive 0.2 ml inocula, containing procaine penicillin and streptomycin sulfate (10,000 units of each/ml inocula). In one-third of the cases, 3 of 4 embryos will die on the first passage, unless VVND is inoculated, in which case all will die in about 50 hr.

Chicks and/or susceptible adult chickens may be inoculated, but the clinical signs are not diagnostic. One-day-old chicks may be used to advantage to assay the pathogenicity of various strains using the intracerebral or intramuscular routes of inoculation.

Suckling mice die 72 hr following intravenous inoculation with $10^{-8.2}$ plaque-forming units of virus (Sanders and Lennette 1968).

NDV can produce changes in many cell culture lines, including cell death and a change in cellular form or function. Plaque formation may also be induced on cell monolayers.

Prevention. Erickson et al. (1980) evaluated three commercial NDV vaccines and eight programs of application in pigeons, using standard doses for chickens. The vaccines were administered as intraocular live LaSota, subcutaneous inactivated aluminum-hydroxide adsorbed, and intramuscular inactivated oil emulsion. Four VVND challenge strains were used. Only one of the vaccine programs protected pigeons against VVND. In that program pigeons were vaccinated with live LaSota vaccine, which was followed in 6 wk with an inactivated oil-emulsion vaccine. This method of control is presently the only one effective against VVND, outside of the use of quarantine and slaughter of visibly affected birds. Good isolation methods are effective in preventing the spread of VVND infection to pigeon lofts. Milder strains, however, do not create a problem for the pigeon but every effort should be made to avoid exposure.

Treatment. No treatment is indicated for mild strains, and none is effective for VVND.

References

Ackermann, W. W. 1964. In *Newcastle Disease Virus: An Evolving Pathogen,* ed. R. P. Hanson, pp. 153-65. Madison: University of Wisconsin Press.

Ahmed, A. A. S., and I. M. Reda. 1967. Avian Dis 11:734-40.

Alexander, D. J. 1982. World Poult Sci J 38:97.

Alexander, D. J., and M. S. Collins. 1981. Arch Virol 67:309-23.

Andrews, C., and H. G. Pereira. 1972. *Viruses of Vertebrates.* 3d ed. Baltimore: Williams & Wilkins.

Beach, J. R. 1942. Proc 46th Annual Meeting of the U S Livestock Sanitary Assoc., pp. 203-23.

Beard, C. W., and R. P. Hanson. 1984. In *Diseases of Poultry,* 8th ed., ed. M. S. Hofstad. Ames: Iowa State University Press.

Beaudette, F. R. 1943. Proc 47th Annual Meeting of the U S Livestock Sanitary Assoc, pp. 122-77.

______. 1949. Proc 53d Annual Meeting of the U S Livestock Sanitary Assoc, pp. 202-20.

______. 1950. Proc 54th Annual Meeting of the U S Livestock Sanitary Assoc, pp. 132-53.

______. 1951. Proc 55th Annual Meeting of the U S Livestock Sanitary Assoc, 108-74.

Beaudette, F. R., and J. J. Black. 1946. Proc 50th Annual Meeting of the U S Livestock Sanitary Assoc., pp. 49-58.

Beaudette, F. R., and C. B. Hudson. 1956. Cornell Vet 46:227.

Brandly, C. A. 1952. In *Diseases of Poultry,* 4th ed., ed. H. E. Biester and L. H. Schwarte, p. 464. Ames: Iowa State University Press.

Brandly, C. A., H. E. Moses, and E. E. Jones. 1944. Spec Rep from Huntington Lab to War Dep, March 27.

Braude, A. I., C. E. Davis, and J. Fierer. 1981. *Medical Microbiology and Infectious Diseases.* Vol. II. Philadelphia: Saunders, p. 1935.

Cunningham, C. H. 1948. Am J Vet Res 9:195-97.

Davis, B. D., et al. 1973. *Microbiology.* 2d ed. Hagerstown, Md.: Harper & Row.

Dobson, N. 1939. Proc 7th World Poultry Congress. Washington, D.C., p. 250.

Doyle, T. M. 1927. J Comp Pathol Therap 40:144-69.

______. 1935. J Comp Pathol 48:1.

Erickson, G. A., M. Brugh, and C. W. Beard. 1980. Avian Dis 24:257.

Hanson, R. P. 1978. In *Diseases of Poultry,* 7th ed., ed. M. S. Hofstad, pp. 513-35. Ames: Iowa State University Press.

Hillbrich, P. 1972. Dtsch Tieraerztl Wochenschr 79:177.

Hitchner, S. B., and E. P. Johnson. 1948. Vet Med 43:525-30.

Iyer, S. G. 1939. Indian J Vet Sci Anim Husb 9:379-82.

Kaschula, V. R. 1951. J S Afr Vet Med Assoc 22:193.

______. 1952. Onderstepoort J Vet Res 25:25.

Kogo, T., et al. 1969. J Jpn Vet Med Assoc 22:600-605.

Konno, T., Y. Ochi, and K. Hashimoto.1929.Dtsch Tieraerztl Wochenschr 37:515-17.

Kraneveld, F. C. 1926. Ned Indische Bl Diergeneeskd Dierent 5:448-50.

Lancaster, J. E., and D. J. Alexander. 1975. Can Dep Agric Monogr 11.

Levinson, S. O., et al. 1944. J Am Med Assoc 125:531-32.

Moses, H. E., C. A. Brandly, and E. E. Jones. 1947. Science 105:477-79.

Nagai, Y., et al. 1979. J Gen Virol 45:263-72.

Olah, P., and Z. Palatka. 1963. Acta Vet Acad Sci Hung 13:37.

Picard, W. K. 1928. Ned Indische Bl Diergeneeskd Dierent 40 Alf 1:1-52.

Placidi, L., and J. Santucci. 1953. Ann Inst Pasteur 84:588.

______. 1956. Ann Inst Pasteur 90:528-29.

Pomeroy, B. S., and R. Fenstermacher. 1948. U S Egg Poult Mag 54:19.

Popovic, B. 1951. Acta Vet Belgr 1:168.

Reuss, U. 1961. Monatsh Tierheilkd 13:153.

Richter, J., and J. Goosens. 1971. Tijdschr Diergeneeskd 96:470.

Rott, R. 1964. In *Newcastle Disease Virus: An Evolving Pathogen,* ed. R. P. Hanson, pp. 133-46. Madison: University of Wisconsin Press.

Russell, P. H., and D. J. Alexander. 1983. Arch Virol 75:243-53.

Sanders, M., and E. Lennette. 1968. *Medical and Applied Virology.* St Louis: Green.

Scheid, A., and P. W. Choppin. 1973. J Virol 11:263-71.

Schloer, G. M., and R. P. Hanson. 1968. Am J Vet Res 29:883-95.

Sharman, E. C., and J. W. Walker. 1973. J Am Vet Med Assoc 163:1089-93.

Stewart, G. H. 1971. Vet Rec 21:225-26.
Ulbrich, F., and U. Sodan. 1965. Monatsh Veterinaermed 20:340-44.
Utterbach, W. W., and J. H. Schwartz. 1973. J Am Vet Med Assoc 163:1080.
Vindevogel, H., et al. 1972. Ann Rech Vet 3:519-32.
Vindevogel, H., et al. 1982. Vet Rec 110:497-99.
Walker, R. V. L., R. Gwatkin, and P. D. McKercher. 1953. Can J Comp Med Vet Sci 17:225-29.
Walker, R. V. L., P. D. McKercher, and G. L. Bannister. 1954. Can J Comp Med Vet Sci 18:244.
Wilson, G. W. C. 1986. World Poult Sci J 42: 143-53.
Winslow, N. S., et al. 1950. Proc Soc Exp Biol Med 74:174-78.

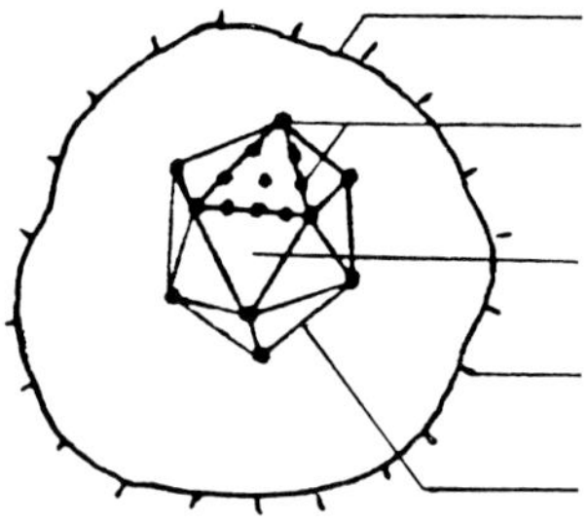

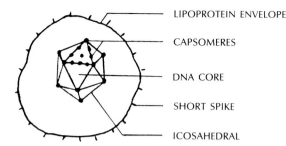

Fig. 3.3. Enveloped icosahedral virion of herpes (from Davis et al. 1973).

Herpesvirus Infections

Definition and Synonyms. Herpes is a common acute, infectious, contagious virus disease of humans, animals, and birds. Depending on which pigeon strain of virus is involved, pigeon herpes is characterized in the mild form by respiratory signs and multiple, focal, degenerative lesions of the liver and in the severe fatal form by paralysis and twisted necks with no respiratory signs. It is also called pigeon herpesvirus (PHV) and pigeon herpes encephalomyelitis virus (PHEV). The latter is the severe nervous form.

Cause and Classification. Pigeon herpesvirus belongs in the family Herpesviridae, but no name has been approved for the genus, which in birds includes Marek's disease and laryngotracheitis of chickens, duck plague, Lake Victoria cormorant herpes, owl and falcon viruses, and Pacheco disease of parrots.

Nature

FORM. The morphology of all herpesviruses is similar. The herpes family of viruses is composed of double-stranded DNA in a core about 760 Å in diameter (Davis et al. 1973). The enveloped virion varies from 120 to 200 nm in diameter. The icosahedral nucleocapsid (See Fig. 3.3) is approximately 100 nm in diameter and composed of 162 hollow elongated hexagonal prisms or capsomeres (Fenner et al. 1987). Early basophilic intranuclear inclusions (INI) contain DNA; older eosinophilic INI do not (Cornwell and Wright 1970). Davis et al. (1973) further indicated that the envelope contained lipid and possessed periodic short projections. In addition, they noted that many human herpesvirus particles do not have envelopes and some have empty capsids. In PHV material examined by Saik et al. (1985), a few nonenveloped virions were present. The capsids in this material were 183 nm and the cores 116 nm in diameter.

METABOLISM AND REPRODUCTION. Herpesvirus DNA replicates in the cell nucleus and the virus particles are assembled within the nucleus and cytoplasm of the infected cell. Most of the members of the family have a special affinity for ectodermal cells and a tendency to produce latent infections as typified by herpes simplex, the human fever blister virus (Davis et al. 1973).

SEROLOGY. The viral envelope may contain both host cell and viral components and is important in adsorption to susceptible host cells (Davis et al. 1973). Lee et al. (1972) reported another finding that has a bearing on serology. PHV has a higher DNA buoyant density than chicken or turkey herpesvirus or laryngotracheitis virus. It ranks with owl and cormorant (Lake Victoria) virus, which also have higher densities. Al Falluji et al. (1979) in their studies found that neither Newcastle nor epidemic tremor antisera neutralized the PHEV and the virus failed to agglutinate red blood cells. The PHV has also been studied by the use of neutralization tests and gel precipitation tests.

Distribution

GEOGRAPHY AND HISTORY. The PHV was first reported in the United States by Smadel et al. (1945) and Lehner et al. (1967). Following these reports it was observed in Denmark by Marthedal and Jylling (1966). The herpesvirus also surfaced in the United Kingdom and Ireland and was identified by Cornwell et al. (1967). Lando and Ryhiner (1969) published on the disease in France. The following year Krupicka et al. (1970) noted the problem in pigeons in Czechoslovakia. Nearby in Hungary the infection was reported by Vetesi and Tanyi (1975). In the same year Vindevogel et al. (1975) started their extensive studies in Belgium. Boyle and Binnington (1973) found the agent in Australia, followed by Thompson et al. (1976) in New Zealand.

The first U.S. case of PHV was undoubtedly observed in New York by Smadel et al. (1943), but it was not established until 1945 when he verified the presence of INI in Army pigeons in southern states. In California Russell and Shetrone (1977) found INI in ulcerative proctitis. A final viral diagnosis was not established. Later Saik et al. (1985, 1986) identified herpesvirus in pigeons in Pennsylvania by electron microscope studies.

In 1978 Tantawi et al. and Mohammed et al. both reported isolations of PHEV in Iraq and later Tantawi et al. (1980) reported the disease in Egypt and possibly Syria. In 1986 Carranza et al. observed PHEV in the Canary Islands.

INCIDENCE. In the United States the PHV appears to be endemic. In the United Kingdom it reoccurs annually in some young flocks (Cornwell and Wright 1970). Vindevogel et al. (1981) conducted a serological survey in Belgium in which 62% of the pigeons had antibodies to PHV.

In Iraq, Al Falluji et al. (1979) reported the PHEV to be slow-spreading and sporadic in certain areas.

SEASON. The disease may occur at any time of the year but is recognized more often in summer months.

AGE, SEX, AND BREED. Birds of any age may be affected, but younger birds 1–6 mo old are more commonly afflicted. Both sexes and all breeds are affected equally.

ARTIFICIALLY INFECTED SPECIES. Pigeons developed paralysis and died following intracerebral inoculation of PHV (Lehner et al. 1967). Surman et al. (1975) observed fatal intravenous infection of cockatiels and budgerigars with INI in liver, pancreas, and small intestine. According to Vindevogel and Duchatel (1977) budgerigars always have a systemic infection with PHV.

NATURALLY INFECTED SPECIES. Pigeons are natural hosts for the viruses.

SPECIES, REFRACTORY TO INFECTION. Mice, rabbits, and guinea pigs do not develop PHV infection (Smadel et al. 1945; Lehner et al. 1967). Tantawi et al. (1979) reported chickens, ducks, turkeys, sparrows, lovebirds, rabbits, guinea pigs, and mice refractory to PHEV.

CARRIERS. The pigeon is the primary host and reservoir. Once infected, a bird is infected for life and will continue to shed virus at various times without clinical signs of infection. Even with high antibody titers PHV may be shed (Vindevogel et al. 1980; Vindevogel and Pastoret 1980). Vindevogel et al. (1980) used cyclophosphamide to cause stress and provoke reexcretion of PHV. After weaning, most squabs are asymptomatic carriers of the virus but are devoid of detectable neutralizing antibodies (Vindevogel and Pastoret 1980).

Transmission. The PHV is spread from bird to bird by discharges from infected birds. Most squabs are infected during the first few days of life while they are protected by parental antibodies. Vindevogel and Pastoret (1980) also reported egg transmission to be unlikely.

Epidemiology

DISTRIBUTION IN BODY. The virus may be spread by contact from one cell to another following localized pharynx inoculation. It may later spread in the blood to lesions outside the upper digestive and respiratory tracts. This appears to be enhanced by stress. PHV has been recovered from the brain but not from genital organs (Vindevogel and Pastoret 1981).

ENTRY AND EXIT. The viruses normally enter the body by way of the mouth. PHV was isolated from 2-day-old clinically negative squabs raised by infected but clinically negative parents. No virus was isolated from squabs from infected clinically negative parents when the squabs were hatched and reared artificially (Vindevogel and Pastoret 1980).

The agent leaves the body in discharges. Immune suppression provokes heavy PHV excretion. In tests by Vindevogel and Pastoret (1980) all squabs excreted infectious PHV particles 24 hr after infection. Excretion persisted off and on regardless of the antibody titer. Pastoret et al. (1978) indicated, however, that the amount of PHV excreted and recovered on a swab was under a titer of 10^{-6}, which was too little for electron microscopy.

PATHOGENICITY. Young birds from noninfected parents are susceptible to clinical infection by both PHV and PHEV. Stressed PHV carrier birds also develop clinical disease. Debilitating factors precipitate infection in clinically normal pigeons (Vindevogel and Duchatel 1979).

IMMUNITY. All mature pigeons tested by Vindevogel and Pastoret (1980) possessed neutralizing antibodies, indicating that they had been exposed to the virus, but eggs from infected birds contained no PHV. Egg yolks, which conferred passive protection to hatched squabs, had neutralizing antibodies but crop milk did not. In addition, squabs lost PHV parental antibodies transmitted by the yolk in 14 days (Vindevogel and Pastoret 1980).

Incubation. Clinical lesions develop 3–4 days after pharynx inoculation of susceptible birds.

Course. The PHV disease may be fatal within a week or may take a prolonged course as in carrier birds. With PHEV, birds often die within a week (Al Falluji et al. 1979).

Morbidity. In PHV outbreaks most of the susceptible birds within a loft will be infected within a short time. Infection of isolated lofts are often delayed. Smadel et al. (1943) observed a 90% infection rate in birds they handled. Over 50% of a loft will be affected when PHEV strikes, but separated lofts may not succumb for some time (Al Falluji et al. 1979).

Mortality. With PHV outbreaks, scattered losses may be expected. These losses occur chiefly in young birds. Mortality largely depends on the degree of susceptibility of the flock. With PHEV infections, 90% of those sick usually die quickly (Al Falluji et al. 1979).

PHV death is caused by systemic infection resulting in paralysis and by upper digestive or respiratory lesions that interfere with eating or drinking and breathing. PHEV usually causes death by paralysis.

Signs. Outbreaks of PHV cause dullness, depression, dehydration, inappetence, inability to fly, reluctance to move, green diarrhea, serous conjunctivitis, and varying degrees of respiratory distress from mild rhinitis to severe dyspnea. Death of affected birds may occur quickly, but a percentage will resist clinical disease. Brain damage will be evident in some birds. This is particularly true in outbreaks of PHEV, where a majority of the birds exhibit acute depression, puffiness, no desire to move or eat, paresis, paralysis, circling, one-sided wing or leg paralysis, trembling, twitching, and occasionally twisted necks. Many have bile-stained droppings and water-filled crops (Tantawi et al. 1978).

Necropsy. Smadel et al. (1945) and Lehner et al. (1967) observed focal necrosis of the liver, spleen, and pancreas in PHV cases they examined. Cornwell and Wright (1970) did not find pancreatic necrosis but did observe varying degrees of necrosis of the liver and spleen. They found the liver, kidneys, and upper respiratory tract to be primarily involved. Peritonitis with liver to airsac adhesions and fibrin over the liver were sometimes noted. In some of the advanced cases, scattered yellow focal liver necrosis reached 1 cm in diameter and in some instances became confluent (Cornwell et al. 1970). Gallbladder distention and congestion of visceral organs were also present. Kidney lesions reported by Cornwell and Wright (1970) were multiple cream-colored, cortical foci about 2 mm in diameter. Diphtheroid lesions were observed on the membranes of the mouth, pharynx, larynx, and trachea (Cornwell et al. 1967; Cornwell and Wright 1970; Vindevogel et al. 1975; Vetesi and Tanyi 1975; Vindevogel and Duchatel 1979; Callinan et al. 1979). Serous conjunctivitis was present in most cases and mild rhinitis was seen at times (Cornwell and Wright 1970). Marthedal and Jylling (1966) and Jylling (1967) found a high percentage of esophageal diphtheritic membrane lesions with INIs. In 20% of 250 birds they examined, lesions were present in the liver, lungs, trachea, esophagus, kidneys, and pancreas. Inoculated chick embryos died in 6–9 days. Kamionokowski (1981) described brain and spinal cord inflammation. This was substantiated by the recovery of PHV virus from brain tissue by Vindevogel and Pastoret (1981). Saik et al. (1986) noted multiple 1-mm yellow-white foci on the epicardium, in addition to multifocal hepatic necrosis and cloudy air sacs.

Micropathology. Eosinophilic INI were first reported by Smadel et al. (1945). Cornwell and Wright (1970) found inclusions in liver and kidney tissue and in the epithelium of the pharynx and larynx. Their further work with PHV made note of leucocytic activity mainly limited to accumulations of heterophils at the margins of necrotic foci. Cellular kidney changes included tubular necrosis, hemorrhage, and lymphocytic infiltrations. Inflammatory cells appeared concentrated at the borders of degenerative lesions. According to Al Falluji et al. (1979) PHEV produces mononuclear infiltrations and perivascular cuffing in the brain. There is also a loss of Purkinje cells with mononuclear accumulations in the cerebellum.

Diagnosis. A diagnosis must be based on signs and necropsy findings typical of the disease. In addition, virus recovery, INIs, and an absence of red blood cell agglutination are diagnostic.

Differential Diagnosis. Velogenic Newcastle disease and paramyxovirus must be excluded as causes. Meningoencephalitis of pigeons as described by Dougherty and Saunders (1953) must be considered, as well as inclusion body hepatitis of chickens (Helmboldt and Frazier, 1963).

Biological Properties

STRAINS. Barling (1977) reported research with several strains of PHV: M3, HPS1 (Cornwell), P5 (Smadel), and VF-74-282LIY (McFerran). Tantawi et al. (1979) characterized two PHEV strains: BVC78, a virulent strain, and BVC78 T/7, a cell-culture-adapted strain.

HEMATOLOGY. The viruses do not agglutinate chicken or pigeon red blood cells.

TISSUE AFFINITY. The virus has been found primarily in tissues of liver, kidney, spleen, pancreas, brain, and respiratory tract. Digestive tract lesions may appear in the esophagus and terminal intestine.

INCLUSION BODIES. Eosinophilic INI with a

clear halo were observed in necrotic liver and kidney foci. Dense basophilic inclusions occupied the entire nucleus, and these had no halo. Inclusions were widely scattered throughout the liver but were primarily near the periphery of the foci. INI were also observed in chick embryo lesions (Cornwell and Wright 1970). Russell and Shetrone (1977) observed INI in the terminal intestinal areas associated with acute ulcerative lesions.

STAINING. Vindevogel et al. (1977) stained chick embryo fibroblasts (CEF) with hemalum eosin for general examination and with acridine orange to demonstrate nucleic acid. At 20 hr fully developed basophilic inclusions fluoresced green with acridine orange and were Feulgen-positive.

STABILITY

Chemical. Cornwell et al. (1970) reported the PHV to be sensitive to ether and Tantawi et al. (1981) reported the envelope of PHEV sensitive to ether and chloroform. The latter workers also found the virus inactivated by 1% cresol and 2% sodium hydroxide in 2 hr and by 2% septol in 24 hr; 2% phenol and 2% formalin decreased the titer significantly in 2 hr.

pH. The PHV infectivity is reduced at a pH of 4 (Cornwell and Wright 1970).

Temperature. The PHV is destroyed at 56°C in 30 min, and in fluid the half-life is 15 hr at 37°C (Cornwell and Wright 1970). Tantawi et al. (1981) reported that PHEV was stable at −70°C for 4 mo but had an 80% loss at −20°C in 4 mo and complete loss of titer at −10°C in 12 wk.

CULTURE MEDIA

Tissue Cell. The virus may be grown in specific cell cultures in artificial maintenance media. PHV grows best in chick embryo fibroblast (CEF) but less well in chick embryo (CE) kidney cells or in pigeon kidney cells (Cornwell and Wright 1970). Vindevogel et al. (1977a) described viral studies using CEF plaques grown in Falcon petri dishes on coverslips in 5% carbon dioxide. Each petri dish was inoculated with 0.2 ml virus containing 10^{-5} plaque-forming units (PFU). The cultures were incubated 1 hr at 37°C to permit adsorption of the virus. The cells were then washed with phosphate-buffered saline (PBS) to remove unadsorbed virus. At harvest the coverslips were rinsed in PBS and fixed in Clarke's solution for 20 min and preserved in 80% ethanol. Vindevogel et al. (1977b) determined that mammalian cell lines were refractory to the virus but duck embryo fibroblast, chicken embryo liver, and chick kidney cells were susceptible but cellular changes varied.

In CEF cultures the earliest change, noted 10 hr after incubation, was chromatin margination and basophilic INI surrounded by a clear zone, or halo. By 14 hr the cell size increased and many round cells had 2–4 nuclei, with some having 5–10. By the 16th hr typical type A Cowdry INI were present. After 36 hr the entire monolayer was involved, with detachment of cells evident by 48–60 hr and destruction complete by 72 hr. The virus concentration of the cultures increased from the 12th hr to a maximum at 36 hr after inoculation. After 10 hr, virus also appeared in the extracellular fluid (Vindevogel et al. 1977a).

Tantawi et al. (1979) grew PHEV on CEF, and duck and turkey EF, with the production of acidophilic and basophilic INI with clear plaques 1–2 mm in diameter 3–4 days postinoculation (PI). Tantawi and Al Sheikhly (1980b), working with PHEV, noted good growth with the virulent strain, BVC78, in all avian fibroblast cultures with the formation of large syncytia 24–48 hr PI whereas the tissue-modified strain, BVC78 T/7, induced only cell rounding 72 hr PI. This cytopathic effect on avian cell lines appears similar to falcon and owl herpesvirus (Lee et al. 1972). Other mammalian cell lines with the exception of swine embryo kidney reacted to PHEV with rounding and clumping (Tantawi and Al Sheikhly 1980a).

Chick Embryo. Tissues from PHV cases are ground in a sterile mortar with sufficient physiological saline to produce a 10% suspension. Antibiotics may then be added to the centrifuged supernatent. Ten-day-old chick embryos are inoculated with 0.2 ml supernatent on the chorioallantoic membrane (CAM). Inoculation by the amniotic and the allantoic sacs do not readily induce hepatic necrosis. Harvest of the CAM may be made at 2–4 days PI. Embryos die 4–5 days PI with PHV. All strains of PHV produce CAM lesions like poxvirus. PHV produces focal, cream-colored, distinctly elevated, pocklike lesions, often with center craters. There is early necrosis of the ectoderm, which extends to the mesoderm (Cornwell and Weir 1970). The PHV also causes focal necrosis of the embryo liver and spleen and INI in these lesions (Cornwell and Wright 1970).

Tantawi et al. (1979) successfully grew PHEV on pigeon, chicken, duck, and turkey CAMs. Duck and turkey embryos died in 72–96 hr PI. Al Sheikhly et al. (1980), also using PHEV BVC78, observed curled, stunted embryos with pinpoint subcutaneous and visceral hemorrhages. Maximum virus yield developed 72 hr PI with BVC78 and 120 hr PI with BVC78 T/7. Most embryos die 3 days PI with BVC78.

Prevention. Vindevogel et al. (1982b) found attenuated and killed vaccine ineffective in preventing carrier birds, but the killed vaccine reduced the primary excretion of virus, which reduces the dissemination of virus.

Tantawi et al. (1980) and Tantawi and Al Sheikhly (1980b) reported using their mild BVC78 T/7 cell-culture-adapted strain to subcutaneously vaccinate pigeons and protect them from an intravenous challenge with their virulent BVC78 strain 28 days later. Killed PHEV vaccines were tested with poor results.

Control. Presently the natural infection of squabs by infected parents while the squabs have parental immunity offers the best means of controlling PHV losses.

Treatment. Schwers et al. (1980, 1981) used trisodium phosphonoformate, an antiherpes agent, at 100 mM/ml on the PHV in tissue culture with some reduction in plaque size. The drug may thus have some value on local lesions of the mouth and throat. The drug did not prevent carrier birds from shedding the virus (Vindevogel et al. 1982a).

References

Al Falluji, M. M., M. M. F. Al Sheikhly, and H. H. Tantawi. 1979. Avian Dis 23:777-84.
Al Sheikhly, M. M. F., H. H. Tantawi, and M. M. Al Falluji. 1980. Avian Dis 24:112-19.
Barling, R. V. 1977. Racing Pigeon Pict 8:200-206.
Boyle, D. B., and J. A. Binnington. 1973. Aust Vet J 49:54.
Callinan, R. B., et al. 1979. Aust Vet J 55:339-41.
Carranza, J., J. B. Poveda, and A. Fernandez. 1986. Avian Dis 30:416-20.
Cornwell, H. J. C., and A. R. Weir. 1970. J Comp Pathol 80:509-15.
Cornwell, H. J. C., and N. G. Wright. 1970. J Comp Pathol 80:221-27.
Cornwell, H. J. C., A. R. Weir, and E. A. C. Follett. 1967. Vet Rec 81:267-68.
Cornwell, H. J. C., N. G. Wright, and H. B. McCusker. 1970. J Comp Pathol 80:229-32.
Davis, B. D., et al. 1973. *Microbiology.* 2d ed. Hagerstown, Md.: Harper & Row.
Dougherty, E., and L. Z. Saunders. 1953. Am J Pathol 29:1165-70.
Fenner, F., et al. 1987. *Veterinary Virology.* New York: Academic.
Helmboldt, C. F., and M. N. Frazier. 1963. Avian Dis 7:446-50.
Jylling, B. 1967. Nord Vet Med 19:415-19.
Kamionokowski, M. 1981. Med Vet J 37:531-32.
Krupicka, V., et al. 1970. Vet Med (Praha) 15:609-12.
Lando, D., and M. L. Ryhiner. 1969. C R Acad Sci III 269:527.
Lee, L., R. L. Armstrong, and K. Nozerian. 1972. Avian Dis 16:799-805.
Lehner, N. D. M., B. C. Bullock, and T. B. Clarkson. 1967. J Am Vet Med Assoc 151:939-41.
Marthedal, H. E., and B. Jylling. 1966. Nord Vet Med 18:565-68.
Mohammed, M. A., S. M. Sokkar, and H. H. Tantawi. 1978. Avian Pathol 7:637-43.
Pastoret, P. P., et al. 1978. Ann Med Vet 122:441-47.
Russell, S. W., and W. R. Shetrone. 1977. University of Calif San Diego Animal Care Facility, Postmortem Record, 9/1/77, 10/12/77.
Saik, J. E. 1985. Proc 14th Pigeon Conference, Cook College, Rutgers University, New Brunswick, N.J.
Saik, J. E., et al. 1986. Avian Dis 30:426-29.
Schwers, A., et al. 1980. J Comp Pathol 90:625-33.
Schwers, A., et al. 1981. Vet Med Univ Liege, Brussels, Belg.
Smadel, J. E., M. J. Wall, and A. Gregg. 1943. J Exp Med 78:189-203.
Smadel, J. E., E. B. Jackson, and J. W. Harman. 1945. J Exp Med 81:385.
Surman, P. G., et al. 1975. Aust Vet J 51:537-38.
Tantawi, H. H., and F. Al Sheikhly. 1980a. Avian Dis 24:595-603.
______. 1980b. Avian Dis 24:455-63.
Tantawi, H. H., M. M. Al Falluji, and M. O. Shony. 1978. 2d Science Conference of the Iraqi Microbiology Society, pp. 2-3.
Tantawi, H. H., M. M. Al Falluji, and F. Al Sheikhly. 1979. Avian Dis 23:785-93.
Tantawi, H. H., M. M. Al Falluji, and H. K. Al Falluji. 1980. Avian Dis 24:1011-15.
Tantawi, H. H., et al. 1981. Avian Dis 25:272-78.
Thompson, E. J., R. C. Gumbrell, and P. R. Watson. 1976. N Z Vet J 25:74.
Vetesi, F., and J. Tanyi. 1975. Magror Allatorv Lapja 3:193-97.
Vindevogel, H., and J. P. Duchatel. 1977. Ann Med Vet 121:193-95.
______. 1979. Ann Med Vet 123:17-27.
Vindevogel, H., and P. P. Pastoret. 1980. J Comp Pathol 90:409-13.
______. 1981. J Comp Pathol 91:415-26.
Vindevogel, H., et al. 1975. Ann Rech Vet 6:431-36.
Vindevogel, H., J. P. Duchatel, and M. Gouffaux. 1977a. J Comp Pathol 87:597-603.
Vindevogel, H., et al. 1977b. J Comp Pathol 87:605-10.
Vindevogel, H., P. P. Pastoret, and G. Burtonboy. 1980. J Comp Pathol 90:401-8.
Vindevogel, H., et al. 1981. Vet Rec 109:285-86.
Vindevogel, H., P. P. Pastoret, and A. Aguilar-Setien. 1982a. J Comp Pathol 92:177-80.
Vindevogel, H., P. P. Pastoret, and P. Leroy, 1982b. J Comp Pathol 92:483-94.

Arbovirus Infections

Definition and Synonyms. Arbovirus infections are infectious, arthropod-borne virus diseases of birds, animals, and humans transmitted by mosquitoes and other vectors. They generally cause subclinical disease in pigeons. Sleeping sickness, encephalitis, encephalomyelitis, and togavirus infections are other names for these infections in pigeons.

Cause and Classification. The viruses are classified as arboviruses. They multiply in both the vertebrate host and in bloodsucking arthropods. There are 21 arboviruses that affect avian species as natural hosts or occasionally as necessary vertebrate hosts (Berge 1975). These are members of the togavirus family and include two subgroup A viruses, eastern and western equine encephalitis (EEE and WEE) and one subgroup B virus, turkey meningoencephalitis, that produce clinical disease in economic bird species (Coleman 1984). St Louis virus, EEE, and WEE tend to produce subclinical disease in pigeons. EEE and WEE are group A togaviruses because they share antigenic components on the basis of hemagglutination inhibition reactions with others of group A but not with members of group B, such as St. Louis virus (Davis et al. 1973).

Nature. Groups A and B togaviruses have the same general spherical appearance, and each contains three proteins. They are single-stranded RNA viruses with icosahedral symmetry and a lipoprotein envelope. They have a dense central nucleocapsid of ribonucleoprotein and an outer membrane covered with fine projections. In size, EEE measures 470 Å and WEE 530 Å. Group B viruses are considerably smaller viruses (Davis et al. 1973).

Distribution

GEOGRAPHY AND HISTORY. Over 350 distinct arboviruses are encountered worldwide (Berge 1975), but few reports involve pigeons. In 1931 Meyer established that EEE in the United States was indeed caused by a virus. In 1933 Giltner and Shahan inoculated four pigeons intracerebrally with 0.2 ml 1:50 dilution of guinea pig brain infected with WEE and produced weakness, ataxia, and marked tremors by the third day and death on day 3 or 4 with paralysis. Traub and Ten Broeck (1935) used pigeons for serial passage of EEE. Fothergill and Dingle (1938) reported the isolation of EEE from the brain of a pigeon presented by a Massachusetts fancier. The bird had been raised on a farm adjacent to where two horses died of the disease. Tyzzer et al. (1938) encountered the first natural outbreak of EEE in ring-necked pheasants in Connecticut. Beaudette (1939) also observed EEE in New Jersey pheasants in 1938. It was not until 1941 that WEE was recovered by Cox et al. (1941) from a prairie chicken in Regly, N. Dak. Later Hetrick (1954) isolated WEE from English sparrows in New Jersey. In 1964 Gainer et al. isolated St. Louis encephalitis from the brain and blood of domestic pigeons during a clinical flock disease, but they could not reproduce the infection.

INCIDENCE. Encephalitis is more common after wet summers when mosquitoes are prevalent.

SEASON. In cooler areas the diseases are seldom experienced until late in the summer or fall after a buildup of infected vectors. The full moon increases marsh flooding and the mosquito population.

AGE, SEX, AND BREED. Birds of any age may contract the viruses. There appears to be no sex or breed difference in incidence.

SPECIES. There is extensive involvement and transmission of EEE and WEE within a broad host and vector spectrum. Encephalitis is a problem because of the great number of infective hosts. It was first reported in 1831 in horses. Campbell (1938) reported infection in Wertenberg, Germany, and Thompson (1931) observed the horse disease in England.

The eastern and western diseases have since been cited by species, with virus isolations and antibody determinations in at least 63 mammalian species, including humans, 1 alligator, 3 frogs, 12 snakes, 12 turtles, 4 lizards, 1 fish, 3 lice, 4 mites, 1 flea, 3 blackflies, 7 ticks, 1 stable fly, 1 assassin bug, 156 birds, and 53 species of mosquitoes. Of 29 reports of EEE or WEE in pigeons, 3 EEE and 4 WEE were antibody-positive but only one had EEE virus isolated (Tudor 1965). During the 1959 outbreak of EEE in New Jersey, 17 pheasant cases were examined and 21 of 33 people died but no pigeon cases were received or reported. This testifies to the lack of clinical pigeon disease.

CARRIERS. To be an effective carrier reservoir of infection, the vertebrate host must be susceptible to infection and must circulate a threshold level of virus sufficient to infect a vector. This means that the virus must survive and multiply within the host so that it is in sufficient concentration to infect a vector. Some hosts develop and circulate so little virus that they are rarely capable of infecting even the most susceptible mosquitoes (Chamberlain 1953). Given an infected host with ample virus, the carrier state and thus the transmission window is essentially limited to the period when virus is present in the bloodstream. This varies with each host species.

Transmission. Transmission of virus from vector to host and host to vector involves the susceptibility of the host, blood virus titers of host and vector, and the feeding habits of the vector. Mosquitoes that feed predominantly on birds such as *Culiseta melanura* and *Culex pipiens,* even though heavily infected, are not potential EEE or WEE hazards for mammals. On the other hand, *Aedes sollicitans, Mansonia per-*

turbans, and *Anopheles quadrimaculatus,* which are mammalian feeders (Crans 1962, 1964), do not contribute to bird-to-bird transmission.

Mosquito infectivity develops as the virus multiplies in the gut for about 2 wk. The virus then collects in the salivary glands of the mosquito and remains infectious for 1 or 2 mo (Davis et al. 1973). In addition to mosquitoes as primary vectors, the recovery of virus from lice, fleas, and assassin bugs and the transmission of virus by mites, blackflies, stable flies, and ticks increases the chances for virus dissemination (Tudor 1965).

Transmission also involves hibernating reptiles or mammals and active year-round residents, such as wild and domestic birds. Mammals permit overwintering of the latent virus and the reinfection of new generations of mosquitoes and other vectors. Some, such as humans and horses, are dead-end hosts. Because of the low titer and short duration of virus in the bloodstream of the host, the virus has virtually no chance of infecting a vector. Aside from vector transmission, virus released from discharges of infected pheasants can be aerosol-transmitted and feather-picking and cannibalism among pheasants may also transmit the virus. This probably would not be a problem with pigeons.

Epidemiology

ENTRY AND EXIT. The viruses enter the body by the wound inflicted by an infected mosquito or other vector. Viremia follows and finally virus enters all tissues including the central nervous system.

The viruses leave the bloodstream of the infected host when the vector bites and extracts a blood meal. This bite must occur, however, during the period when the virus is in the circulating blood.

PATHOGENICITY. Each host species has a minimum threshold above which infection can become well enough established to stimulate an antibody response. The dose may be calculated in chick LD_{50} (the level of virus required to kill 50% of the embryos inoculated). If the salivary glands of the mosquito contain sufficient virus and the host has few neutralizing antibodies, infection becomes established.

As a general statement, EEE usually produces serious disease with high mortality and severe neurological damage in a few individuals. In contrast, WEE usually causes widespread disease but with less severe consequences. Most individuals recover completely or have an abortive disease. This contrast is not observed clinically in pigeons.

IMMUNITY. Antibodies develop in the bloodstream about 7 days after inoculation and persist for some time. The level of hemagglutination-inhibiting antibodies is increased notably following a second encounter 3–6 wk later with the same or a group-related virus (Davis et al. 1973).

Incubation. Clinical disease appears in 3 days following intracranial inoculation of pigeons, but clinical signs seldom occur with natural infection.

Course. Pigeons can die in 2–4 days, but the subclinical disease is seldom recognized.

Morbidity. Most pigeons raised near high vector populations will carry antibody titers.

Mortality. Mortality seldom occurs in pigeon flocks.

Signs. A pigeon with brain infection may show varying clinical signs: walking in circles; stumbling; falling over; retraction of the head over the back; depressing of the head; twisting of the neck; inability to eat, drink, or fly; and lack of sight or recognition.

Necropsy. Gross examination of the viscera of infected birds seldom reveals any notable changes. The spleen may be enlarged and pale. Brain coverings may be engorged.

Micropathology. Tyzzer and Sellards (1941) found changes in inoculated chickens in the brain, heart, gizzard, and sometimes liver. In young chickens cardiac lesions were the most important. These changes could also occur in pigeons, depending on the extent of the infection.

Diagnosis. Diagnosis is based on the recovery of the virus in 10-day-old chicken embryos or in newborn mice. A brain sample may be collected by Pasteur pipette after the skin has been carefully removed from the head and the bony skull is seared with a hot table knife or spatula. The sample is ground under sterile conditions with about 2 ml broth. Embryos are each injected in the chorioallantoic sac with 0.2 ml centrifuged suspension. Acute injection and death of embryos occurs in 24–48 hr. In subclinical cases noncontaminated tissue should also be collected from the liver, spleen, and heart.

The newly isolated virus can be identified by serum neutralization titrations and by hemagglutination inhibition tests with known antisera. Complement fixation tests permit an evaluation of unknown sera collected during the acute phase of the infection in a bird. Such antibodies do not persist; thus the test is not used for flock surveys.

Differential Diagnosis. Botulism, paramyxovirus, and herpesvirus must be excluded.

Biological Properties

HEMATOLOGY. The hemagglutinating component of the viruses probably resides in the glycoprotein spikes, which agglutinate red blood cells. Maximum agglutination occurs at pH 6.4 and 37°C for group A and at pH 6.5–7.0 at 4° or 22°C for group B (Davis et al. 1973).

TISSUE AFFINITY. The virus has an affinity for nerve tissues, including the brain and spinal cord.

INCLUSION BODIES. Inclusion bodies are not observed.

STABILITY. The viruses survive only a short time outside the body or in a dead carcass, but live birds and mammals permit overwintering of the virus in a latent state.

The viruses are resistant to glycerin and weak phenol solutions but are inactivated by lipid solvents, formalin, chloroform, and chlorine. Cold preserves for several months at −20°C or indefinitely at −70°C or if lyophilized. Heat of 60°C for 30 min inactivated the virus (Kelser 1948).

Proteolytic enzymes disrupt the hemagglutinating activity and infectivity of group B togaviruses, but not group A, and these reactions are most stable at pH 8.5 (Davis et al. 1973).

CULTURE MEDIA. Young mice and guinea pigs permit study of the viruses. Animals die in 2–5 days from EEEV and 4–7 days from WEEV.

Tissue culture using chicken and duck embryo fibroblasts and hamster kidney cells also yield good viral growth (Coleman 1984).

Prevention and Control. Lofts should be mosquito-screened to exclude mosquitoes. Vector numbers are reduced by spraying with insecticides. Sick birds should be isolated and dead birds buried under 2 ft of soil. Killed horse vaccines used on pheasants have not been successful in preventing losses but may have reduced mortality in naturally infected flocks.

Treatment. No treatment is available.

References

Beaudette, F. R. 1939. Proc 43d Annual Meeting of the U S Livestock Sanitary Assoc, p. 185.

Berge, T. O., ed. 1975. *International Catalogue of Arboviruses.* Washington, D.C.: U.S. Department of Health, Education and Welfare, Public Health Service.

Campbell, J. N. 1938. North Am Vet 19:31–36.

Chamberlain, R. W. 1953. Proc 40th Meeting of the New Jersey Mosquito Extermination Assoc, pp. 49–52.

Coleman, P. H. 1984. In *Diseases of Poultry,* 8th ed., ed. M. S. Hofstad et al., pp. 576–80. Ames: Iowa State University Press.

Cox, H. R., W. L. Jellison, and L. E. Hughes. 1941. U S Pub Health Rep 56:1905.

Crans, W. J. 1962. Proc 49th Meeting of the New Jersey Mosquito Extermination Assoc, pp. 121–25.

______. 1964. Proc 51st Meeting of the New Jersey Mosquito Extermination Assoc, pp. 51–58.

Davis, B. D., et al. 1973. *Microbiology.* 2d ed. New York: Harper & Row, pp. 1377–97.

Fothergill, L. D., and J. H. Dingle. 1938. Science 88:549–50.

Gainer, J. H., et al. 1964. Am J Trop Med Hyg 13:472–74.

Giltner, L. T., and M. S. Shahan. 1933. Science 78:63.

Hetrick, F. M. 1954. Ph.D. diss., University of Pittsburgh.

Kelser, R. A., and H. W. Schoening. 1948. *A Manual of Veterinary Bacteriology.* 5th ed. Baltimore: Williams & Wilkins.

Meyer, K. F., C. M. Haring, and B. F. Howitt. 1931. J Am Vet Med Assoc 79:376–89.

Thompson, A. 1931. N Engl Farmer 10:108.

Traub, E., and C. Ten Broeck. 1935. Science 81:572.

Tudor, D. C. 1965. Proc 69th Meet of the U S Livestock Sanitary Assoc, pp. 210–53.

Tyzzer, E. E., and A. W. Sellards. 1941. Am J Hyg 33:69.

Tyzzer, E. E., A. W. Sellards, and B. L. Bennett. 1938. Science 88:505.

Avian Influenza

Definition and Synonyms. Avian influenza is a subacute or acute,highly infectious, often fatal virus disease of chickens. It may be characterized by a fatal systemic viremia with edema and hemorrhages or by a mild transient respiratory infection. Other wild birds and pigeons may show no evidence of infection. In pigeons it is considered as an incidental, subclinical disease. Avian influenza is also called fowl plague or fowl peste.

Cause and Classification. Avian influenza is caused by an RNA virus and is classified in the family Orthomyxoviridae. Antigenically there are three types of virus: A, B, and C. Type A viruses have been isolated from humans, swine, and horses and on occasion from birds and other mammals. Subtypes are identified by antigenic activity of both the hemagglutinins (HA) presently designated as HA1–HA13 and the neuraminidase (N) antigens specified as N1–N9.

Types B and C have been isolated from only humans except for one report, in which C was found in swine (Easterday and Beard 1984).

Nature. The single-stranded RNA virus particles are generally ovoid with a diameter of 80–120 nm. Filamentous forms also occur. The entire surface of the particles is covered with evenly spaced glycoprotein spikes or rods 10–12 nm in length that bear HA and N antigens. Beneath this outer zone is a membrane-like lipid envelope that covers the helical nucleocapsid (Easterday and Beard 1984).

Distribution

GEOGRAPHY AND HISTORY. Type A avian influenza viruses (AIV) have been found throughout the world (Alexander 1982). The first report, which called the disease fowl plague, was made by Perroncito in Italy in 1878 (Stubbs 1965). It was observed in Great Britain in 1922 (Alexander 1982). In 1924 it appeared in the New York poultry market (Mohler 1924). In 1924, 1925, and 1929 it was diagnosed in New Jersey chickens by Beaudette et al. (1929). In 1952 Walker and Bannister (1953) recovered a filterable agent in ducks (A/duck/Can/52) that was later identified as type A influenza (Mitchell et al. 1967). In 1955 Schafer showed that fowl plague virus had the same RNA protein as Type A influenza virus. In 1962 infected Canadian turkeys were first found by Lang and Wills (1966). Since then turkeys have been shown to be the most frequently infected domestic species. In 1983 Pennsylvania chickens were infected with a highly pathogenic H5N2 strain. In 1986 two lots of pigeons, one in a New York market and another in Florida, were each found to have a strain of low pathogenicity. Numerous avian influenza serological studies and reports of outbreaks have indicated widespread dissemination of the viruses with migratory ducks as the primary reservoir (Pearson 1987).

INCIDENCE. Between 1964 and 1984 there were at least 65 serotype isolations in the United States alone, involving 22 states (Bankowski 1984).

SEASON. The virus may appear at any time of the year, but most avian cases are observed during the warmer months.

AGE AND SEX. There is a clear indication of increasing resistance with age (Beard et al. 1984). Virus is more readily isolated from juvenile ducks than from adults. Sex does not appear to influence the development of the disease in the various species.

NATURALLY INFECTED SPECIES. Pigeons are seldom infected, but in 1987 virus isolations were reported (Pearson 1987). Chickens were first reported infected in Italy in 1878. Since then influenza has been recognized in turkeys, wild and domestic ducks and geese, guinea fowl, pheasants, coturnix quail, a partridge, seagulls, budgerigars, myna birds (Easterday and Beard 1984), a sparrow, a crow, a jackdaw, sandpipers, wagtails, jays, doves, starlings, a rock partridge, swallows, herons, and terns (Alexander 1982). In addition, influenza viruses that resemble those of avian origin have been recovered from seals (Lang et al. 1981) and humans (Hinshaw and Webster 1982).

ARTIFICIALLY INFECTED SPECIES. Experimentally cats, swine, ferrets, mink, and humans can be infected with avian influenza (Alexander 1982). Narayan et al. (1969) inoculated an Ontario turkey isolate into pigeons, chickens, ducks, and geese. The pigeons, ducks, and geese remained healthy but had serological evidence of infection.

CARRIERS. AIV was isolated from ducks in 1953 and 1956. More isolations have been made from ducks throughout the world than from any other species. Carrier ducks are asymptomatic intermittent shedders of virus, but most chickens are not carriers (Pearson 1987).

Transmission. Direct close contact with infected birds and their discharges is necessary for the spread of some strains of virus, but others do not require close contact (Webster and Laver 1975). This is supported by Pearson (1987), who reported air transmission of 45 m. Wildlife such as crows, blackbirds, starlings, house sparrows, pigeons, mice, and rats are not significant spreaders of AIV (Nettles 1984). Even though pigeons are not considered significant disseminators of AIV, they can carry the virus mechanically. Narayan et al. (1969) has shown that turkey eggs can carry the virus, but there is presently no evidence of this in chicken eggs (Beard et al. 1984).

Epidemiology

DISTRIBUTION IN BODY. The virus may be found throughout the body.

ENTRY AND EXIT. The virus enters by the mouth and respiratory system and leaves in the droppings and respiratory discharges.

PATHOGENICITY. The characteristics of AIV cannot be specified on the basis of HA and N components or the host of origin. Infection or virulence resides in the HA polypeptide, and this is a polygenic character (Rott et al. 1979). Trypsin, which is a proteolytic enzyme when combined with the HA antigen, can cause cleavage or separation of the polypeptide into two nucleotide segments. The virulent virus produces plaques in chicken fibroblast cultures and undergoes cleavage of the HA precursor into HA1 and HA2 in the presence or absence of trypsin. The nonvirulent virus produces plaques, and the HA is cleaved only in the presence of trypsin (Webster et al. 1984). There are amino acid differences between virulent and nonvirulent viruses, and it appears that basic amino acids cluster at the cleavage line (Pearson 1987). This partially explains the presence of virulent and nonvirulent virus in the same area or on the same farm, as in the 1983 Pennsylvania outbreak.

Influenza virus may lack virulence for one species but be virulent for another. The poultry pathogenic HA5N2 virus isolated in Pennsylvania would not reproduce in ducks (Buisch et al. 1984). The same HA5 virus appears to have the same relationship with pigeons, where it produces only a mild or inapparent disease. From the standpoint of infectivity it has also been shown that ducks can frequently carry two or more distinctly different influenza viruses (Hinshaw and Webster 1982).

Incubation. The incubation period is variable, as little as a few hours up to 2–3 days.

Course. The course depends on the strain of virus. Virulent strains kill chickens in just a few days.

Morbidity and Mortality. The extent of the disease and the losses largely depend on the virulence of the virus.

Signs. Respiratory signs predominate in acute cases. Coughing, edema of the head and face, nervous evidence, diarrhea, lowered feed consumption, a definite drop in egg production, and increased mortality are typical of the highly pathogenic strains. With strains of low pathogenicity, the signs are not pronounced. Slightly lowered production, diarrhea, and perhaps an increase in mortality may be noted.

Necropsy. Pigeons, as well as chickens and other birds, may show no evidence of infection. Pathogenic strains in chickens may produce edema of the face and straw-colored fluid in the subcutaneous tissues. Blotch hemorrhages may be present on the shanks of the legs. Other hemorrhages may appear in the trachea, proventriculus, beneath the gizzard lining, and in the fat around the heart.

Diagnosis. Tracheal or cloacal swabs from live or dead birds are useful in virus isolation. Strain identification is best established by hemagglutination inhibition (HAI) and neuraminidase inhibition (NI) tests. Other tests include the agar gel precipitin test, the double diffusion test, the complement fixation test, the serum neutralization test (Beard 1970), and the single radial diffusion test (Dowdle and Schild 1975).

Differential Diagnosis. Because of the variability of diagnostic signs Newcastle disease, herpes, paramyxovirus and chlamydial infections, and mycoplasmosis must be considered.

Biological Properties

STRAINS. All the traditional fowl plague strains are recognized as HA7 strains and include Alexandria, Brescia, Dutch, and Rostock. Neuraminidase activity of these strains differs, however. Brescia has N1 and Dutch has N7 antigens. All of the avian influenza virus strains have been Type A (Easterday and Beard 1984).

Alexander (1982) also reported other virulent HA7 strains. Of the 13H serotypes only HA5 and HA7 viruses are considered highly pathogenic for the avian species (Bosch et al. 1979). Bankowski (1984) identified the following hemagglutinin antigens between 1964 and 1984 (* = incomplete data): turkeys, 1, 2, 3, 4, 5, 6, 7, 8, 9, 10; chickens, 1, 4, 5, 6; ducks,* 3, 4, 5, 6, 11; geese,* 5; guinea fowl,* 5; pheasants, 3, 7; quail, 5; and pigeons, 5.

The HA5 antigen was identified in chickens in Maryland in 1983, along with ducks, geese, and guinea fowl. This was during the 1983 Pennsylvania outbreak in chickens, when virus was recovered from a chukar partridge and a pheasant. At this time antibodies to HA5N2 were detected in wild and domestic ducks, geese, and seagulls but not in 309 pigeons tested in Pennsylvania (Davison 1986).

HEMATOLOGY. Chicken red blood cells are agglutinated by hemagglutinins in serum and egg yolks (Palmer et al. 1975). Egg yolk testing has provided a reliable surveillance tool.

TISSUE AFFINITY. The virus is commonly found in respiratory tissues but may be found in blood, spleen, bone marrow, and cloaca.

STABILITY. AIV can survive in moist feces for several days at 25°C (Beard et al. 1984) and as long as 30 days at 4°C (Webster et al. 1978). It is relatively stable at pH 7–8 but is destroyed by acid (Lang et al. 1968). It is destroyed by drying (Pearson 1987).

CULTURE MEDIA. The virus can be isolated in 10-day-old chicken embryos by the allantoic route. Some embryos may die in 48 hr; others survive. Fluids from dead embryos and chilled surviving embryos must be harvested and tested for red cell hemagglutinins.

Highly pathogenic chicken strains have been identified by inoculation of 4- to 8-wk-old chickens by intravenous, intramuscular, or caudal air sac methods and by observing mortality in excess of 75%.

Cell cultures of chicken embryo fibroblasts and monkey kidney cells have been used to distinguish between pathogenic and highly pathogenic strains. Plaque formation in the absence of trypsin is characteristic of highly pathogenic strains (Bosch et al. 1979), but there are considerable strain differences.

Prevention and Control. Eradication is impossible. The disease exists in the migratory bird population. It is thus difficult to interrupt the cycle of infection. Infected live domestic and migratory birds are the primary source of infection. Swine may also be a possible source of infection for turkeys (Hinshaw et al. 1983).

Pigeons can mechanically carry live virus on their feet, but presently pigeons do not appear to contribute to the spread of the disease. The movement of live birds (including wild birds), personnel, vehicles, equipment, animals, flies, and rodents serve as a means of transmission. Control efforts must be directed toward limiting such movement. Disinfecting vehicles and premises with standard disinfectants following cleaning is essential. Premise security must be a part of any control program. Pet shops, live poultry markets, supply houses, race headquarters, and pigeon transport vehicles all contribute to the spread of infection. Contamination of clothing, crates, and birds at these locations can introduce infection.

Present plans have generally excluded the use of vaccines because antibodies derived from a vaccine cannot be distinguished readily from those generated by the natural disease. This makes it difficult to evaluate a vaccination program by testing. Secondly, an effective vaccine must contain antigenic components that are the same or similar to the field strain of virus. There are many possible antigenic combinations. Therefore, it is almost impossible to match a field strain before it appears.

Successful tests have been conducted, however, using a killed polyvalent vaccine (Brugh et al. 1979). Beard and Easterday (1973) also vaccinated chickens with a live turkey strain.

Treatment. No treatment is indicated.

References

Alexander, D. J. 1982. Vet Bull 52:341-59.

Bankowski, R. A. 1984. Proc 88th Annual Meeting of the US Animal Health Assoc, p. 282.

Beard, C. W. 1970. Bull WHO 42:799-86.

Beard, C. W., and B. C. Easterday. 1973. Avian Dis 17:173-81.

Beard, C. W., M. Brugh, and D. C. Johnson. 1984. Proc 88th Annual Meeting of the U S Animal Health Assoc, pp. 462-73.

Beaudette, F. R., C. B. Hudson, and A. H. Saxe. 1929. J Agric Res 49:83-92.

Bosch, F. X., et al. 1979. Virology 95:197-207.

Brugh, M., C. W. Beard, and H. D. Stone. 1979. J Am Vet Med Assoc 171:1105.

Buisch, W. W., A. E. Hall, and H. A. McDaniel. 1984. Proc 88th Annual Meeting of the U S Animal Health Assoc, pp. 430-46.

Davison, S. 1986. Personal communication.

Dowdle, W. R., and G. C. Schild. 1975. In *The Influenza Viruses and Influenza,* ed. E. D. Kilbourne, pp. 243-64. New York: Academic.

Easterday, B. C., and C. W. Beard. 1984. In *Diseases of Poultry,* 8th ed., ed. M. S. Hofstad et al., pp. 482-96. Ames: Iowa State University Press.

Hinshaw, V. S., and R. G. Webster. 1982. In *Basic and Applied Influenza Research,* ed. A. S. Beare, pp. 79-104. Boca Raton: CRC.

Hinshaw, V. S., et al. 1983. Science 220:206-8.

Lang, G., and C. G. Wills. 1966. Arch Gesamt Virusforsch 19:81-90.

Lang, G., A. Gagnon, and J. R. Geraci. 1981. Arch Virol 68:189-95.

Lang, G., et al. 1968. Can Vet J 9:22-29.

Mitchell, C. A., L. F. Guerin, and J. Robillard. 1967. Can J Comp Med Vet Sci 31:103-5.

Mohler, J. R. 1924. USDA, BAI Rep.

Narayan, O., G. Lang, and B. T. Rouse. 1969. Arch Gesamt Virusforsch 26:266-82.

Nettles, V. F. 1984. Southeastern Cooperative Wildlife Disease Study, University of Georgia, Athens.

Palmer, D. F., et al. 1975. *Immunology, Ser 6, Procedural Guide.* Atlanta, Ga.: Centers for Disease Control.

Pearson, J. E. 1987. Proc 59th Avian Disease Conference, Atlantic City, N.J. p. 38.

Rott, R., M. Orlich, and C. Scholtissek. 1979. J Gen Virol 14:471-77.

Schafer, W. 1955. Z Naturforsch 106:81-91.

Stubbs, E. L. 1965. In *Diseases of Poultry,* 5th ed. H. E. Biester and L. H. Schwarte, pp. 813-22. Ames: Iowa State University Press.

Walker, R. V. L., and G. L. Bannister. 1953. Can J Comp Med Vet Sci 17:248-50.

Webster, R. G., and W. G. Laver. 1975. In *The Influenza Viruses and Influenza,* ed. E. D. Kilbourne, pp. 270-310. New York: Academic.

Webster, R. G. et al. 1978. Virology 84:268-76.

Webster, R. G., et al. 1984. Proc 88th Meeting of the U S Animal Health Assoc, pp. 447-55.

Avian Adenovirus Infections

Definitions and Synonyms. Avian adenovirus has been recovered from pigeons with clinical respiratory or liver diseases, but the true relationship of the virus to the disease condition has not been established. This infection has also been called pigeon inclusion body hepatitis and pigeon bronchitis.

Cause and Classification. The virus is listed as family Adenoviridae, genus *Aviadenovirus.*

There are at least 12 known avian serotypes (Cowan and Naqi 1982). Individual avian enteric viruses within the serotype have been isolated from healthy, as well as diseased, chickens and more than one agent may be present in a flock (Burke et al. 1959). Actually, serologically related isolants have differed in their ability to produce disease (Fadly and Winterfield 1975). Yates et al. (1976) found avian adeno-associated viruses (AAV) with adenovirus isolants. The defective single-stranded DNA AAV agents, 200–250 Å in diameter, are dependent on the replication of unrelated adenoviruses for their multiplication. Further, when the AAV is assisted by an adenovirus, the multiplication of the adenovirus itself is inhibited (Davis et al. 1973). This may mediate the pathogenicity of the adenovirus and explain the presence of silent infections in flocks.

Only serotype 8 adenovirus has been reported isolated from pigeons (McFerran et al. 1976; McFerran and Adair 1977). Taylor and Calnek (1962) studied fowl virus 1359A-4, later called A2, which was classified as serotype 8 by cross-neutralization (Calnek and Cowan 1975). This is mentioned because further studies of this virus may yield more information concerning type 8 viruses and pigeon isolates. Future studies, however, may reveal that pigeons contract serotypes other than type 8.

Nature. The CELO adenovirus is a double-stranded DNA virus that is an isometric, naked (nonenveloped), cubic icosahedral capsid 69–76 nm in diameter having 252 polygonal capsomeres (Dutta and Pomeroy 1963). The outer capsid cover encloses a dense protein central core 40–45 nm in diameter called a nucleoid (Yates 1975). Twelve of the capsomers situated at the vertices or corners of the icosahedron bear *pentons* (filamentous fibers with terminal knobs). Pentons detach as the virus penetrates the host cell wall, and virus replication occurs after the viral core enters the nucleus (Davis et al. 1973). The virus particles isolated from pigeons were 68–77 nm in diameter (McFerran et al. 1976).

Distribution

GEOGRAPHY AND HISTORY. Adenoviruses of birds appear to be ubiquitous. McFerran et al. (1972) found that 6 of 7 serologically distinct viruses recovered from birds in Northern Ireland were identical to 6 of 8 serotypes reported by Kawamura et al. (1964) in Japan. Calnek and Cowan (1975) noted that the serotype spectra worldwide was quite similar and 9 of 10 of the then-described serotypes were isolated in the United States.

Type 2 virus causing hemorrhagic enteritis (HE) was first recognized in turkeys by Pomeroy and Fenstermacher (1937) in Minnesota. In 1949 Olson (1950) first isolated type 1 avian adenovirus from West Virginia quail with bronchitis. Yates et al. (1954) and Yates and Fry (1957) made the first isolation of type 1 chicken embryo lethal orphan (CELO) virus from endogenously infected chick embryos. This virus is now considered the cause of quail bronchitis (Du Bose and Grumbles 1959). CELO virus was also reported as the cause of pancreatitis in guinea fowl (Pascucci et al. 1970) and a systemic infection in geese (Csontos 1967).

Fontes et al. (1958) in Michigan recovered the type 2 unrelated gallus adeno-like (GAL) virus from a case of lymphomatosis in chickens. Inclusion body hepatitis and anemia syndrome were first described by Helmboldt and Frazier (1963) and later by Stein and Wills (1974) as hemorrhagic aplastic anemia. Domermuth et al. (1975) and Domermuth and Gross (1975), working with type 2 turkey hemorrhagic enteritis, recognized this virus as the cause of marble spleen of pheasants and splenomegaly of chickens. In 1976 Van Eck et al. described egg drop syndrome, which has been recovered from healthy ducks in the United States by Villegas et al. (1979) and in Great Britain by Baxendale (1978). It appears to be unrelated to 11 fowl and 2 turkey adeno prototypes (Adair et al. 1979) but has 7 of 13 polypeptides found in the Phelps strain of CELO type 1 adenovirus (Todd and McNulty 1978).

McFerran et al. (1976) and McFerran and Adair (1977) in Ireland isolated fowl adenovirus serotype 8 from pigeons. This appeared to be the first recorded pigeon isolation. However, a report by Goryo et al. (1988) takes issue with this statement. They report the isolation of adenovirus particles and liver inclusion bodies in four white pigeons in Japan in 1986. AGP antibody against adenovirus was only detected in the serum from pigeon number 4, which had severe kidney damage. Virus particles recovered from these birds were 80–100 nm in diameter and had a round or hexagonal outline. The virus failed to produce cytopathic effects in chick embryo fibroblasts or pock lesions on chorioallantoic membranes. Both McFerran et al. (1976) and Coussement et al. (1984) previously reported adenovirus as the cause of inclusion body hepatitis in pigeons.

INCIDENCE. The clinical disease in pigeons is uncommon. Thus the details of its distribution in this species need further study.

SPECIES. The work by McFerran et al. (1976) indicates that serotype 8 virus was found in pigeons, chickens, ducks, and budgerigars. The strain of virus was not delineated, and the host of origin was not determined.

CARRIERS. The latency of the virus in pigeons is unknown.

Transmission. Airborne dust transmission of fecal and respiratory discharges and mechanical transfer of the virus may serve to transmit infection. McFerran et al. (1976) noted pigeon infection following close contact with chickens, but there was no proof that infection actually came from the chickens. McFerran and Adair (1977) also commonly observed transmission through the egg in chickens. Even though pigeons are susceptible to fowl serotypes, they are probably not a major disseminator of the virus.

Epidemiology

DISTRIBUTION IN BODY. Once infected, the adenovirus spreads throughout the body but appears to stop at the blood-brain barrier, except for the Phelps strain of CELO, which is proposed as the type strain (Winterfield and Du Bose 1984).

ENTRY AND EXIT. The digestive and respiratory tracts permit viral infection and shedding.

PATHOGENICITY. Most strains of adenovirus do not initiate disease by themselves. They are more likely to participate as an intercurrent disease agent (McFerran and Adair 1977). From the standpoint of pathogenicity, it is of interest that Sarma et al. (1965) inoculated CELO virus into newborn hamsters with the development of fibrosarcomas. Further, at least nine human adenoviruses produce tumors in newborn hamsters, rats, and mice (Davis 1973). Although serotype 8 was isolated from pigeons, the role of this adenovirus in disease remains to be established (McFerran and Adair 1977). Bergmann and Kiupel (1982), however, suggest that parvovirus may serve as an adeno-associated virus.

Signs. Adenoviruses affect domestic birds with a wide spectrum of pathogenic manifestations. Respiratory signs were recorded for strain VF71-327 in pigeons (McFerran et al. 1976).

Necropsy. Inclusion body hepatitis with death in 24 hr was noted by McFerran et al. (1976) for strain VF74-282. Diarrhea was also present. Bergmann and Kiupel (1982) reported duodenal and jejunal inflammation with INI in enterocytes.

Diagnosis. Immunofluorescent antibody, virus neutralization, complement fixation (CF), immunodiffusion, and agar gel precipitin tests have been used in diagnosis.

Differential Diagnosis. The infection must be distinguished from herpes, mycoplasmosis, paramyxovirus infection, vibrionic hepatitis, and drug toxicities. Mixed infection with reovirus must also be considered.

Biological Properties

STRAINS. McFerran et al. (1976) isolated adeno strains VF71-327 and VF74-282 from separate pigeons. Both were type 8.

HEMATOLOGY. The agglutination of red blood cells (RBC) varies with the strain of virus. For example, fowl adenovirus type 1 CELO causes agglutination of rat RBC but not those of sheep or chicken (Burke et al. 1968). Indiana C type 1, on the other hand, agglutinates RBC of sheep and rat (Winterfield et al. 1973).

SEROLOGY. Avian group antigens are distinct from mammalian group antigens in that complement-fixing antigens are not shared. Chicken adenoviruses share group antigens that are involved in CF and immunodiffusion tests (Fenner et al. 1974).

INCLUSION BODIES. Intranuclear liver cell inclusions were found in pigeons by McFerran et al. (1976) and Goryo et al. (1988). Coussement et al. (1984) described them as Cowdry type A inclusions. Bergmann and Kiupel (1982) and Coussement et al. (1984) found inclusions in enterocytes in pigeons they examined.

STAINING. Burke et al. (1968) studied 7 avian isolates and observed basophilic intranuclear inclusions with hematoxylin-eosin stain. Feulgen and acridine orange also stained the inclusions. Goryo et al. (1988) reported both basophylic and eosinophilic inclusions in liver and kidney cells.

STABILITY. Adenoviruses are resistant to lipid solvents and are fairly stable at 56°C and pH 3 (Yates 1975). In addition, the A2 virus studied by Calnek and Cowan (1975) was unaltered by chloroform.

CULTURE MEDIA. Chick embryos are resistant to all but avian adenoviruses (Davis 1973). Tissue culture using chicken kidney cells permits satisfactory growth, but round refractile cells develop with the pigeon virus (McFerran et al. 1976).

Prevention and Control. No avian adenovaccines are available for the prevention of the disease in pigeons. Pet shop pigeons in contact with ducks, budgerigars, and chickens should be avoided.

Treatment. Only supportive treatment is considered helpful.

References

Adair, B. M., et al. 1979. Avian Pathol 8:249–64.

Baxendale, W. 1978. Vet Rec 102:285–86.

Bergmann, V., and H. Kiupel. 1982. Arch Exp Vet Med 36:445–53.

Burke, C. N., R. E. Luginbuhl, and E. L. Jungherr. 1959. Avian Dis 3:412–18.

Burke, C. N., R. E. Luginbuhl, and L. F. Williams. 1968. Avian Dis 12:483–505.

Calnek, B. W., and B. S. Cowan. 1975. Avian Dis 19:91–103.

Coussement, W., et al. 1984. Vlaam Diergeneeskd Tijdschr 53:277–83.

Cowan, B. S., and S. Naqi. 1982. J Am Vet Med Assoc 181:283.

Csontos, L. 1967. Acta Vet Hung 17:217–19.

Davis, B. D., et al. 1973. *Microbiology.* 2d ed. Hagerstown, Md.: Harper & Row, pp. 1222–37.

Domermuth, C. H., and W. B. Gross. 1975. Avian Dis 19:657–65.

Domermuth, C. H., et al. 1975. J Wildl Dis 11:338–42.
Du Bose, R. T., and L. C. Grumbles. 1959. Avian Dis 3:321–44.
Dutta, S. K., and B. S. Pomeroy. 1963. Proc Soc Exp Biol Med 114:539–41.
Fadly, A. M., and R. W. Winterfield. 1975. Am J Vet Res 36:532–34.
Fenner, F., et al. 1974. *The Biology of Animal Viruses.* 2d ed. New York: Academic.
Fontes, A. K., et al. 1958. Proc Soc Exp Biol Med 94:712–17.
Goryo, M., et al. 1988. Avian Pathol 17:391–401.
Helmboldt, C. F., and M. N. Frazier. 1963. Avian Dis 7:446–450.
Kawamura, H. F., F. Shimizu, and H. Tsubahara. 1964. Natl Inst Anim Health Q 4:183–93.
McFerran, J. B., and B. Adair. 1977. Avian Pathol 6:189–217.
McFerran, J. B., J. K. Clarke, and T. J. Connor. 1972. Arch Gesamt Virusforsch 39:132–39.
McFerran, J. B., T. J. Connor, and R. M. McCracken. 1976. Avian Dis 20:519–24.
Olson, N. O. 1950. Proc 54th Annual Meeting of the US Livestock Sanitary Assoc, pp. 171–74.
Pascucci, A., et al. 1970. Atti Soc Ital Sci Vet 24:667–68.
Pomeroy, B. S., and R. Fenstermacher. 1937. Poult Sci 16:378–82.
Sarma, P. S., R. J. Huebner, and W. Thone. 1965. Science 149:1108.
Stein, G., and C. R. Wills. 1974. J Am Vet Med Assoc 165:742.
Taylor, P. J., and B. W. Calnek. 1962. Avian Dis 6:51–58.
Todd, D., and M. S. McNulty. 1978. J Gen Virol 40:63–75.
Van Eck, J. H., et al. 1976. Avian Pathol 5:261–72.
Villegas, P., et al. 1979. Avian Dis 23:507–14.
Winterfield, R. W., and R. T. Du Bose. 1984. In *Diseases of Poultry,* 8th ed., ed. M. S. Hofstad et al., pp. 508–10. Ames: Iowa State University Press.
Winterfield, R. W., A. M. Fadly, and A. M. Gallina. 1973. Avian Dis 17:342–44.
Yates, V. J. 1975. In *Isolation and Identification of Avian Pathogens,* ed. S. B. Hitchner et al. Ithaca, N.Y.: Arnold.
Yates, V. J., and D. E. Fry. 1957. Am J Vet Res 17:657–60.
Yates, V. J., D. E. Fry, and B. Wasserman. 1954. Proc 26th Northeast Conference of Laboratory Workers, Pullorum Control [Abstr].
Yates, V. J., et al. 1976. Avian Dis 20:146–52.

Reovirus and Rotavirus Infections

Definition. Reoviruses and rotaviruses have been important causes of disease in domestic animals, birds, and humans and have also been identified in pigeons, but information is incomplete.

Cause and Classification. The name *reovirus* (respiratory enteric orphan) was proposed in 1959 for a group of viruses affecting the intestine and respiratory tract. They are listed as family Reoviridae, and genera *Reovirus, Orbivirus* (ring), and *Rotavirus* (wheel) (Fenner et al. 1987).

Reoviruses are respiratory and intestinal viruses hosted by many animals and humans. Orbiviruses are actually arboviruses of animals in which the wood ticks serve as vectors and reservoir. Rotaviruses are enteric pathogens of domestic animals and children (Kucera and Myrvik 1985).

Eleven avian reovirus serotypes have been reported by Wood et al. (1980). Kawamura et al. (1965) recognized 5 avian serotypes. One serotype was identified in pigeons by McFarren et al. (1976). Infectious tenosynovitis (viral arthritis) (Olson 1984), infectious bursal disease (IBD) (Lukert and Hitchner 1984), and malabsorption syndrome (Van der Heide et al. 1981), all reoviruses of chickens, have not been reported in pigeons. Vindevogel et al. (1981), however, have reported pigeon antibodies to *Rotavirus.*

Nature

FORM. The virus family is characterized by a virion composed of 10 segments of double-stranded RNA and protein. The virions have icosahedral symmetry, are nonenveloped, spherical, and about 70 nm in diameter with inner and outer concentric capsids (Davis et al. 1973). The outer capsid of IBD is composed of 32 capsomers (Hirai et al. 1979). The core within the capsid contains the viral nucleic acid. McFerran et al. (1976) determined the pigeon reovirus to be 70 nm in diameter with an inner capsid 50 nm in diameter.

REPRODUCTION. Reovirus reproduces in various tissues of the body. The principal site of rotavirus replication occurs in the cytoplasm of epithelial cells of the small intestine (McNulty 1984).

Distribution

GEOGRAPHY AND HISTORY. The first isolation of a reovirus was made from chickens by Fahey and Crawley in 1954. In West Virginia Olson et al. (1957) recognized the first case of viral arthritis (VA) in chickens, in which *Mycoplasma synoviae* was also isolated. Ruptured hemorrhagic gastrocnemius tendons have often been observed in VA infections of chickens. Phillips et al. (1970) reported negative results in attempts to establish reovirus infection in pigeons, but McFerran et al. (1976) in Ireland isolated strain VF72-106 from a pigeon with diarrhea. This appears to be the only isolation report involving pigeons.

Bergeland et al. (1977) first found rotavirus in watery droppings of poults in the United States, and McNulty (1984) in Britain recovered the virus from chickens. In a serological study of pigeons in Belgium, Vindevogel et al. (1981) used a bovine strain of rotavirus as an antigen to detect pigeon antibodies and found 10.7% of 75 pigeons positive.

INCIDENCE. The economic significance of reo- and rotavirus infection in pigeons has not been evaluated. Generally once a virus is identified in a species, continued awareness by diagnosticians usually detects the extent of infection. It would appear that both viruses are more prevalent than reports indicate.

SPECIES. Reoviruses can be isolated from diseased and healthy birds (Takehara et al. 1987). They have been recognized in turkeys with signs of bluecomb (Wooley and Gratzek 1969), in geese with influenza signs (Csontos and Csatari 1967), and in muscovy ducks (Gaudry and Tektoff 1973). Attempts to establish VA reovirus in other than chickens and turkeys have failed (Olson 1984). Other reoviruses are widespread in domestic and wild animals and humans.

Chickens, turkeys, ducks, guinea fowl, and pigeons are naturally infected with rotavirus (McNulty 1984). In addition, foals, lambs, calves, piglets, monkeys, infant mice, and humans may host rotaviruses, but whether the strains cross to avian species and vice versa remains for further study.

CARRIERS. Since VA reovirus persists in chickens for at least 289 days, the carrier state is a factor in transmission (Olson and Kerr 1967), but there is no indication that birds are carriers of rotavirus or that biologic vectors occur (McNulty 1984).

Transmission. Direct and indirect contact serve to spread VA infection in chickens (Olson 1984). Transmission of chicken rotavirus, on the other hand, is enhanced because it survives in the feces for months (Woode and Bridger 1975).

Epidemiology

DISTRIBUTION IN BODY. Reovirus of pigeons has been found in the intestine. The actual distribution of pigeon rotavirus is unknown, but in chickens the virus may be found in the intestine and the liver.

ENTRY AND EXIT. Specific information concerning infection in pigeons is not available.

IMMUNITY. Even though circulating neutralizing antibodies develop in both diseases, protection may not occur because the antibodies may not be in contact with the virus growing in locations such as the lumen of the intestine.

Incubation. Olson (1984), in describing VA in chickens, indicated an incubation period of 24 hr for footpad inoculation and longer for intramuscular or intravenous methods. Chicken rotavirus infection is detectable in the feces in 48 hr (Pascucci et al. 1981). The incubation period for both viruses is short.

Signs. Both reo- and rotaviruses can produce enteritis and watery diarrhea in pigeons. This may result in loss of body weight and thin breasts. The pigeon from which the VF72-106 strain was isolated had diarrhea.

Micropathology. Chicken rotaviruses thicken and blunt the villi and dilate the crypts of the intestine (McNulty 1984).

Diagnosis. Identification of reoviruses may be made by complement fixation (CF) and hemagglutination inhibition or virus neutralization tests (Davis et al. 1973). According to Wood et al. (1980), CF and agar gel precipitin tests failed to distinguish between the three mammalian serotypes, but virus neutralization tests were effective. Hsiung (1982) used the enzyme-linked immunosorbent assay (ELISA) and the immunofluorescent antibody test to demonstrate the presence of reovirus.

McNulty (1984), however, reported the use of electron microscopy of feces in the identification of rotavirus. He also demonstrated antibodies by the ELISA, immunofluorescence, and neutralization tests, and counterimmunoelectric osmophoresis. Since avian rotaviruses are antigenically not directly related to mammalian rotaviruses, pigeon rotavirus antigens should be used in serum surveys involving pigeons. Rotavirus antibody appears to be widespread; thus serology alone can not identify the cause of a disease. The agent must be isolated and identified.

Differential Diagnosis. Other causes of diarrhea must be considered.

Biological Properties

STRAINS. McFerran et al. (1976) isolated pigeon reovirus VF72-106.

SEROLOGY. The pigeon reovirus shares a common antigen with fowl reoviruses (McFerran et al. 1976).

McNulty et al. (1979) showed that some but not all avian rotaviruses share common antigenic determinants with mammalian rotaviruses.

TISSUE AFFINITY. The viruses may be found in pigeon feces or intestinal tissues.

INCLUSION BODIES. Reovirus cytoplasmic inclusions stain greenish yellow with acridine orange stain (Davis et al. 1973).

STABILITY. Since the reovirus family lacks lipid the individual viruses are resistant to lipid solvents. They are also fairly stable at pH 3 (Davis et al. 1973).

CULTURE MEDIA. Reovirus VA may be isolated in chicken kidney cell cultures. VA virus also grows readily in embryos following yolk sac inoculation. Blind passages may be necessary for initial isolations (Olson 1984).

Chick kidney cell cultures have been used with some difficulty in the growth of rotaviruses. Since most rotaviruses on primary isolation are not pathogenic for cells, immunofluorescence must be used to identify viral growth (McNulty 1984).

Prevention. Routine sanitary precautions and avoidance of live bird traffic are essentially the only means of prevention for either virus.

Treatment. No treatment is indicated except supportive therapy.

References

Bergeland, M. E., J. P. McAdaragh, and I. Stotz. 1977. Proc 26th Western Poultry Disease Conference, pp. 129–30.

Csontos, L., and M. M-K. Csatari. 1967. Acta Vet Acad Sci Hung 17:107–14.

Davis, B. D., et al. 1973. *Microbiology.* 2d ed. Hagerstown, Md.: Harper & Row, pp. 1399–1408.

Fahey, J. E., and J. F. Crawley. 1954. Can J Comp Med 18:13–21.

Fenner, F., et al. 1987. *Veterinary Virology.* New York: Academic.

Gaudry, D., and J. Tektoff. 1973. Proc 5th International Congress, Muench World Veterinary Poultry Assoc, 2:1400–2.

Hirai, K., et al. 1979. J Virol 32:232.

Hsiung, G. D. 1982. *Diagnostic Virology.* 3d ed. New Haven, Conn.: Yale University Press.

Kawamura, H., et al. 1965. Natl Inst Anim Health Q Tokyo 5:115–24.

Kucera, L. S., and O. N. Myrvik. 1985. *Fundamentals of Medical Virology.* 2d ed. Philadelphia: Lea & Febiger.

Lukert, P. D., and S. B. Hitchner. 1984. In *Diseases of Poultry,* 8th ed., ed. M. S. Hofstad et al., pp. 566–76. Ames: Iowa State University Press.

McFerran, J. B., T. J. Connor, and R. M. McCracken. 1976. Avian Dis 20:519–24.

McNulty, M. S. 1984. In *Diseases of Poultry,* 8th ed., ed. M. S. Hofstad et al., pp. 580–85. Ames: Iowa State University Press.

McNulty, M. S., et al. 1979. Arch Virol 61:13–21.

Olson, N. O. 1984. In *Diseases of Poultry,* 8th ed., ed. M. S. Hofstad et al., pp. 560–66. Ames: Iowa State University Press.

Olson, N. O., and K. M. Kerr. 1967. Avian Dis 11:578–85.

Olson, N. O., D. C. Shelton, and D. A. Munro. 1957. Am J Vet Res 18:735–39.

Pascucci, S., M. E. Misciattelli, and G. Giovanetti. 1981. Abstr 8th International Congress of the World Veterinary Poultry Assoc, p. 57.

Phillips, P. A., N. F. Stanley, and M. Walters. 1970. Aust J Exp Biol Med 48:277–84.

Takehara, K., et al. 1987. Avian Dis 31:730–34.

Van der Heide, L., D. Lutticken, and M. Horzinek. 1981. Avian Dis 25:847–56.

Vindevogel, H., et al. 1981. Vet Rec 109:285–86.

Wood, G. W., et al. 1980. J Comp Pathol 90:29–38.

Woode, G. N., and J. C. Bridger. 1975. Vet Rec 96:85–88.

Wooley, R. E., and J. B. Gratzek. 1969. Am J Vet Res 30:1027–33.

Rubella

Definition and Synonym. Rubella causes a red rash of the face and neck in humans. Women in their first trimester of pregnancy may give birth to babies with congenital defects. In pigeons rubella is considered a common subclinical disease. Rubella is also called German measles.

Cause and Classification. The virus tentatively resides in the family Togaviridae.

Nature. The RNA virion, 500–850 Å in diameter has a spherical core of 300 Å and a loose envelope. Small spikes project from the envelope (Davis et al. 1973).

Distribution

GEOGRAPHY AND HISTORY. In 1941 Gregg, an Australian, observed the congenital defects in babies born by infected women. In 1976 Plissier and Andre in France identified rubella antibodies in pigeons. Heffels (1980), in a broad study, also detected antibodies in pigeons in Germany and suggested that pigeons may be a reservoir for human infection. In 1981 Fritzsche et al. included rubella in a review of pigeon viruses.

INCIDENCE. Plissier and Andre (1976) reported hemagglutination antibodies to the RNA virus in 70% of 126 captive pigeons in France.

SPECIES. Rubella has been reported in humans, primates, and pigeons. Further research may reveal other hosts.

CARRIERS. A carrier state is considered likely.

Transmission. It is highly contagious, and nasal discharges may be airborne for short distances (Davis et al. 1973).

Epidemiology

ENTRY AND EXIT. Nasal discharges probably spread the virus and inhalation introduces the infection.

PATHOGENICITY. Rubella is less contagious than measles.

IMMUNITY. One attack protects people for life.

Incubation. The period is 14–21 days in people (Lyght 1966).

Course. In humans, transmission occurs for 7 days after the rash is noted (Davis et al. 1973).

Signs. The disease is often inapparent in humans. This is also true of pigeons, as reported by Plissier and Andre (1976), who observed no signs of infection after experimental inoculation.

Diagnosis. A complement fixation test provides a reliable serologic diagnosis.

Differential Diagnosis. Rubella has been confused with measles, scarlet fever, and picornavirus infections.

Biological Properties

STRAINS. Only a single antigenic type of virus has been observed (Davis et al. 1973).

SEROLOGY. In people, neutralizing and hemagglutinating antibodies develop along with the rash. Immunity also forms and lasts for years (Davis et al. 1973). Plissier and Andre (1976) reported rubella-specific antibodies following experimental subcutaneous infection of pigeons.

STABILITY. Rubella virus is rapidly inactivated by ether and chloroform. It is fairly stable at −60 to −70°C but sensitive at 4°C (Davis et al. 1973).

CULTURE MEDIA. Initial virus isolation is best conducted in grivet monkey kidney cell cultures or in rabbit or hamster cells (Davis et al. 1973).

Prevention, Control, and Treatment. Since the disease in pigeons does not manifest itself clinically, no recommendations are indicated.

References

Davis, B. D., et al. 1973. *Microbiology.* 2d. ed. Hagerstown, Md.: Harper & Row, pp. 1353–60.

Fritzsche, von K., U. Heffels, and E. F. Kaleta. 1981. Dtsch Tieraerztl Wochenschr 88:72–76.

Gregg, N. M. 1941. Trans Ophthalmol Soc Aust (BMA) 3:35.

Heffels, U. 1980. Ph.D. diss., Veterinary College of Hannover, Hannover, Germany.

Lyght, C. E. 1966. *The Merck Manual.* 11th ed. Rahway, N.J.: Merck.

Plissier, M., and M. Andre. 1976. Bull Acad Nat Med (Paris) 160:224–27.

Rabies

Definition and Synonym. Rabies is an acute infectious viral disease that causes signs of central nervous disturbance, paralysis, and death in most animals. Spontaneous natural infection seldom occurs in pigeons. Hydrophobia is a synonym.

Cause and Classification. The virus is placed in the family Rhabdoviridae.

Nature. The ether-sensitive, single-stranded, bullet-shaped RNA virus has a helical nucleocapsid (Davis et al. 1973).

Distribution

HISTORY. In 1880 Pasteur in France recognized the infectious, nonbacterial nature of the agent. In 1903 Remlinger reported that rabies virus passed bacteria-retaining filters.

SPECIES. All animals are considered susceptible to rabies. Zurn (1882) cited Spinola and Halot, who indicated that by the end of the century rabies had been observed in turkeys, ducks, and chickens. Kraus and Clairmont (1900) found that falcons, ravens, and old pigeons could not be infected, but old pigeons could be infected after a period of starvation. Young pigeons were susceptible. Lote (1904) infected a mouse hawk (*Buteo vulgaris*) with a rabid rabbit brain and transmitted rabies to eagle owls. He considered chickens and pigeons to be more resistant than birds of prey. Remlinger and Bailly (1929) induced infection in a chicken by the bite of a rabid dog. They (1936) also inoculated *street virus* (naturally occurring rabies virus found in the saliva of animals) intracerebrally in a stork (*Ciconia ciconia*) and produced paralysis. In 1938 Jacotot transmitted rabies to a pheasant. Fritzsche et al. (1981), citing Gough and Jorgenson (1976), reported that pigeons could be experimentally infected.

Transmission. Bite wounds caused by dogs, bats, or other animals infected with rabies may establish infection in birds or animals. This is the way that pigeons usually become infected. Rubber gloves must be worn when examining or handling such specimens.

Epidemiology

IMMUNITY. Marie (1904) unsuccessfully attempted to hyperimmunize mature pigeons. More recent research has involved the Flury strain in vaccine production for humans.

Incubation. The period is variable. Clinical disease may occur within 2 wk in owls and geese and longer in chickens (Schwarte 1965). In animals the incubation may take 6 days or not be observed for a year (Davis et al. 1973).

Signs. Incoordination, tremors, convulsions, paralysis, and death may be observed. Kraus and Clairmont (1900) noted slow recovery in some chickens.

Diagnosis. Diagnosis depends on the identification of Negri bodies in brain tissue. Fluorescent antibody techniques enable confirmation.

Biological Properties

STRAINS. After 50 intracerebral serial passages in rabbits the virus becomes fixed in its virulence, producing death of rabbits in 6–7 days.

SEROLOGY. All rabies viruses appear to be of a single immunological type (Davis et al. 1973).

TISSUE AFFINITY. The virus primarily invades nerve tissue.

INCLUSION BODIES. Cytoplasmic inclusions (Negri bodies) may be found in the hippocampus major of the brain of infected animals and birds.

STABILITY. Lye 1-2% is rapidly viricidal (Kelser and Schoening 1948).

CULTURE MEDIA. The virus may be cultivated in chick embryos and in laboratory animals such as mice.

Prevention. Pigeons should not be housed near colonies of free-flying bats, which may serve to transmit infection.

Control and Treatment. No control or treatment is suggested.

References

Davis, B. D., et al. 1973. *Microbiology.* 2d ed. Hagerstown, Md.: Harper & Rowe, pp. 1368–75.
Fritzsche, v. K., U. Heffels, and E. F. Kaleta. 1981. Dtsch Tieraerztl Wochenschr 88:72–76.
Gough, P. M., and R. D. Jorgenson. 1976. J Wildl Dis 12:392–95.
Jacotot, H. 1938. C R Soc Biol (Paris) 127:131.
Kelser, R. A., and H. W. Schoening. 1948. *A Manual of Veterinary Bacteriology.* 5th ed. Baltimore: Williams & Wilkins. pp. 544–64.
Kraus, R., and P. Clairmont. 1900. Hyg 34:1–30.
Lote, J. 1904. Zentralbl Bakteriol [Orig A] 35:741.
Marie, M. A. 1904. C R Soc Biol (Paris) 56:573.
Pasteur, L. 1885. C R Seances Acad Sci [III].
Remlinger, P. 1903. Ann Inst Pasteur 17:834.
Remlinger, P., and J. Bailly. 1929. Ann Inst Pasteur 43:153.
______. 1936. C R Soc Biol (Paris) 123:383.
Schwarte, L. H. 1965. In *Diseases of Poultry,* 5th ed., ed. H. E. Biester and L. H. Schwarte, pp. 832–33. Ames: Iowa State University Press.
Zurn, F. A. 1882. Venag Bernhard Friedrich Voigt Weimar S118.

Infectious Bronchitis

Definition. Avian infectious bronchitis (AIB) is a common acute, highly contagious, viral respiratory disease of chickens but an uncommon disease of pigeons.

Cause and Classification. AIB virus, the type species for the genus, was placed in the family Coronaviridae and in the genus *Coronavirus* (Cunningham 1970). Since 1956 other species were identified (Hofstad 1984). Wadey and Faragher (1981) and Faragher (1987) listed serum subtypes A–I. Clewley et al. (1981) listed 11 glycoprotein subtypes with M and C protein patterns and Johnson and Marquardt (1975) classified 10 distinct isolates.

Nature. Avian coronavirus particles tend to be circular, but some are pleomorphic. They are single-stranded RNA viruses (Cunningham 1970). Club-shaped envelope projections 20 nm long are uniformly widely spaced (McIntosh et al. 1967). The virus particles with projections are 80–120 nm in diameter (Berry et al. 1964). MacNaughton and Davies (1980) observed two types of particles. One contained all the polypeptides and the genome; the other lacked the ribonucleoprotein polypeptide and the genome.

Distribution

GEOGRAPHY AND INCIDENCE. In 1930 Schalk and Hawn (1931) observed the infection in chickens in North Dakota. Since then numerous reports of the disease in chickens have appeared in literature. In 1962 Winterfield and Hitchner noted nephrosis in chickens affected with AIB. This condition was also observed in chickens in Australia (Cumming 1963, 1967). In 1985 Barr et al. (1988) identified AIB in racing pigeons. This is the first confirmed report of the agent in other than chickens.

SPECIES. AIB is reported only in chickens and pigeons. Other members of the family infect humans, dogs, cats, cattle, horses, swine, rats, and mice and primarily produce gastroenteritis (Green 1984).

CARRIERS. Infected chickens shed and carry the virus for 49 days, but no permanent carrier state has been demonstrated (Cunningham 1970). This is probably true for pigeons.

TRANSMISSION. A stray pigeon was observed to join the Australian flock a week before the disease developed and undoubtedly introduced the infection. The virus is spread rapidly as an aerosol and is carried by air. In work conducted by the author, the highly infectious AIB virus was not transmitted to control chickens a distance of 50 ft away.

Epidemiology

DISTRIBUTION IN BODY. The virus may be found throughout the body but is largely concentrated in the respiratory tract.

ENTRY AND EXIT. The virus is spread in body discharges, including tracheal exudate and feces.

PATHOGENICITY. Pigeon serum samples were taken 26 days after the disease onset in the affected flock. Similar samples were taken from a so-called "unaffected" flock 400 m away. Hemagglutination inhibition antibody levels of 2–2.5 were present in both flocks. Four 8-wk-old pigeons and four 4-week-old specific-pathogen-free chickens housed together were inoculated by intranasal, intraocular, and oral routes with 10^{-3} units of AIB virus. The pigeons remained healthy for 18 days, but the chickens had marked evidence of infection (Barr et al. 1988). The pigeon infection may have developed as a result of stress or intercurrent infection with trichomonads, which were observed.

INCUBATION. In chickens signs of respiratory infection are evident in 18–48 hr.

COURSE. In the case reported by Barr et al. (1988) the affected pigeons recovered in 2–3 wk.

MORBIDITY AND MORTALITY. All birds in a loft become infected within a short time. Losses in the Australian pigeons amounted to 15%.

SIGNS. As reported by Barr et al. (1988) a loft of 150 racing pigeons became ill and 11 died in the first 24 hr and 11 more during the next 2 days. They had ruffled feathers, difficult breathing, and an excessive mucus discharge.

NECROPSY. Mucoid pharangitis and tracheitis were noted in the pigeons. AIB in chickens usually produces serous or catarrhal tracheal exudate and some air sac frothiness or cloudiness.

DIAGNOSIS. The pigeon tracheal isolate was inoculated into 9-day-old chicken embryos and allantoic fluids were harvested 72 hr after inoculation. An aliquot was passed three times in chicken embryos at 48-hr intervals according to Barr et al. (1988), who also reported that Endo and Faragher demonstrated AIB virus fluorescence. Hemagglutination was then observed after virus treatment with a phospholipase C type 1. Wadey and Faragher (1981) also used plaque reduction serum neutralization tests for each of the 9 Australian AIB virus subtypes and placed the virus in type B.

DIFFERENTIAL DIAGNOSIS. Other respiratory diseases must be considered, such as Newcastle, paramyxovirus infection, and mycoplasmosis.

Biological Properties

STRAINS. The strain of AIB recovered from the pigeon flock was shown to be the same serotype, type B, as used in Australian AIB vaccine. Faragher (1987) classified 10 serologically different strains by the neutralization test. In 1981 Clewley et al. used oligonucleotides and small glycoproteins to identify 11 distinctly separate strains. Massachusetts, Massachusetts-Connaught, Georgia SE17, Beaudette, and 927 from England were designated protein pattern M. Connecticut, Holte Delaware, Iowa 97, Australian T, and New Zealand A were designated C. Massachusetts C, Beaudette, and 927 were virtually the same, but each of the strains were distinct. Even the Australian T and New Zealand A were unrelated.

Hematology and Serology. In vitro test methods for the detection of AIB humoral immunoglobulins include hemagglutination inhibition (HI), agar gel precipitation, serum neutralization, complement fixation, plaque reduction, indirect fluorescent antibody, and ELISA. Faragher (1987) describes an HI test that modifies methods of Corbo and Cunningham (1959) and Bingham et al. (1975). Prior work with untreated AIB virus did not reveal its innate ability to agglutinate avian red blood cells.

TISSUE AFFINITY. AIB virus is concentrated in respiratory and splenic tissues.

STABILITY. AIB lyophilized virus sealed under 5 μm vacuum and stored in a Revco freezer at −45°C or under remains viable for 30 yr or more. Cunningham (1970) noted that AIB virus was most stable at pH 7.8. Cowen and Hitchner. (1975) found considerable variation in stability of 17 strains at pH 3.

Heat of 56°C for 15 min inactivated most strains (Hofstad 1984). Most disinfectants destroy AIB virus, but 1% phenol at room temperature for 1 hr was not effective (Quiroz and Hanson 1958).

CULTURE MEDIA. Chicken embryos 9–11 days old may be inoculated with 0.2 ml 1–5 dilution of virus in the chorioallantoic (CA) sac and virus harvested in 24–30 hr from the CA membrane or fluid. Dead embryos yield less virus. Gross lesions include embryo stunting, dwarfing, and curling after two to three passages. The Beaudette embryo-killing strain is lethal in 36 hr. Embryo mortality from other strains increases with serial passages. Mortality is late in early passes but is usually reduced to 2 days in later passages. Microscopic perivascular cuffing occurs in one-third of the embryos by the 6th day, and extensive kidney nephritis and congestion may be noted.

Prevention and Control. Since the disease appears uncommon in pigeons, vaccination is not considered necessary. However, the fact that the nearby "unaffected" Australian flock had an HI titer equal to the known infected flock perhaps indicates a more widespread distribution in pigeon flocks.

Treatment. There is no specific treatment indicated for AIB.

References

Barr, D. A., et al. 1988. Aust Vet J 65:228.
Berry, D. M., et al. 1964. Virology 23:403-7.
Bingham, R. W., M. H. Madge, and D. A. J. Tyrrell. 1975. J Gen Virol 28:381-90.
Clewley, J. P., et al. 1981. Infect Immunol 32:1227-33.
Corbo, L. J., and C. H. Cunningham. 1959. Am J Vet Res 20:876.
Cowen, B. S., and S. B. Hitchner. 1975. J Virol 15:430-32.
Cumming, R. B. 1963. Aust J Sci 25:314-15.
______. 1967. Ph.D. diss., University of New England, New South Wales, Australia.
Cunningham, C. H. 1970. Adv Vet Sci Comp Med 14:105-48.
Faragher, J. T. 1987. Aust Vet J 64:250.
Green, C. E. 1984. *Clinical Microbiology and Infectious Diseases of the Dog and Cat.* Philadelphia: Saunders.
Hofstad, M. S. 1984. In *Diseases of Poultry,* 8th ed., ed. M. S. Hofstad et al., pp. 429-43. Ames: Iowa State University Press.
Johnson, R. B., and W. W. Marquardt. 1975. Avian Dis 19:82-90.
MacNaughton, M. R., and H. A. Davies. 1980. J Gen Virol 47:365.
McIntosh, K., et al. 1967. Proc National Academy of Sciences 57:933-40.
Quiroz, C. A., and R. P. Hanson. 1958. Avian Dis 2:94-98.
Schalk, A. F., and M. C. Hawn. 1931. J Am Vet Med Assoc 78:413-22.
Wadey, C. N., and J. T. Faragher. 1981. Res Vet Sci 30:70-74.
Winterfield, R. W., and S. B. Hitchner. 1962. Am J Vet Res 23:1273-79.

Spongioform Brain Disease

A prion is a proteinaceous infectious particle totally devoid of genetic DNA or RNA. The agent, now the apparent cause of spongioform brain disease of cows in Great Britain, may have originated from sheep infected with the disease scrapie. Infected sheep carcasses rendered at a recently lowered temperature and included in dairy feed appear to be the source of the cattle disease. Dr. C. Gibbs at the National Institute of Health showed that scrapie of sheep, and kuru, and Creutzfeldt-Jacob diseases of man can be transmitted to monkeys. In addition, Dr. R. Barlow of the Royal Veterinary College at Hertfordshire demonstrated oral transmission of the cow prion to mice. A Siamese cat and exotic antelopes in Great Britain are reported infected. In the United States, goats, mule deer, Rocky Mountain elk, and mink have succumbed. It is speculated that the mink contracted the infection from slaughtered dairy cows in Wisconsin (Reeve 1990).

To date birds have not contracted the disease even though meat and bonemeal has been included in their feed. Pigeons, as grain eaters, are not considered candidates for infection.

Reference

Reeve, Mary P. 1990. *Harvard News Letter* 16:1-3.

4

Bacterial Diseases

Bacterial organisms exhibit characteristics of both plants and animals. They grow on artificial media and can be observed under a light microscope. Sixteen bacterial diseases are noted in pigeons: salmonellosis, avian chlamydiosis, tuberculosis, omphalitis, vibriosis, staphylococcosis, streptococcosis, colibacillosis, listeriosis, erysipelas, fowl cholera, ulcerative enteritis, avian pseudotuberculosis, *Pseudomonas* infection, avian arizonosis, and mycoplasmosis.

Salmonellosis

Definition and Synonym. Salmonellosis is an enteric bacterial infection in pigeons, other birds, animals, and humans, distinguished by the causative agents as pullorum, fowl typhoid, or paratyphoid.

Pullorum is an uncommon infectious bacterial disease of pigeons and humans but more commonly a disease of chickens, primarily transmitted by parent stock in eggs or droppings with resultant mortality.

Fowl typhoid is also an uncommon infectious bacterial disease of pigeons but a common disease of chickens, somewhat similar to pullorum.

Paratyphoid, on the other hand, is a common major infectious bacterial disease of pigeons, characterized as a subclinical infection in adult pigeons and often as an acute, fatal disease of young squabs.

Food poisoning is often used as a synonym for salmonellosis.

Cause and Classification. These organisms are classified as class Schizomycetes, order Eubacteriales, family Enterobacteriaceae.

Pullorum disease, named by Rettger (1900) is caused by Salmonella pullorum. Fowl typhoid (Curtice 1902) is caused by *Salmonella gallinarum* (Klein 1889).

Pigeon paratyphoid is caused by many species of the genus *Salmonella.* The second name is always the species or serotype name. Originally different names were mistakenly given the same organism. Following are synonyms for *S. typhimurium* var. *copenhagen,* which was named by Kauffman (1934). These names include *S. typhimurium* var. *binns,* named by Schuete (1920); *S. typhimurium* var. *storrs,* named by Jungherr and Wilcox (1934) and Edwards (1935). Earlier *S. aertrycke* was named by Castellani and Chalmers (1919). This organism was the primary species causing paratyphoid in pigeons (Edwards and Bruner 1940).

Other reported paratyphoid serotypes isolated from pigeons include *S. california* B, *S oranienburg* Cl, *S. anatum* E1, and *S. meleagridis* E1 (Edwards et al. 1948a); *S. thompson* C1 (Scott 1926); *S. schottmulleri* (Emmel 1929); *S. enteritidis* D (Schutt 1931; Shirlaw and Iver 1937); *S. dublin* D (Panwitz and Pulst 1972); *S. hadar* C2 (Pomeroy 1987); *S. blockley* C2 and *S. montevideo* C1 (Moran 1961); *S. london* E1 (Blackburn 1976); *S. heidelberg* B (Blackburn and Harrington 1978); *S. braenderup* C1, *S. bredeney* B (Blackburn and Harrington 1979); *S. indiana* B and *S. manhattan* C2 (Sutch and Harrington 1982); *S. havana* G2 (Sutch et al. 1985); *S. typhimurium* B (Ferris et al. 1986);

and *S. tennessee* C1 (Frerichs 1987). (Letters following the species indicate the Kauffmann-White grouping.)

Nature

PULLORUM. *S. pullorum* is a long, slender, gram-negative rod with slightly rounded ends that occurs singly and in pairs but seldom in longer chains. It is nonmotile, forms no capsule or spores, and grows as a facultative anaerobe. It ferments dextrose with acid and gas but not lactose, maltose, or sucrose. Edwards and Bruner (1946) typed *S. pullorum* as Group D, O (IX, XII1, [XII2], XII3); H phase 1,2 negative (no flagella). The standard strains contain only a small amount of XII2, but variant strains contain a large amount of this antigen (Van Roekel 1965).

FOWL TYPHOID. The organism is a relatively short, plump, gram-negative rod that occurs singly and occasionally in pairs. It is nonmotile, forms no capsules or spores, and grows as a facultative aerobe. It ferments dextrose and maltose with acid and no gas but fails to ferment lactose or sucrose. *S. gallinarum* was typed Group D, O (1,9,12); H phase 1,2 negative (no flagella) (Hall 1965).

PARATYPHOID. This is a large group of bacteria composed of over 2000 serotypes, each with a specific type designation (Davis 1973). The organisms are gram-negative rods that occasionally form short filaments. They are generally motile by peritrichous flagella but form no capsules or spores. They are facultative anaerobes.

Paratyphoid organisms ordinarily ferment dextrose and maltose with acid and gas but not lactose or sucrose; however, pigeon isolates of *S. typhimurium* var. *copenhagen,* which are common in pigeons, usually ferment only dextrose not lactose, maltose, or sucrose.

Most salmonella organisms possess two antigens, the somatic or body O (from the German "ohne Hauch") antigen and the alcohol and heat-labile flagellar or H (for "mit Hauch") antigen. The H antigen is further subdivided into phase 1 and phase 2. Both somatic and flagellar antigens often consist of several component factors, which may be shared by other types in the genus. *S. typhimurium* var. *copenhagen,* the major cause of the disease in pigeons, is typed Group B, O (1,4,-,12); H phase 1 (i); phase 2 (1,2). *S. typhimurium* is typed Group B, O (1,4,5,12); H phase 1 (i); phase 2 (1,2). The latter contains somatic factors 4 and 5 but var. *copenhagen* lacks the 5 factor (Williams 1972). Since there are flagella on paratyphoid organisms they carry the H antigen but pullorum and fowl typhoid organisms do not.

Distribution

GEOGRAPHY AND HISTORY

Pullorum. Even though the disease is widespread in chickens, there are very few reports in literature suggesting that pigeons contract pullorum. In 1929 Van Heelsbergen in Germany reported the disease in a textbook of bird diseases, and in 1936 Reis and Nobrega in Brazil provided a treatise on diseases of birds and included pullorum as a disease of pigeons.

Fowl Typhoid. Several conflicting reports pertain to fowl typhoid in pigeons. Klein (1889) in Germany had no success in infecting pigeons subcutaneously. In France, Lucet (1891) had no success with a 1-ml dose. Moore (1895) in the United States killed pigeons in 8 days with 2 ml of a broth culture. Pfeiler and Reopke (1917) in Germany considered pigeons not naturally susceptible. Kraus (1918) killed pigeons in 4 days with 1 ml of a 24-hr broth culture. Te Hennepe and van Straaten (1921) in Holland indicated that pigeons were not always susceptible when injected. Kaupp and Dearstyne (1924) in the United States showed that pigeons became sick in 3–4 days with recovery in 15 days. El-Dine in Egypt (1939) reported no evidence of the disease in pigeons. Hall (1946, 1965) reported on tests conducted in 1946 in which pigeons died in 4.5 days from intraperitoneal or intramuscular inoculation using 1 ml of a 5-hr broth culture.

Paratyphoid. From the sparcity of these reports it is evident that neither pullorum nor fowl typhoid is likely to occur naturally in pigeons. Paratyphoid, on the other hand, is a common widespread disease of pigeons, other birds, animals, and humans, and numerous reports attest to its prevalence. Moore (1895b), with the USDA Bureau of Animal Industry (BAI), recorded an outbreak in Vineland, N.J., in which several hundred pigeons died. This appears to be the first investigation of the pigeon disease. He called the causal organism *Bacillus suipestifer* and considered it a variant of the hog cholera organism. Mohler, also with the BAI, made the first authentic report in 1904, dealing with seven outbreaks in Indiana, Missouri, New Jersey, and Washington, D.C. Zingle (1914) noted the wet appearance of carrier pigeons with paratyphus at the military station in Strassburg. In Rotterdam, Reitsma (1924) published an account of *B. paratyphi* in pigeons. The same year Elbert and Schulgina found paratyphus in three pigeons. In 1925 Meder and Lund described the wet infection in young pigeons. Also in Germany, Beck and Meyer (1926) observed an outbreak. In New Jersey Beaudette (1926) identified the organism as *B. aertrycke.* In 1927 Lahaye and Willems called a pigeon isolate in Belgium *Salmonella aertrycke.* Pfaw (1928) also noted the disease in pigeons. *S. schottmuller* was isolated by Emmel (1929) from a wing joint. In Germany, Berge (1929) described the infection. In 1930 Brunett, at Cornell, called the organism *S. aertrycke.* Cash and Doan reported spontaneous infection of pigeons with *B. aertrycke* in 1931. The same year Schutt found *B. enteritidis* in 25% of pigeons examined in Germany. Lesbouyries and Verge (1932) observed the infection in France. In 1932 Clarenburg and Dornickx described a carrier pigeon outbreak in the Haag. *S. aertrycke* was considered the cause of infection in Rumania by Cernainu and Popovici (1933). In 1934 Cernea and Berdicek each reported pigeon isolations. The same year Jungherr and Wilcox recovered an organism, *S. aertrycke* var. *storrs,* from spontaneously infected squabs that were negative to the tube agglutination test. Kauffmann (1934) designated the cultures lacking the V (5) antigen of the Kauffmann-White schema as *S. typhimurium* var. *copenhagen.* Khalifa (1935) and Morcos (1935) in Egypt reported the disease in pigeons. Infection was also observed by Gaede (1935), Niemeyer (1939), and Marthedal (1951). In 1961 Moran reported that *S. typhimurium* var. *copenhagen* was essentially host-specific. Most of the pigeon cultures that she examined were *S. typhimurium* and of these nearly all were var. *copenhagen.* Later Faddoul et al. (1965) in Massachusetts reported 88% of their pigeon cultures as variants (var. *copenhagen*).

INCIDENCE. Edwards et al. (1948b) tabulated their *S. typhimurium* pigeon cultures and reported 200 of 205 were var. *copenhagen.* The occurrence of this serotype in other animals (2.4%) was due to direct or indirect contact with pigeons. Actually 97.5% of the cultures isolated from pigeons were *S. typhimurium* var. *copenhagen.* Harms (1965) in Hanover found 8% of feral pigeons infected. Pannwitz and Pulst (1972) in East Germany examined 353 wild pigeons and found 1.43% infected with *S. dublin* or *S. typhimurium* var. *copenhagen.* Since then Frerichs (1987) in the United States reported pigeon isolates to be 4% *S. typhimurium,* 90% *S. typhimurium* var. *copenhagen,* and 4% other serotypes, with 2% unknown.

SEASON. The diseases may occur at any time of the year.

AGE, SEX, AND BREED. Clinical salmonellosis occurs most frequently in very young individuals and in adult debilitated birds or animals (Edwards et al. 1948a). Older pigeons seldom show evidence of the clinical disease. Within the pigeon family there appears to be no relation to sex or breed.

SPECIES. Lerche (1939) found IV variant in isolations from horses, cattle, swine, pigeons, geese, pheasants, and wild birds. Edwards and Bruner (1940, 1943) reported this IV variant in pigeons, rabbits, turkeys, chickens, ducks, horses, a calf, a fox, and a goose. Edwards et al. (1948c) added a canary, cattle, swine, guinea pigs, and humans. Sato et al. (1975) also reported infection in calves. However, practically all cultures isolated from pigeons lacked the 5 factor (Edwards and Bruner 1940).

CARRIERS. Adult subclinically infected birds remain as chronic carriers that serve to infect young birds.

Transmission. There are many sources of infection aside from infected carrier birds. Fecal contamination of feed, water, and equipment is first in importance. Other sources include rodents, flies, roaches, free-flying birds, nesting materials, baskets, crates, training trucks, animals, and human feet. Clarenburg and Dornickx (1932) reported transmission in the pigeon egg. This was confirmed by Gauger et al. (1940), who isolated paratyphoid from unhatched eggs from serologically positive birds and from blighted ova of autopsied birds 6 mo after the infected eggs were laid. Eggshells may even become contaminated during laying. Washing eggs removes the bloom, which opens the pores to the bacteria. Hoffman and Edwards (1937) reported the spontaneous transmission of the pigeon variant to rabbits, indicating that close contact with infected pigeons can spread the infection to other animals and to humans.

Epidemiology

DISTRIBUTION IN BODY. Paratyphoid organisms may be readily recovered from most organs of infected squabs. Liver, spleen, and heart blood are usually infected. Adult birds may have the infection walled off in a wing or leg joint or in a blighted ova.

ENTRY AND EXIT. The bacteria enter the body by ingestion and leave in the feces. The reproductive tract may also permit entrance of the infection in semen.

PATHOGENICITY. Only *S. typhimurium* var. *copenhagen* is an important continuing cause of salmonellosis in pigeons. This organism is quite infectious for chicks (Seuna 1979) and is also very infectious for squabs.

Human Susceptibility and Public Health Concerns. *S. typhimurium* var. *copenhagen* does not present a significant public health hazard. The 1980 CDC annual salmonella summary (Porcher) listed 26,844 salmonella isolations from humans for 1974 through 1980. Of these isolations 326 were *S. typhimurium* var. *copenhagen,* which is 0.012%, or 1.2 human infections per 10,000 people. In addition, less than 3% of other serotypes occur in pigeons. This means that pigeons are not an important source of salmonellae for people.

Franssen (1962) and Scholtens and Caroli (1971) reported pigeons as an insignificant cause for salmonellae in food. This was further investigated by van Oye and Borghijs (1973) in Holland.

IMMUNITY. Adult birds develop a degree of immunity, but squabs infected under 5 wk of age may be incapable of producing antibodies (Gauger et al. 1940).

Incubation. Death may occur in 1–2 days, but clinical signs of the disease may be evident in 1–5 days.

Course. The disease varies from a rapidly fatal disease of young birds to a chronic prolonged infection in adults.

Morbidity. Only 10–20% of a flock may show infection at one time.

Mortality. Young squabs suffer most with losses up to 20% by the third week of infection.

Signs. Adult pigeons usually show no evidence of infection, but affected birds may have a watery eye or nasal discharge, loose greenish bowel discharges, soiled feet and feathers, stiff leg or wing joints, wing joint nodules, and respiratory rales. Eyes are often glassy. Squabs have greenish, loose intestinal discharges and become emaciated. Many squabs have respiratory trouble, which increases losses. Fertile eggs may not hatch. If they do hatch, squabs are weak and often die in a short time.

Necropsy. Swelling and congestion of the spleen, liver, and kidneys are primary signs. Necrotic foci may also be found in the liver and lungs. Urates may be found in the ureters and kidneys. The mucosa of the intestine is usually congested and covered with catarrhal exudate.

Diagnosis. Diagnosis is based on the isolation of the organism on suitable artificial media and its identification in fermentation tubes. Cultures from several sick and several dead birds and/or unhatched eggs may be required to establish a diagnosis. Birds to be cultured should be free of medication for a week prior to examination. Nielsen (1965) found that repeated individual fecal cultures were necessary to identify carriers.

Tube agglutination tests may be tried using *copenhagen* antisera prepared in rabbits or chickens and diluted to 1:25, 1:50, and 1:100. This sera may be combined with an unknown culture to determine the degree of agglutination, or pigeon sera may be combined with a phenolized autogenous paratyphoid culture using the same dilutions. Gauger et al. (1940), however, found that live antigens were agglutinated in higher dilutions than killed antigens and more suspicious readings were observed with live antigens. They also recovered the organisms from some birds that tested negative with live and killed antigens. Their blood studies likewise could not be depended on to identify chronic carriers. There was considerable variation in blood counts, but the white blood count averaged 24,270. The normal range is 13,000–18,500/mm^3 (Sturkie 1954).

Differential Diagnosis. Pullorum and fowl typhoid must be excluded as causes of disease. To avoid errors in diagnosis, colonies should grow for 3 days, because *Proteus* and *Pseudomonas* spp. provide similar fermentations in 24 hr before spreading growth on agar can be observed. *Campylobacter* sp., another intestinal pathogen, may also cause confusion.

Biological Properties

STRAINS. Within the species var. *copenhagen* individual cultures vary in their production of component antigens. Thus strain differences occur.

SEROLOGY. White (1926) recognized the importance of considering bacterial variation in relation to the antigenic analysis of salmonellae. This work was extended by Kauffmann (1941) and established the Kauffman-White classification, which made possible the uniform methods of typing now in world use. This was simplified by Edwards and Kauffmann (1952). Kauffmann (1934) noted that certain cultures of *S. typhimurium* lacked antigen V, found in the somatic complex of typical strains. Edwards (1935, 1936a, 1938) observed that 13 cultures of *S. typhimurium* isolated from pigeons lacked the heat-stable V antigen. Hohn and Herrmann (1937) also described a lack of antigen V in pigeon cultures. Lerche (1939) found IV variants among pigeon cultures. Edwards (1936b) and Edwards and Bruner (1939) also found that all their pigeon-derived *S. typhimurium* var. *copenhagen* cultures were IV variants and suggested that the lack of the V antigen was characteristic of pigeon cultures more so than the failure to utilize carbon from ammonium agar, as Hohn and Hermann suggested in 1937.

TISSUE AFFINITY. The organism may be recovered from the liver, spleen, gallbladder, ovary, lungs, heart blood, kidney, bursa, and sometimes wing or leg joints.

STAINING. Salmonellae are gram-negative and stain with the usual aniline dyes.

STABILITY. Environmental factors such as

temperature, humidity, moisture, and pH must be considered in the survival and destruction of salmonellae. These bacteria may survive as saprophytes or pathogens. They can live in the soil or environment above a pH of 4.0 and below 8.0. Delage (1961) determined that *S. abortusovis* (*S.* var. *storrs,* Fenstermacher 1952) survived over a year in soil. *S. typhimurium* survived 47 days at 2°C and 1.2 min at 57°C (134.6°F) (Diesch 1978). The endotoxin is heat-resistant, however, and no soluble exotoxin is formed. Further, the usual disinfectants destroy the organism.

CULTURE MEDIA. A loop culture taken from a seared liver or tissue surface can be seeded on plate media, such as trypticase soy agar, Endo's agar, eosin methylene blue agar, MacConkey's agar, salmonella shigella agar, or sodium ammonium phosphate agar. Intestinal or crop contents can be added to selective growth media, such as selenite broth or tetrathionate-enriched broth containing brilliant green 1/100,000. Galton (1961) suggested laboratory procedures for the isolation of salmonellae.

Following growth on agar plates, typical moist, grayish white, circular, entire individual colonies can be transferred by loop to Dunham's peptone water containing 1% Andrade's indicator and 1% fermentable substance. Cultures are usually tested for acid and gas production in dextrose, lactose, maltose, sucrose, xylose, rhamnose, trehalose, inositol, salicin, dulcitol, adonitol, arabinose, and mannitol. If acid and gas develops in 24 hr at 37°C in dextrose, mannitol, and maltose, but not in malonate, lactose, sucrose, salicin, and adonitol, and if indole is not formed and gelatin is rarely liquefied, the culture is considered a paratyphoid. It must be remembered that fermentation of maltose and xylose may be very slow and dulcitol is usually positive. Nonpigeon strains of *S. typhimurium* var. *copenhagen* usually ferment maltose with acid and gas (Smyser and Snoeyenbos 1972). Jungherr and Wilcox (1934), Edwards and Bruner (1940), and Smyser and Snoeyenbos (1972) each encountered pigeon isolations that were maltose-negative. Virtually all the New Jersey pigeon cultures were also maltose negative. Van Dorssen (1953) noted a pigeon culture that was rhamnose-negative. All cultures that Gauger et al. (1940) isolated were doubtful rhamnose fermenters.

Prevention. Several management procedures are an essential part of prevention. Breeding stock should come from lofts free of the disease. All imports are isolated 30–60 days away from other birds. Feed and water containers are shielded to keep out manure and birds, using good sanitary practices and feeding only in clean hoppers. Proper disposal of dead birds is important. Litter and manure must not be spilled when cleaning. Free-flying birds should be screened out and rodents, roaches, and flies controlled. Sick birds should not fly or be shown. Good nutrition, exercise, and sunshine are essential. If the disease strikes, breeding should be suspended and disease control and treatment measures instituted.

Control. To control salmonellae, single, rapid, whole blood tests with antigens prepared from autogenous cultures are not satisfactory. Repeated testing may be successful in identifying carriers since Gauger et al. (1940) determined that antibody titers may last 8 mo in pigeons.

Chemical control is effective. The premises and loft should be routinely acidified with sodium acid sulfate (Du Pont) at the rate of 1 lb salt/100 ft² on a weekly basis or as often as is necessary to maintain pH 3.5 or below. Test Hydrion paper may be used periodically to determine the pH of litter or soil solutions. The salt converts to sulfuric acid when in contact with water, so caution must be used in the application of the chemical; rubber gloves and eye protection are essential. The dry salt can be raked into litter or soil about the loft. If no litter is used on the floor, kitty litter will work. Small quantities will not hurt the feet of the pigeons and ingestion of a small amount is not considered harmful. Chicks were held on litter with pH 2.5 for 1 mo with no visible harm. A wet solution (1 lb salt/8 q water) may be applied by broom about the floor areas (boots should be worn during application). The acid burns and corrodes metal, so it should not be applied to metal or sprayed.

Acidification is effective in controlling salmonellae because it prevents replication on the floor or ground and thus prevents recycling bacteria from the droppings. Pigeons are grain eaters and thus seldom contract serotypes other than *S. typhimurium* var. *copenhagen* because uncontaminated grains usually do not carry these bacteria. Animal and poultry feed containing meat scrap and bonemeal, which pigeons seldom eat, provides a continuing source of other salmonellae for animals and birds.

Another control avenue was investigated by Hauser and Robl-Furst (1972) in Germany. Heat-inactivated vaccine was given for 9 days in drinking water with little value in the presence of infection. Marcus (1964) also used a killed vaccine, which was given by injection and per os without success. In addition, Horter (1971) used a different product for immunization and Marx (1987) prepared an injectable killed bacterin in an effort to prevent the disease.

Treatment. After the premise has been acidified, treatment can be successful. Acidification prevents reinfection. Treatment is designed to destroy salmonellae in the bird and consists of chlortetracycline (Aureomycin, American Cyanamid) soluble powder used at the rate of 2 0.25-lb packages (25.6 grams [g] per package) per 25 gal water or 100 g per ton of grain. Grain can be coated with corn oil so the powder will stick to the grain. In addition, furazolidone (NF180, Hess and Clark) should be added at 2 lb pure drug per ton of grain. The drug treatments should be given for 2 wk, but acidification should continue for at least 1 mo.

Other reports of treatments are as follows. Devos et al. (1965) used Furaltadone 1/10,000 in drinking water for 10 days for pigeons. Schoop and Zettl (1956) treated pigeons with 160 mg streptomycin orally on day 1 and 85 mg on days 2 and 3. Klingler and Morgenstern (1971) also reported treating army pigeons in Bern. Batril (Mobay Corp.) has been used on individual birds at 2.5 mg/kg body weight twice daily for 5 days.

References

Beaudette, F. R. 1926. J Am Vet Med Assoc 68:644-52.

Beck, A., and E. Meyer. 1926. Z Infektionskr Parasitol Kr Hyg Haustiere 30:15.

Berdicek, V. 1934. Zantralbe Bakteriol I Abt Orig. 115:12.

Berge, R. 1929. Dtsch Tieraerztl Wochenschr 1:247.

Blackburn, B. O. 1976. Proc US Animal Health Assoc, p. 367.

Blackburn, B. O., and R. Harrington. 1978. Proc US Animal Health Assoc, pp. 460-62.

______. 1979. Proc US Animal Health Assoc, pp. 400-402.

Brunett, E. L. 1930. Cornell Vet 20:169-76.

Cash, J. R., and C. A. Doan. 1931. Am J Pathol 7:373-98.

Castellani and Chalmers. 1919. Man Trop Med, p. 939.

Cernainu, C., and I. Popovici. 1933. C R Soc Biol (Paris) 112:829.

Cernea, J. 1934. Arch Vet 26:119.

Clarenburg, A., and C. G. J. Dornickx. 1932. Ned Tijdschr Geneeskd 76:1579.

Curtice, C. 1902. R I Exp Stn Bull 87.

Davis, B. D., et al. 1973. *Microbiology.* 2d ed. Hagerstown, Md.: Harper & Row.

Delage, B. 1961. Arch Inst Pasteur Maroc 6:139-42.

Devos, A., N. Viaene, and M. Staelens. 1965. Vlaam Diergeneeskd Tijdschr 34:209-20.

Diesch, S. L. 1978. Proc National USDA Salmonella Seminar, Washington, D.C., pp. 1-10.

Edwards, P. R. 1935. J Bacteriol 30:465-71.

______. 1936a. J Hyg 36:229.

______. 1936b. J Hyg 36:348.

______. 1938. J Bacteriol 35:123-28.

Edwards, P. R., and D. W. Bruner. 1939. Am J Hyg 29:24-31.

______. 1940. Ky Agric Exp Stn Bull 400, pp. 41-70.

______. 1943. J Inf Dis 72:58-67.

______. 1946. Cornell Vet 36:318.

Edwards, P. R., and F. Kauffmann. 1952. Am J Clin Pathol 22:692-97.

Edwards, P. R., D. W. Bruner, and A. B. Moran. 1948a. Cornell Vet 38:247-56.

______. 1948b. J Inf Dis 83:220-31.

______. 1948c. Ky Agric Exp Stn Bull 525.

Elbert, B., and O. Schulgina. 1924. Zentralbl Bakteriol I Abt Orig 91:496.

El-Dine, H. S. 1939. Proc 7th World Poultry Congress, p. 229.

Emmel, M. W. 1929. J Am Vet Med Assoc 75:369-70.

Faddoul, G. P., and G. W. Fellows. 1965. Avian Dis 9:377-81.

Fenstermacher, R. 1952. In *Diseases of Poultry,* 3d ed., ed. H. E. Biester and L. H. Schwarte, pp. 261-300. Ames: Iowa State University Press.

Ferris, K., C. D. Murphy, and B. O. Blackburn. 1986. Proc US Animal Health Assoc, pp. 381-96.

Franssen, J. G. 1962. Tijdschr Diergeneeskd 87:1576-77.

Frerichs, W. M. 1987. Personal communication.

Gaede, H. 1935. Zeitschr Fleisch Milchhyg 45:182.

Galton, M. M. 1961. Proc US Animal Health Assoc, pp. 434-40.

Gauger, H. C., R. E. Greaves, and F. W. Cook. 1940. N C Agric Exp Stn Tech Bull 62:71.

Hall, W. J. 1946. USDA Circ 755, p. 1-9.

______. 1965. In *Diseases of Poultry,* 5th ed., ed. H. E. Biester and L. H. Schwarte, pp. 329-58. Ames: Iowa State University Press.

Harms, F. 1965. Dtsch Tieraerztl Wochenschr 72:232.

Hauser, K. W., and H. Robl-Furst. 1972. Tieraerztl Umsch 27:444-50.

Hoffmann, H. A., and P. R. Edwards. 1937. Am J Hyg 26:135-37.

Hohn, J., and W. Herrmann. 1937. Zeitschr Hyg 119:369.

Horter, R. 1971. Vet Bull. no. 6127.

Jungherr, E., and K. S. Wilcox. 1934. J Infect Dis 55:390-401.

Kauffmann, F. 1934. Zeitschr Hyg 116:368.

______. 1941. Acta Pathol Microbiol Scand 18:351.

Kaupp, B. F., and R. S. Dearstyne. 1924. Poult Sci 3:119.

Khalifa, I. A. B. 1935. J Am Vet Med Assoc 86:24.

Klein, E. 1889. Zentralbl Bakteriol 5:689.

Klingler, L., and R. Morgenstern. 1971. Schweiz Arch Tierheilkd 113:159-63.

Kraus, E. J. 1918. Zentralbl Bakteriol [Orig A] 82:282.

Lahaye, J., and R. Willems. 1927. Ann Med Vet 72:241-60.

Lerche, M. 1939. Z Hyg 122:72.

Lesbouyries, M. M., and J. Verge. 1932. J Bull Acad Vet Fr 5:294.

Lucet, A. 1891. Ann Inst Pasteur 5:312.

Marcus, I. 1964. Berl Muench Tieraerztl Wochenschr 77:385–89.
Marthedal, H. E. 1951. 9th World Poultry Congress, pp. 149–53.
Marx, D. 1987. Personal communication.
Meder, E., and L. Lund. 1925. Dtsch Tieraerztl Wochenschr, p. 377.
Mohler, J. R. 1904. 21st Annu Rep BAI, USDA, p. 29.
Moore, V. A. 1895a. 12–13th Annu Rep BAI, USDA.
______. 1895b. USDA BAI Bull 8, p. 71.
Moran, A. B. 1961. Proc US Animal Health Assoc, pp. 442–45.
Morcos, Z. 1935. Br Vet J 91:11.
Nielsen, B. B. 1965. Nord Vet Med 17:156–63.
Niemeyer, W. E. 1939. J Am Vet Med Assoc 94:434–35.
Pannwitz, E., and H. Pulst. 1972. Monatsh Veterinaermed 27:573–75.
Pfaw, M. 1928. J Vet Med 48:1318.
Pfeiler, W., and W. Roepke. 1917. Zentralbl Bakteriol 79:125.
Pomeroy, B. S. 1987. Proc Northeast Avian Disease Conference, Atlantic City, N.J., p. 50.
Porcher, F. H. 1980. U S Dept Health Hum Serv Salmonella Surveillance Rep 1974–1980.
Reis, J., and P. Nobrega. 1936. Ed Inst Biol, Sao Paulo, Braz, p. 109.
Reitsma, K. 1924. Tijdschr Vgl Geneeskd 10:6.
Rettger, L. F. 1900. N Y Med J 71:803.
Sato, G., et al. 1975. AJVRAH 36:839–42.
Scholtens, R. T., and G. Caroli. 1971. Antonie van Leeuwenhoek J Microbiol Serol 37:473–76.
Schoop, G., and K. Zettl. 1956. Tieraerztl Umsch 11:256–57.
Schuete, H. 1920. Lancet 93.
Schutt, G. 1931. Dtsch Tieraerztl Wochenschr I, p. 401.
Scott, W. M. 1926. J Hyg 25:398.
Seuna, E. 1979. Avian Dis 23:392–400.
Shirlow, J. F., and S. G. Iver. 1937. Indian J Vet Sci Anim Husb 7:231.
Smyser, C. F., and G. H. Snoeyenbos. 1972. Avian Dis 16:270–77.
Sturkie, P. D. 1954. *Avian Physiology.* New York: Comstock.
Sutch, K., and R. Harrington. 1982. Proc US Animal Health Assoc, pp. 460–62.
Sutch, K., C. D. Murphy, and B. O. Blackburn. 1985. Proc US Animal Health Assoc, pp. 383–97.
te Hennepe, B. J. C., and H. van Straaten. 1921. Trans 1st World Poultry Congress 1:259.
Van Dorssen, C. A. 1953. Tijdschr Diergeneeskd 78:161–63.
Van Heelsbergen, T. 1929. *Handbuch der Geflugelkrankheiten und der Geflugelzucht.* Stuttgart: Ferdinand Enke.
van Oye, E., and J. Borghijs. 1973. Vlaam Diergeneeskd Tijdschr 42:446.
Van Roekel, H. 1965. In *Diseases of Poultry,* 5th ed., ed. H. E. Biester and L. H. Schwarte, pp. 220–59. Ames: Iowa State University Press.
White, P. B. 1926. Med Res Counc Spec Rep Ser 103.
Williams, J. E. 1972. In *Diseases of Poultry,* 6th ed., ed. M. S. Hofstad. Ames: Iowa State University Press.
Zingle, M. 1914. Z Infektionskr Parasitol Kr Hyg Haustiere 15:268.

Avian Chlamydiosis

Definition and Synonyms. Avian chlamydiosis is a clinical or nonapparent subclinical, acute or chronic, sometimes fatal, naturally occurring, systemic, infectious, transmissible, respiratory, bacterial disease caused by *Chlamydia psittaci.* It infects a wide variety of birds and mammals, as well as humans. In pigeons it is a cosmopolitan, enzootic, usually chronic, subclinical, widespread infection. In humans it is observed as interstitial pneumonia.

Avian chlamydiosis is also called ornithosis, psittacosis, parrot fever, and bedsonia, miyagawanella, or Levinthal-Cole-Lillie infection.

The term *avian chlamydiosis* has been used for some time (Storz 1971; Page 1978; Eugster 1980), but the term *Chlamydia* was first proposed by Jones et al. (1945) before Moulder (1964) showed it to be of bacterial nature.

Cause and Classification. The avian disease is classified under the bacterial class Schizomycetes, order Chlamydiales, family Chlamydiaceae, genus *Chlamydia,* and species *psittaci* (Group B). Other species, *C. pneumoniae trachomatis* (Group A), are largely limited to humans and do not involve birds (Cutting 1990). The work of Gordon and Quan (1965) enabled the separation of the groups and permitted this classification.

According to Cutting (1990), "Each strain of *C. psittaci* preferentially infects one host species," and monoclonal antibody studies indicate that "strains of *Chlamydia* from different host species may actually be separate species." Certainly the nature of *C. psittaci* infection in individual species of humans, other mammals, and birds suggests this conclusion.

Nature

FORM AND DEVELOPMENTAL CYCLE. The chlamydial organisms have a special developmental cycle that alternates between (1) spherical elementary bodies 300 nm in diameter adapted for extracellular survival and (2) large vegetative, intracellular, spherical or irregular reticulated initial bodies 600–1500 nm in diameter that reproduce in the host cell only. The infectious, small, dense elementary body with a limiting wall or membrane is taken into the host cell by phagocytosis. There it becomes a large, flexible-walled, reticulated spheroid (inclusion) 0.6–1.5 μm in diameter within the cytoplasmic vacuole, which is bounded by the host cell cytoplasmic membrane. The initial or reticulate body containing nuclear fibrils and ribosomal elements multiplies by fission or binary divisions until reticulate microcolonies are formed. The vesicle wall of the microcolony ruptures, releasing the organisms into the cytoplasm of the cell. Other larger particles, intermediate bodies, appear to be precursors of infective particles and also reproduce by binary fission. The growth, multiplication, and maturation occurs over a period of 30 hr or less. As the numbers of daughter reticulate bodies increase, their size is reduced; they form small compact elementary bodies that are released when the cell wall ruptures, permitting infection of other cells (Page 1978; Page and Grimes 1984).

METABOLISM. Chlamydiae survive phagocytosis and reproduce within cells designed to destroy them. Compared with other bacteria they have limited enzymatic ability outside the host cell. Isolated chlamydiae may have some ability to make RNA. They survive by using the ATP energy of the host cell and by negating the action of the host's lysosomal enzyme activity. They inhibit the synthesis of DNA by the host but initiate their own RNA, DNA, and protein synthesis within the host cells (Davis et al. 1973).

CHARACTERISTICS. These agents can be identified by family-specific and type-specific antibody response. They have a common family-specific, heat-stable antigen in lysates of the organism. The type-specific antigen is related to the cell wall protein. Pigeon chlamydiae are classified in Group B by type-specific antigens, which are heat-labile.

Group B chlamydiae are antibiotic but not sodium sulfadiazine-sensitive. They also have nonrigid inclusions, which do not contain iodine-staining glycogen (carbohydrate) or lipid. They share other bacterial properties with Group A chlamydiae. These properties include (1) the ability to be observed under an ordinary light microscope, (2) cytoplasmic inclusions that lie close to the cell nucleus, and (3) ribosomes similar in size to those of bacteria (Davis et al. 1973).

Distribution

GEOGRAPHY AND HISTORY. The disease is encountered wherever pigeons are grown. It has been known in pigeons since 1939 when it was first reported by Coles (1940) in Johannesburg, South Africa, and by Pinkerton and Swank (1940) in the United States. Andrews and Mills (1943) described an outbreak in England. This was followed by Labzoffsky (1947) in Ontario, Canada; Lepine (1950) in Paris; Meenan et al. (1950) in Ireland; Rugiero et al. (1950) in Argentina; Dekking and Ruys (1951) in Holland; Komarov and Goldsmit (1952) in Israel; Mohr (1954) in Germany; Varella (1955) in Mexico; Grubb (1955) in Sweden; Olivo and Badiali (1956) in Italy; Michael (1957) in Greece; Schmidtke (1957) in Norway; Meyer and Genewein (1957) in Switzerland; Terzin (1958) in Yugoslavia; Kalra (1958) in India; Miles (1959) in New Zealand; Inaba et al. (1959) in Japan; Kuen (1962) in Moscow; Fries (1966) in Denmark; Ahmed and Elsisi (1966) in Cairo; and Dane and Beech (1955) in Australia.

INCIDENCE. Numerous studies have shown that most pigeon flocks contain a percentage of infected birds. Levinson et al. (1944) examined 14 pigeons in Philadelphia and 42% had positive complement-fixing (CF) titers. Labzoffsky (1947) found the incidence in Ontario to be 2%, 20%, and 45% in pigeons in three cities and he isolated chlamydiae from 58% of the sero-positive birds. Davis and Ewing (1947) isolated the agent from 15% of 100 pigeons in Baltimore. Davis (1950) found 3 of 22 infected adult pigeons in Washington, D.C., in 1949. In the same year he isolated the agent from 5 of 120 in Birmingham, Alabama. Sixty percent of this group were CF-positive. In California and Iowa, Page (1978) noted that half of the loft flocks he tested were positive serologically but in feral pigeons he found 48 of 50 groups positive. Kapitancik and Vrtiak (1965) in Slovakia examined feral pigeons and 58% were CF-positive. Meyer and Eddie (1951) found only 1% serologically positive in 100 pigeons in Mexico. In Utrecht, Holland, Jansen (1955) found infection in 16.4% of 2637 pigeons examined. Monreal (1958) observed that only 17% of the flocks were free of infection in Germany. In 1965 Meyer reviewed the results of 50 investigators in 24 countries, and of 16,539 feral and loft pigeons tested, 26.9% were CF-positive and 19.9% were active carriers. Meyer and Eddie (1951) ascertained that pigeons alone, which were probably domestic birds, were responsible for 69 human cases and three deaths in the United States from 1945 to July 1950. A discussion by Shaughnessy (1955) indicates that 4–26% of feral pigeons harbor chlamydiae and serve as a reservoir for people. A more recent U.S. review from the CDC by Gregg (1983), covering 1975–1981, reports 54 cases in domestic pigeons, 27 in feral pigeons, and 41 human cases related to pigeons.

According to Durfee and Moore (1975) and Durfee (1975), 21% of all human cases in the United States between 1953 and 1960 (excluding abattoir turkey cases) were caused by pigeons. In Czechoslovakia in 1949–1960 there were 1072 human cases recorded and 3% were associated with pigeons and pheasants (Strauss 1967). Meyer (1965) surveyed the number of human cases of psittacosis worldwide between 1931–1963. Of 5390 cases, 680 (12.5%) were associated with pigeons. In Czechoslovakia 32% of 63 pigeon fanciers were serologically positive compared with 35% of 169 fanciers in Germany (Mohr 1954; Wohlrab 1955; Siegmund 1960). There appears to be no evidence of widespread human infection of pigeon origin, but the uncommonly high incidence of significant human serum titers in pigeon fanciers points to pigeons as a public health hazard (Meyer et al. 1942; Dekking 1950; Davis 1955; Shaughnessy 1955).

SEASON. The disease may occur at any time of the year.

AGE, SEX, AND BREED. Old birds often have latent, nonapparent infections, but young birds are more likely to develop clinical illness and die. Both sexes are susceptible and all pigeon breeds are equally susceptible.

SPECIES. Meyer (1965), by the end of 1963, had listed 127 species of birds, belonging to 10 orders, that had been found infected. The list included psittacines, pigeons, seagulls, herons, egrets, ducks, geese, chickens, and turkeys, and all showed latent infections. In the 1975 turkey outbreaks in Texas, Page (1978) reported as seropositive 65% of blackbirds, 44% of killdeer, 27% of sparrow, 100% of goats, and some cattle. The organism produces pneumonitis in cats, sheep, cattle, goats, horses, rabbits, and swine; arthritis in cows, sheep, and swine; and conjunctivitis in cats, sheep, and guinea pigs. In humans interstitial pneumonia may develop. In addition, mice and hamsters used as laboratory animals can be infected.

CARRIERS. Shaughnessy (1955) reviewed pigeon cases in which the agent was isolated from feral pigeons and concluded that up to 26% of wild pigeons harbor chlamydiae. Panigrahy et al. (1982) examined 695 pigeons and 29.6% tested positive with the CF test. A titer of 1:8 or greater was considered positive. They quoted Meyer's surveys of 1941 and 1962 in which 29.5% of 1183 were positive.

Antibodies do not appear to inactivate organisms already localized within the cells. Individuals free of apparent disease and relatively immune to reinfection may continue to shed virulent chlamydiae (Davis et al. 1973). Also pigeons with or without CF antibodies may harbor and shed the organism (Meyer 1965). Human infection does occur following contact with carrier birds (Meyer 1941). Meyer and Eddie (1951) showed that infected humans may serve as carriers and Bruu (1981) and Nagington (1984) both reported transmission between humans.

Transmission. Transmission between pigeons occurs as a result of direct or indirect contact with the agent. Aerosol nasal discharges undoubtedly play an important role. Clinically recovered birds continue to pass the organism intermittently for long periods. An infected parent may transmit the disease in regurgitated crop milk. Indirect contact with contaminated discharges, droppings, equipment, water, and feed also provides a means of transfer. Dried droppings become dust that can remain infectious for feed and water for many months. The disease may thus be established by ingestion or inhalation. Even lice and mites have been shown to harbor chlamydiae, but they apparently serve only as mechanical carriers (Eddie et al. 1962).

Transovarian infection (congenital transmission of infection through the egg) also remains a possibility. Experimental studies have shown that chicken embryos inoculated with a sublethal dose of chlamydiae will hatch and the organism can be recovered from apparently normal chicks for up to 22 days (Davis and Vogel 1949). The agent has been transmitted through eggs of ducks (Illner 1962) and of black-headed gulls (Lehnert 1962). Two attempts to recover the agent from pigeon eggs from infected lofts failed. Davis (1955) inoculated mice intracerebrally (IC) and intraperitoneally with material from 70 embryonated eggs and Fritzche et al. (1956) inoculated material from 55 eggs from infected pigeons without success.

Monreal (1958) was successful in establishing nasal mucosal infection but experienced difficulty by the oral route. Aerosol or dust transmission from pigeons to turkeys on a turkey ranch was shown to occur by Page (1960).

Transmission of the disease from squabs to humans was first reported by Coles (1940). Levinson et al. (1944) reported 6 human cases with pigeon contacts. Cohen et al. (1946) also observed that a man and his wife developed chlamydiosis 1 wk after two pigeons were kept in their bedroom. Meyer (1941) observed infection in a zoologist working with doves and also in a retired individual whose son had racing pigeons. In addition, he (1965) reported the risk to laboratory workers. Between 1929 and 1956 there were 108 U.S. human cases with 12 deaths.

Epidemiology

DISTRIBUTION IN BODY. The organism may be found in all parts of the body but is more easily recovered from the spleen and air sac or pericardial exudate. The presence of the organism can also be demonstrated in the intestinal discharges.

ENTRY AND EXIT. Inhalation or ingestion of the agent in sufficient quantities can establish infection. Conjunctival infection may occur from contaminated dust or water. The agent leaves the body in droppings and discharges.

PATHOGENICITY. The natural disease producing ability of strains of *C. psittaci* for domestic fowl are classified into two groups. Highly virulent or toxigenic strains that may cause up to 30% mortality have been recovered from turkeys, psittacines, and wild birds. Strains of low virulence, which may produce 5% losses, have been isolated from pigeons and ducks and occasionally from turkeys and sparrows. Strains of low and high virulence spread rapidly and with equal ease within a flock. Serologic studies indicate over 90% infection rate by the time there is clinical evidence. Titrations of virulent strains in laboratory animals indicate that they are quite infectious and lethal for mice, guinea pigs, turkeys, and parakeets but do not bother pigeons and sparrows. On the other hand, strains of low virulence, such as pigeon isolates, have a high infectivity but low mortality rate for mice, pigeons, sparrows, parakeets, and turkeys. These strains seldom are infectious for guinea pigs (Page 1978). Two turkey isolates studied by Page (1960) were equally virulent in chicken embryos and mice but differed in their effect on pigeons and turkeys. The C/2 Woodland strain was 30 times more severe in turkeys than the C/4 Modesto strain, which was 525 times more lethal for pigeons by intraperitoneal inoculation. Meyer and Eddie (1942) evaluated a chicken isolate from New Jersey and found it to be of low virulence for mice but lethal for pigeons, parakeets, and rice birds by intramuscular and IC inoculation. Pigeons are inherently not susceptible to IC inoculation of large numbers of highly toxigenic chlamydial strains from turkeys and domestic herbivores (Page and Grimes 1984), but IC inoculation of pigeons with pigeon isolates usually produces meningitis (Pinkerton and Moragues 1942). Pigeons are, however, immune to the same pigeon strains given intramuscularly (Meyer 1965). The chlamydial agent in pigeons is usually of low virulence, suggesting that the organism has been associated with pigeons for a long time.

Incubation. The incubation period varies from a few days to several weeks depending on the extent of exposure, resistance of the individual, and the strain involved.

Course. The course is indefinite. A somewhat resistant yet susceptible bird may carry the infection throughout life, whereas other resistant birds may be exposed but present neither serologic nor clinical evidence of infection. Young or weakened birds, on the other hand, may succumb to infection. Concurrent infections usually intensify the chlamydial infection and shorten the course, resulting in death.

Morbidity. The number of sick birds in an infected flock will vary depending on the age and resistance of the birds and the strain to which they are exposed. It is often difficult to determine if the disease is present. Positive serology or recovery of the organism may be the only clue.

Mortality. Adult birds seldom die, but young pigeons often do, particularly if the strain is a virulent one.

Signs. Older pigeons may present no outward evidence of infection. Birds may appear to be healthy but have latent disease. Sick birds have no typical syndrome. Young birds under 4 mo of age often die; they become thin and feeble, often having a greenish diarrhea with matted vent feathers. Adult carrier birds may be listless and inactive and show weakness, inappetence, loss of weight, and diarrhea. A mucopurulent ocular or nasal discharge may be present. Any pigeon with a conjunctivitis must be considered as suspicious.

Necropsy. Necropsy findings vary from relatively few abnormalities to extensive air sac involvement with semipurulent exudate over the serous membranes and pericardium. The lungs are not consistently involved. Visibly enlarged, elongated, friable, dark spleens and large, congested livers and kidneys may be noted. Fibrinous exudate may cover the heart and pericardium. Catarrhal enteritis of the intestine with mild swelling may be present.

Diagnosis. Recovery of the organism is the best method of establishing an unequivocable diagnosis. Isolation attempts may be made from both living and dead specimens. Those made on living birds must be made when organisms are actively shed. To isolate chlamydiae from postmortem tissues, a 20% (w/v) suspension of ground spleen, pericardium, and heart in cell culture growth medium that contains streptomycin, kanamycin, and vancomycin 100 μg/ml of each is prepared and. centrifuged for 15 min at 800 times gravity. The supernatent (0.2 ml) is injected intraperitoneally into young mice or into 6- to 7-day-old chicken embryos via the yolk sac. Blind passages are not made in mice, but two blind passages are suggested in chick embryos. Mice are observed for 2 wk. Embryos are harvested after a week.

Chlamydial inclusions may be identified in peritoneal fluid from touch impressions stained with Giemsa or by the Gimenez method. Stained yolk sac impressions made from dead embryos are also necessary for a definitive diagnosis.

Cell cultures inoculated from dilutions of the original supernatent may be examined for inclusions at 72 hr. Air-dried coverslip mounts are lightly heat-fixed and Gimenez-stained or Bouin's-fixed for 5 min and Giemsa-stained in Giemsa diluted 1:20 with phosphate buffer pH 6.0 (Grimes 1985b).

The direct complement-fixation test (DCFT) and the latex agglutination test (LAT) are presently used in the detection of antibody activity in avian sera. In the DCFT sera are tested at dilutions 1:8 to 1:256. Reagents include diluted serum and antigen, guinea pig complement, hemolysin, and washed sheep red blood cells. DCFT titers of psittacines of 8 or under are considered negative, 8 probably negative, 16 and 32 possibly infected, and over 64 as currently infected or in the recovery stage, except for cockatiels where over 16 is considered positive. A CF titer of 32 or greater in a pigeon suggests current infection, but according to Meyer (1941) and Terzin et al. (1957), chlamydiae have been recovered from pigeons with no CF antibodies. This could occur in early stages of infection before a titer rise. Erratic serological test results may occur in exposed pigeons because they are immunologically immature or genetically nonsusceptible or have recovered from a latent infection. Grimes (1984) also indicates that it may take several months for this titer to decrease following an initial infection. Because of this titer duration it is difficult to determine if a treated bird is free of the infection when cultures are negative.

In the LAT (Grimes 1985a), reactions develop in 2 min when antigen/latex and antibody-positive serum mixtures are rotated to facilitate reagent contact. LAT titers of $1/64$ are possibly infected, $1/128$ probably infected, and $1/256$ infected.

For the detection of chlamydiae in tissues, Grimes (1985a) has used the enzyme linked immunosorbent assay (ELISA), but many false positives have been noted. Spencer (1989) quoted Grimes, who indicated that the LAT detects IgM, which is a sign of current infection. Longer-standing infections may produce both IgM and IgG. He further observed that a high complement fixation titer followed by a negative LAT suggests past infection.

Page (1974) used the agar gel precipitin test for serodiagnosis but obtained only a 35% response with pigeon sera, compared with 77% with CFT or 56% with the capillary tube test. A California pigeon number 3 strain was used in the comparison. Moore and Petrack (1985) used a peroxidase-antiperoxidase technique with success using parrot monoclonal antibodies directed against chlamydiae organisms in parrot necropsy tissues. Woods (1986) reported good correlation with antemortem direct fluorescent antibody tests of fecal and nasotracheal smears when compared with other current postmortem diagnostic tests in psittacines.

STRAINS. Zichis and Shaughnessy (1945) reported the isolation of an Illinois strain of chlamydiae from two human patients, one of whom died. They also studied the cross-protection in mice of 6 unidentified Chicago pigeon chlamydiae isolates. These were tested against the following test strains: psittacosis 6BC, ornithosis 207, and an Illinois strain. Two pigeon isolates protected against 207 and 2 against Illinois. This finding suggests that the Chicago pigeons were probably carrying a highly fatal strain for people. A later study (Zichis et al. 1946) revealed 45% of 200 feral pigeons were positive for the Illinois strain. Lepine (1950) has stated, however, that the disease in humans that is of pigeon origin is inclined to be less severe and have a lower fatality rate than strains from other sources. Strain variation occurs between animal species and between bird species. One pigeon isolate injected into mice did not transmit to uninfected cagemates (Weyer and Lippelt 1956). Cross-immunization experiments failed to reveal differences between duck and pigeon isolates, but toxin neutralization tests with a California duck agent did not neutralize toxin from a pigeon strain (Page 1959a). Since then Spencer (1989) quoted reports of the isolation of 12 serotypes, 6 from avian species and 6 from mammalian species. The 3 avian groups of major importance were pigeon, psittacine, and turkey.

SEROLOGY. Serologically the CFT is useful in the detection of flock infection, but it is advisable to test pigeon serum by both the direct and indirect methods (Miles 1954). Karrer et al. (1950) and Hilleman et al. (1951) found the indirect CFT satisfactory, however. Piraino (1965) at the Milwaukee Department of Health examined caged pigeons associated with cases of human psittacosis and 31–57% of the birds were serologically positive depending on the test.

Neutralization tests with chicken antiserum (Hilleman 1945) and the toxin neutralization technique (Manire and Meyer 1950) indicate antigenic differences. Differences also occur between pigeon strains and those from other sources (Davis 1955).

Chlamydial hemagglutination activity has been demonstrated by Hillemen et al. (1951).

TISSUE AFFINITY. The agent can be isolated from the spleen and pericardium or from cloacal contents. White mice may be inoculated IC or intranasally (IN) and chick embryos by yolk sac inoculation (Davis 1955).

INCLUSION BODIES. *C. psittaci* forms intracytoplasmic, nonrigid inclusions that contain no glycogen. These lie free in the cytoplasm without relation to the nucleus (Davis et al. 1973). Inclusions may be observed in wet mounts with the aid of a phase contrast microscope.

STAINING. *C. psittaci* produces no iodine-staining carbohydrate and thus retains a tan color when microcolonies are stained for several hours with a 5% solution of iodine–potassium iodide in cold methanol. Lugol's solution also stains well in 10 min. All forms of the organism may be stained in impression smears of air sacs, spleen, or pericardium with Giemsa's, Castaneda's, and Macchiavello's stain but Gimenez's method stains best. The latter is, however, not good for cloacal swabs. Methyl green–neutral red is also reported effective for inclusions (Woodland et al. 1982). Chlamydiae stain negatively with Gram stain.

STABILITY. Tetracycline, erythromycin, and chloramphenicol inhibit multiplication of the organisms by inhibiting the protein formation on chlamydial ribosomes. Penicillin acts on cell wall synthesis. It is less effective than broad spectrum antibiotics. D-cyclo-serine inhibits some strains by its action on the cell wall (Page 1978). Streptomycin sulfate, kanomycin sulfate, and vancomycin at 1 mg each/ml solution have little or no effect on organism growth. Bacitracin, neomycin, and gentamycin also have no effect.

Quaternary ammonium compounds and fat solvents quickly inactivate chlamydiae. Anything that affects the cell walls or fat within alters the organism (Page 1978). Ethyl alcohol 70%, hydrogen peroxide and silver nitrate 3%, and tincture of iodine will destroy the organism within minutes. Lime and cresyl compounds are not effective disinfectants (Tarizzo and Nabli 1967). According to Davis et al. (1973), "Some drugs that inhibit multiplication penetrate the host cell with difficulty and are thus required in high concentrations. The drugs act reversibly. They stop multiplication without killing."

Freezing at −20°C or below preserves infectious tissue elementary bodies indefinitely. The reticulated bodies are inactivated at −70°C. Six freeze-thaw cycles also destroy the organism. Tissue suspensions (20%) are inactivated by heat of incubation for 5 min at 56°C and 48 hr at 37°C (Page 1959b). The family heat-stable antigen component can withstand boiling or autoclaving at 135°C whereas the type antigen is destroyed by 60°C (Meyer 1965).

CULTURE MEDIA. White mice 3–4 wk of age may be used for isolation purposes using IC, IN, or IP routes. Guinea pigs are not recommended for study or isolation of avian strains.

Chick embryo cell cultures may be used for isolation purposes. Mammalian cell lines are also suitable (Page 1978). Yolk sac inoculation of 6-day-old chicken embryos with 0.2 ml ground liver or spleen in broth to which 20 mg streptomycin has been added per ml inocula yields gray opaque lesions in 3–4 days. Embryo death may result in 4–7 days on the first passage, 3–4 days on the 2nd passage (Hudson et al. 1955).

Prevention. Introductions of new birds and the mingling of birds at shows or races encourage transfer of infection. Thirty days of pen isolation and CF testing of new arrivals may be helpful. Antibiotic treatment may be of value in reducing shedding of the organism and in curing most birds in infected lofts.

Unfavorable environmental conditions reduce resistance to infection. Good ventilation that avoids drafts and extreme temperature changes is essential. Clean, dry, sanitary quarters and avoidance of crowding are measures that improve a flock's resistance.

Soiled cages and equipment should be immersed in 2% phenolic disinfectant or 2 tbsp sodium hypochlorite (bleach)/pt water or a residual disinfectant solution of 400 ppm quaternary ammonium compound. Thirty min contact time is recommended. Pans for feed and water pans should be scrubbed in hot water and rinsed with clean running water after disinfection.

Control. At present bird vaccination is not a practical method of control. Bacterin immunization shows promise, and pigeons may eventually be protected by this method (Harris 1983). However, research has shown that pigeons vaccinated with bacterins and exposed 4 mo later were susceptible (Hughes 1947). It is of interest that immunity in turkeys is related to specific stimulation of cell-mediated immune responses rather than humoral antibody response (Page 1978).

Treatment. Pigeon strains are susceptible to broad spectrum antibiotics, but studies have shown that they may fail to prevent the carrier state (Meyer and Eddie 1951; Davis 1955). A chlamydial strain isolated from ducks in the United Kingdom was shown to possess resistance to tetracyclines, erythromycin, and tylosin (Johnson and Spencer 1983). It is notable that Page and Bankowski (1959) found that they could eliminate infection in turkey flocks with 400 g chlortetracycline/t feed for 7–14 days when they were dealing with a strain of low virulence and low mortality but they were rarely successful when the chlamydial strain was of high virulence and produced high mortality.

No treatment is totally effective in eradicating the disease, but if drug treatment is initiated it must be given continuously for 45 days at the level specified. Resistance develops readily with indiscriminate drug use. Arnstein et al. (1964) suggested the treatment of infected pigeons with readily absorbed tetracycline at 100 mg/kg body weight. His recommendation for infected flocks are as follows.

Infected pigeons can be treated with 100 g chlortetracycline (CTC)/lb ration, consisting of milo, cracked corn, wheat, and kafir with S F Mix 66 and water. S F Mix 66, available from the American Cyanamid Company, N.J., contains 100 g pure chlortetracycline/lb soybean meal base). The S F Mix 66 is added in an amount equal to 4% of the weight of the ration. Water or cooking oil must be added at the rate of 15 ml/100 g grain to make the mix stick to the grain. The final concentration of CTC must be at least 0.89%. The feed must be prepared fresh daily and fed for 45 days, as much as the birds will eat. They will normally consume 50 g/day after they become adjusted to the feed.The above procedure may be avoided if commercially pelleted CTC feed is purchased or medicated seed containing 500 ppm CTC is used (Wachendorfer 1973).

Present human treatment suggests doxycycline hyclate (Danbury Pharmacal, Conn.) at 100 mg orally twice a day for 7 days (Gregg 1985). Weight reduced dosages have been used in pigeons but results are unconfirmed. Rifamycin and rifampicin (Merrell Dow, Cincinnati) have also been used in humans (Davis et al. 1973).

No medication is perfect. While medication reduces mortality and infection in a flock, it fails to totally eradicate the infection and carrier birds remain. To obtain best results, the local veterinarian should be consulted.

References

Ahmed, A., and M. A. Elsisi. 1965. Vet Med J Giza 12:329–30.

Andrews, C. H., and K. C. Mills. 1943. Lancet 1:292.

Arnstein, P., D. H. Cohen, and K. F. Meyer. 1964. J Am Vet Med Assoc 145:921–24.

Bruu, A. L., S. Aasen, and S. Tijaland. 1981. Scand J Infect Dis 16:145–52.

Cohen, L., I. Gray, and S. London. 1946. N Y State J Med 46:1132.

Coles, J. D. A. 1940. Onderstepoort J Vet Sci Anim Ind 15:148.

Cutting, J. B. 1990. Solvay Anim Health Vet Rep 3:5.

Dane, D. S., and M. Beech. 1955. Med J Aust 1:428.

Davis, B. D., et al. 1973. *Microbiology.* 2d ed. Hagerstown, Md.: Harper & Row, pp. 916–27.

Davis, D. J. 1950. J Am Vet Med Assoc 126:220–23.

______. 1955. In *Psittacosis,* ed. F. R. Beaudette. New Brunswick, N.J.: Rutgers University Press.

Davis, D. J., and C. L. Ewing. 1947. Publ Health Rep 62:1484.

Davis, D. J., and J. E. Vogel. 1949. Proc Soc Exp Biol Med 70:585.

Dekking, F. 1950. Psittacosis en Ornithosis in Nederland. Thesis, Amsterdam.

Dekking, F., and A. Ruys. 1951. Rev Belge Pathol 21:92.

Durfee, P. T. 1975. J Infect Dis 132:604.

Durfee, P. T., and R. M. Moore. 1975. J Infect Dis 131:193.

Eddie, B., et al. 1962. J Infect Dis 110:231.

Eugster, A. K. 1980. In *Chlamydiosis Handbook,* ed. J. Steele, vol. 2, pp. 357–417. Cleveland: CRC.

Fries, N. F. 1966. Nord Vet Med 18:44–45.

Fritzsche, F., H. Lippelt, and F. Weyer. 1956. Berl Muench Tieraerztl Wochenschr 4:61.

Gordon, F. B., and A. L. Quan. 1965. J Infect Dis 115:186–96.

Gregg, M. B. 1983. Psittacosis 1975–1981. Atlanta: Centers for Disease Control.

______. 1985. *Chlamydia Trachomatis Infections,* CDC U S Morb Mort Suppl 34 #3s.

Grimes, J. E. 1984. Personal communication.

______. 1985a. Avian Dis 30:60–66.

______. 1985b. J Am Vet Assoc 186:1075–79.

Grubb, R. 1955. Sven Lak Tidn 52:26.

Harris, J. W. 1983. World Poult Sci J 39:5–23.

Hilleman, M. R. 1945. J Infect Dis 76:96.

Hilleman, M. R., D. A. Haig, and R. J. Helmold. 1951. J Immunol 66:115.

Hudson, C. B., et al. 1955. J Am Vet Med Assoc 126:111.

Hughes, D. L., 1947. J Comp Pathol Ther 57:67–76.

Illner, F., 1962. Monatsh Veterinaermed 17:141.

Inaba. et al. 1959. Vet Bull 29:130.

Jansen, J. 1955. Zooprofilassi 10:495–501.

Johnson, F. W. A., and W. N. Spencer. 1983. Vet Rec 112:208.

Jones, H., G. Rake, and B. Stearns. 1945. J Infect Dis 76:55.

Kalra, S. L. 1958. Indian J Med Sci 12:162.

Kapitancik, B., and J Vrtiak. 1965. Veterinarstvi 15:462–65.

Karrer, H., K. F. Meyer, and B. Eddie. 1950. J Infect Dis 87:13-36.
Komarov, A., and L. Goldsmit. 1952. Harefuah 43:3.
Kuen, C. 1962. J Microbiol, Moscow 3:115-17.
Labzoffsky, A. 1947. Can Publ Health J 38:187.
Lehnert, C. 1962. Berl Muench Tieraerztl Wochenschr 75:151.
Lepine, P. R. 1950. Sem Hop Paris 26:3376.
Levinson, D. C., J. Gibbs, and J. T. Beardwood. 1944. J Am Vet Med Assoc 126:1079-84.
Manire, G. P., and K. F. Meyer. 1950. J Infect Dis 86:226.
Meenan, P. N., M. Clarke, and D. S. Breen. 1950. J Med Assoc Eire 26:70.
Meyer, K. F. 1941. Schweiz Med Wochenschr 71:1377.
______. 1965. In *Diseases of Poultry,* 5th ed., ed. H. E. Biester and L. H. Schwarte, pp. 675-770. Ames: Iowa State University Press.
Meyer, K. F., and B. Eddie. 1942. Proc Soc Exp Biol Med 49:522.
______. 1951. Bull Hyg 26:1.
Meyer, K. F., and R. J. Genewein. 1957. Helv Med Acta 24:427.
Meyer, K. F., B. Eddie, and H. Y. Yanamura. 1942. Proc Soc Exp Bio Med 49:609.
Michael, K. P. 1957. Acta Microbiol Hell 2:14.
Miles, J. A. R. 1954. Aust J Exp Biol Med Sci 32:57.
______. 1959. N Z Med J 58:506.
Mohr, W. 1954. Zeitschr Gesamte Inn Med 9:1005.
Monreal, G. 1958. Zentralbl Veterinaermed 5:273.
Moore, F. M., and M. L. Petrak. 1985. Avian Dis 29:1036-42.
Moulder, J. W. 1964. *The Psittacosis Group of Bacteria: Ciba Lectures in Microbial Biochemistry.* New York: Wiley.
Nagington, J. 1984. J Hyg (Camb) 92:9-19.
Olivo, R., and C. Badiali. 1956. G Mal Infect Parassit 8:145.
Page, L. A. 1959a. Avian Dis 3:23.
______. 1959b. Avian Dis 3:67.
______. 1960. Am J Vet Res 21:618.
______. 1974. Proc Am Assoc Vet Lab Diagn, pp. 51-61.
______. 1978. In *Diseases of Poultry.* 7th ed., ed. H. S. Hofstad et al., pp. 337-66. Ames: Iowa State University Press.
Page, L. A., and R. A. Bankowski. 1959. Am J Vet Res 20:941.
Page, L. A., and J. E. Grimes. 1984. In *Diseases of Poultry,* 8th ed., ed. H. S. Hofstad et al., pp. 283-308. Ames: Iowa State University Press.
Panigrahy, B., et al. 1982. J Am Vet Med Assoc 181:4.
Pinkerton, H., and V. Moragues. 1942. J Exp Med 75:575-80.
Pinkerton, H., and R. L. Swank. 1940. Proc Soc Exp Biol Med 45:704-6.
Piraino, F. F. 1965. J Immunol 95:1107-10.
Rugiero, H. R., et al. 1950. Prensa Med Argent 37:2593.
Schmidtke, L. 1957. Ubers Zentralbl Bakteriol [Orig A] 165:1
Shaughnessy, H. J. 1955. In *Psittacosis in Wild Pigeons,* ed. F. R. Beaudette. New Brunswick, N.J.: Rutgers University Press.
Siegmund, I. 1960. Zeitschr Gesamte Inn Med 15:622.
Spencer, L. M. 1989. J Am Vet Med Assoc 195:855.
Storz, J. 1971. *Chlamydia and Chlamydia Induced Diseases.* Springfield, Ill.: Charles C Thomas.
Strauss, J. 1967. Am J Ophthalmol 63:1246.
Tarizzo, M. L., and B. Nabli. 1967. Am J Ophthalmology 63:1550-57.
Terzin, A. L. 1958. J Hyg Epidemiol Microbiol Immunol 2:129.
Terzin, A. L., M. R. Fornazaric, and D. M. Hlaca. 1957. Acta Virol 1:203.
Varella, G. 1955. D F Rev Inst Salubr Enferm Trop Mex 15:221.
Wachendorfer, J. G. 1973. J Am Vet Med Assoc 162:298-303.
Weyer, F., and H. Lippelt. 1956. Z Hyg Infektionskr 143:223.
Wohlrab, R. 1955. Desinfekect Gesundh 47:1.
Woodland, R. M., J. Malam, and S. Darougar. 1982. J Clin Pathol 35:642-44.
Woods, L. W. 1986. Proc American Association of Avian Veterinarians, pp. 75-79.
Zichis, J., and H. J. Shaughnessy. 1945. Science 102:301.
Zichis, J., H. J. Shaughnessy, and C. Lemke. 1946. J Bacteriol 51:55.

Tuberculosis

Definition and Synonyms. Tuberculosis in pigeons is an old sporadic, widespread, granulomatous contagious bacterial disease of birds, animals, and humans caused by an acid-fast bacillus. It is characterized by insidious chronicity with unthriftiness and finally death.

Synonyms. *Mycobacterium avium* infection, *M. avium* complex, Battey avian complex (Thoen et al. 1972a) and *M. intracellulare-M. avium* infection (Feldman 1938).

Cause and Classification. The bacteria are listed as class Schizomycetes, family Mycobacteriaceae, order Actinomycetales, and genus and species *Mycobacterium tuberculosis avium.* Thoen et al.(1972a) listed 20 serotypes in the *M. avium* complex.

Two other type species of tuberculosis, *M. tuberculosis* var. *bovis* and *M. tuberculosis* var. *hominis* are not a problem in chickens or pigeons (Burrows 1954).

Nature. The tubercle bacteria are slender, nonmotile, nonsporulating, often beaded, slightly curved rods with rounded ends, 1-3 μm in length. They have a waxy capsule and are acid-fast but young cultures may have non-acid-fast organisms. They occur singly and in parallel pairs or groups. As a rule, they are gram-positive and are obligate aerobes.

On solid media, avian colonies are small, slightly raised, grayish, moist, and glistening. Growth requires 10–21 days or longer. On nutrient broth with 2% glycerin, the organism forms a soft pellicle on the surface (Burrows 1954). Karlson (1962) observed bright yellow colonies.Details of the biochemical properties of *M. avium* are provided by Thoen et al. (1972b) and Thoen et al. (1984).

Distribution

GEOGRAPHICAL HISTORY. Tuberculosis is worldwide in distribution. Villemin in 1865 showed that the disease could be transmitted by inoculation of human tuberculosis material into rabbits. In 1872 Paulicki noted the similarity between the human and bovine types and what was later shown to be avian tuberculosis. Robert Koch (1882), a German bacteriologist in Posen, discovered the bacillus that caused tuberculosis but considered it the same in all species. In two reports (1884) Cornil and Megnin first recognized avian tuberculosis in chickens as a separate disease. Maffucci (1889, 1892) confirmed that chicken and bovine tuberculosis were distinct entities. In 1890 Koch developed tuberculin, a sterile culture filtrate that contains the growth products of the organism. This he used as a test for the disease (Koch 1890, 1891, 1902). During this period Koch (1901) accepted the fact that tuberculosis of chickens was an infection unlike the bovine or human types. In 1898, while with the Department of Agriculture, Theobald Smith also demonstrated differences between bovine and human tubercle bacteria. It was about 1900 that tuberculosis was first recognized in poultry in the United States. By 1920 it was widely distributed on farms of the northern midwestern states. A survey reported by Johnson and Ranney (1956) centered the avian disease in that area. Jones (1911) reported an outbreak in pigeons in New York state. This was followed by Riddle (1921), who gave an account of 940 cases in pigeons and doves. Zschernitz (1923) experimented with the oral transmission of human and bovine tuberculosis to pigeons. Later in 1926 Schalk in North Dakota conducted transmission tests between tuberculosis-infected pigeons and chickens and noninfected pigeons and chickens. He also found 9% of pigeons from various sources infected. Lahaye (1928) found infrequent infection in Belgium and Van Heelsbergen (1929) observed pigeon infection in Holland. Eber (1924) noted it in Germany. In 1936 Stiles found 58.5% of the pigeons in Denver, Colo., infected.

INCIDENCE. Pigeons in close association with infected chickens have a higher incidence of infection. The disease is often present in small backyard flocks of older chickens. The organism has been isolated from many sources.

SEASON. Tuberculosis may appear at any time of the year, but the clinical disease is observed more often in the northern temperate zone during the cold, rainy, spring months.

AGE, SEX, AND BREED. The disease often starts in young birds but usually only birds over 6 mo are recognized clinical cases because the disease has had a longer time to become established. The incidence does not appear to be altered by sex or breed.

SPECIES. All species of birds may be infected with avian tuberculosis. The avian type readily infects swine, rabbits, mink, and sheep. It is also reported in cattle, deer, and marsupials. Dogs, cats, monkeys, and humans are highly resistant but can contract avian tuberculosis. Guinea pigs are relatively resistant but hamsters are susceptible (Thoen and Karlson 1984).

CARRIERS. Infected adult pigeons serve as carriers.

Transmission

The mode of transmission and the sources of the serotypes of the *M. avium* complex are not completely known. Even though rats and mice, dogs, and cats are resistant to infection, they can mechanically carry the organism. Earthworms in an infected barnyard may also serve as mechanical carriers. Schalk et al. (1935) reported soil and litter infected after 4 yr, but it is unknown how much longer *M. avium* will live in the soil. A contaminated premise and infected birds, however, are considered the chief sources of infection.

In pigeons the infection may be ingested in dropping-contaminated soil, food, grit, or water. Organism-laden loft dust may also infect the equipment and premises. Fritzsche and Allam (1965) found avian tubercle bacilli in 3.5% of chicken eggs from infected flocks. Pigeons that break and eat infected eggs may thus contract infection, but squabs are unlikely to hatch with the infection contracted from eggs from infected breeders. Fitch and Lubbenhusen (1928) and Schalk et al. (1935) were unable to demonstrate transfer of infection by chickens hatching eggs.

Hosty and McDurmont (1975) found *M. avium* serotypes 9 and 16 in oysters and serotypes 9, 13, and 19 in raw milk. Songer (1980), quoting others, reported *M. avium* serotypes 6, 7, 8, 9, 12, 14, 16, 17, 19, and 20 in house dust and 3, 4, 6, 8, 9, 10, 14, and 16 in sawdust.

Epidemiology

DISTRIBUTION IN BODY. Tubercles are often found along the intestine and in the liver and spleen, but avian tuberculosis may be a generalized disease with lesions in many tissues including the skin.

ENTRY AND EXIT. Avian tuberculosis organisms are usually ingested in contaminated feed or soil. Infection by way of the respiratory tract undoubtedly takes place, although this is not the common portal of entry. The bacteria may enter through skin abrasions or other traumatic injuries, but penetration of intact skin is uncertain (Burrows 1954). Because the disease involves the digestive tract, droppings carry the bacteria from the body.

PATHOGENICITY. A relationship appears to be present between colony type and virulence. Older, dark, rough- or smooth-domed colonies lack virulence for chickens, whereas transparent, smooth colonies are virulent (Thoen 1979). Infectivity also depends on the serotype of the organism and the resistance of the host. Malnutrition, overcrowding, and stress decrease resistance to the disease.

IMMUNITY. Each species of animal varies in its susceptibility to *M. avium.* Birds also vary in their resistance to the different serotypes; serotypes 1 and 2 predominate.

Incubation. The length of the period preceding clinical disease is largely a factor of exposure and bird resistance. Clinical signs in injected birds can be noted in 1 wk.

Course. The disease runs a prolonged course ending in death.

Morbidity. Gradually all birds in a loft will become unthrifty and show signs of infection.

Mortality. Eventually all birds in an infected loft will die.

Signs. The feathers become rough and dull and diarrhea develops along with lameness, sunken eyeballs, and emaciation. Hard tumorlike masses may form at leg or wing joints.

Necropsy. Early infection may produce bile stasis, a green liver, an enlarged spleen, and brownish kidneys. Joint lesions often occur and ulcerate, presenting yellow cheesy material when opened. In most cases the liver, spleen, and intestine bear grayish white, thickly studded, irregular, firm, circumscribed, pinpoint to walnut-sized nodules that can be easily peeled out with little capsule left behind. Older lesions may have a capsule and are often hard and gritty from calcium deposits. Pulmonary lesions seldom occur in birds.

Micropathology. Following ingestion, tubercle organisms may be transported to any part of the body by lymph or blood channels. White blood cells, called polymorphonuclear leukocytes, followed by mononuclear epithelioid cells derived from histiocytes, are attracted as phagocytes. The mononuclear cells engulf the organisms, but the bacteria with their waxy capsules resist destruction. Cells accumulate around the area and form a protective wall to stop the spread of the infection. This tissue reaction results in localized encapsulated nodules called *tubercles.* The lesion is characterized by a core of coagulative caseous necrosis bounded by a cluster of mononuclear cells, some of which fuse to form giant cells with multiple nuclei. This nodule is surrounded by a border of lymphocytes. If the tubercle remains stationary and the wall is dense on all sides, it is considered arrested. Calcium salts may become deposited in the lesion over a period of time, transforming it into a gritty calcified nodule. This is typical of avian-type tuberculosis infections.

Diagnosis. The use of tuberculin for intradermal skin testing is not dependable in pigeons or chickens. Likewise, the whole-blood agglutination test is unreliable. The enzyme linked immunosorbent assay (ELISA) appears to have value in detecting mycobacterial antibodies in infected chickens but the use of the test has not been reported for pigeons.

A positive acid-fast smear from a pigeon with a typical liver or spleen tubercle is sufficient for diagnosis. Other acid-fast organisms may be found but seldom occur in conjunction with typical lesions. Culturing may be required in less advanced cases.

Differential Diagnosis. Bovine and human tuberculosis must be distinguished from *M. avium.* These organisms will not grow at temperatures in excess of 41°C and can be recognized from *M. avium* on this basis (Kelser and Schoening 1948). For other biochemical characteristics see previous references. Coligranuloma and other nodules observed on necropsy must be considered suspicious.

Biological Properties

STRAINS. Thoen et al. (1972a) identified *M. avium* serotypes 1 and 2 as the ones most often isolated from chickens and swine in the United States. These also occur in pigeons.

SEROLOGY. Schaefer (1965) demonstrated culture-stable *M. avium* complex serotypes. This enabled the identification of 20 serotypes, with serotypes 1, 2, and 3 corresponding to the USDA system (Thoen 1972a). Serotype 3 has been isolated from birds in Europe but not in the United States (Schaefer et al. 1973).

TISSUE AFFINITY. The organism is generally localized in the liver, spleen, and intestine, but joint infection often occurs. Lung or skin involvement is not common.

STAINING. Because of the fatty or waxy capsule, the tubercle bacteria will not stain readily with staining methods that are effective with other bacteria. The microorganisms are acid-fast, which means that once stained with red carbol-fuchsin they retain the stain and are not easily decolorized by acid alcohol. They can not be restained with a second blue counterstain. Tuberculosis organisms retain the red color.

The Ziehl-Neelsen method of staining is commonly used. A thin film or impression smear of suspected material is gently fixed by heat. Carbol-fuchsin, which is a red stain, is applied to the slide and steamed over a flame (but not boiled) for 3–5 min. This is washed with water and decolorized with acid-alcohol (3 ml hydrochloric acid/100 ml 95% ethyl alcohol) until most of the red color dissipates. The slide is again washed and counterstained with Loeffler's methylene blue for 1 min.

According to Wolinsky (Davis 1973), when mycobacterium are gram-stained they may appear to be positive but they take up the stain weakly and irregularly without requiring iodine to retain it. Burrows (1954) suggests warming the aniline gentian violet for 2–3 min to establish the Gram status.

STABILITY. The sun destroys the organism in a few hours, but it can live in putrefying material for weeks. Water and sewage also permit growth for some time (Burrows 1954). Johnson and Ranney (1956) reported that the tubercle bacillus can remain alive and infectious in a moist, dark, protected site for up to 2 yr. Schalk et al. (1935) found *M. avium* viable in soil after 4 yr. Many serotypes of *M. avium* have been isolated from soil and water, but only serotype 1 of the chicken serotypes has been found in water (Songer 1980).

The heat of pasteurization, 62°C for 30 min, was designed to kill tuberculosis bacilli. Moist heat at 65°C kills in 15 min, at 85°C in 2 min, and at 100°C in 1 min. Tuberculosis organisms are highly resistant to drying and chemical disinfectants. They are usually resistant to acid and alkali and relatively insensitive to cationic detergents (Wolinsky 1973).

CULTURE. Most avian strains readily adapt to artificial media, but others are slow-growing. Jensen's modification of Lowenstein's medium, which contains bone marrow infusion, potato meal, citrate, glycerol, asparagine, and malachite green, is an effective culture medium. Glycerated potato is used in Corper's medium. Egg yolk and potato medium is recommended by the American Trudeau Society (Burrows 1954). Christensen and Leach (1966) recommended egg agar for primary isolation, but Wolinsky (Davis 1973) suggested oleic acid-albumin agar medium.

The avian tubercle bacillus grows at temperatures from 25 to 45°C (Thoen and Karlson 1984). Carbon dioxide 5–10% increases growth (Stafseth et al. 1934).

Prevention. Pigeons should be kept away from other birds and animals and new additions should be isolated for 60 days.

Control and Treatment. If the disease is diagnosed in a loft, all birds must be killed and burned and the loft cleaned up and disinfected. No treatment or vaccination is suggested or approved for pigeons. Streptomycin, isoniazid, aminosalicylic, ethambutol, pyrazinamide, ethioamide, viomycin, and BCG vaccine have been used for the control of tuberculosis in humans and primates (Lyght 1966), but these products should not be used on pigeons. Disinfectants alone are not effective in rendering a premise tuberculosis-free. Any fiber litter should be sprayed with kerosene and burned. The top 6 in. of soil should be separated from dirt floor pens and buried under 2 ft of soil away from a water supply or a high-water-table area. Old wood equipment, boxes, and crates should be burned. If possible the clean loft should be idle for 6 mo, then tested with six birds for another 6 mo. The test birds should be necropsied and cultured at the end of the test period. To achieve the goal of total elimination of *M. avium* from flocks, bird disposal is essential.

References

Burrows, W. 1954. *Textbook of Microbiology,* 16th ed., Philadelphia: Saunders, pp. 542–57.

Christensen, W. B., and R. E. Leach. 1966. Health Lab Sci 3:39–43.

Cornil, V., and P. Megnin. 1884. C R Soc Biol 36:617.

Davis, B. D., et al. *Microbiology.* 2d ed., Hagerstown, Md.: Harper & Row, pp. 844–69.

Eber, A. 1924. Infektionskr 25:145–75.

Feldman, W. H. 1938. *Avian Tuberculosis Infections.* Baltimore: Williams & Wilkins.

Fitch, C. P., and R. E. Lubbenhusen. 1928. J Am Vet Med Assoc 72:636–49.

Fritzsche, K., and M. S. A. M. Allam. 1965. Arch Lebensmittelhyg 16:248–50.

Hosty and McDurmont. 1975. Proc US Animal Health Assoc.

Johnson, H. W., and A. F. Ranney. 1956. *Yearbook of Agriculture,* House Doc 344, pp. 213–21.

Jones, F. S. 1911. Rep N Y State Vet Coll 1911–1912, pp. 159–64.

Karlson, A. G. 1962. In *Advances in Vet Sci,* vol. 9, ed. C. A. Brandly and E. L. Jungherr, pp. 147–81. New York: Academic.

Kelser, R. A., and H. W. Schoening. 1948. *A Manual of Veterinary Bacteriology.* Baltimore: Williams & Wilkins.

Koch, R. 1882. Berl Klin Wochenschr 19:221–30.

______. 1890. Wien Med B1 13:531–35.

______. 1891. Dtsch Med Wochenschr 17:101–2.

______. 1901. Dtsch Med Wochenschr 27:829–34.

______. 1902. Trans Br Congr Tuberc 1:23.

Lahaye, J. 1928. *Maladies des Pigeons: Anatomie Hygiene Alimentation.* 3d ed. Remouchamps, Belg.: Steinmetz Haenen.
Lyght, C. E. 1966. *The Merck Manual.* 11th ed. Rahway, N.J.: Merck.
Maffuci, A. 1889. Centralbl Bakteriol 5:237-41.
______. 1892. Z Hyg Infektionskr 11:445-86.
Paulicki, M. 1872. Mag Gesamte Tierheilkd.
Riddle, A. 1921. J Infect Dis 29:544-52.
Schaefer, W. B. 1965. Am Rev Respir Dis 92:85-93.
Schaefer, W. B., et al. 1973. J Hyg (Camb) 71:549-57.
Schalk, A. F. 1926. Ag Exp Stn Bull 194, Fargo, N.D.
Schalk, A. F., et al. 1935. Ag Exp Stn Bull 279, Fargo, N.D.
Smith, T. 1898. J Exp Med 3.
Songer, J. G. 1980. Proc U S Animal Health Assoc, pp. 528-35.
Stafset, H. J., et al. 1934. J Am Vet Med Assoc 85:342-59.
Stiles, G. W. 1936. Am Pigeon J 25:212-13.
Thoen, C. O. 1979. In *American Society for Microbiology,* ed. R. Schlessinger. Washington, D.C.
Thoen, C. O., and A. G. Karlson. 1984. In *Diseases of Poultry,* 8th ed., ed. M. S. Hofstad et al., pp. 165-77. Ames: Iowa State University Press.
Thoen, C. O., C. M. Himes, and A. G. Karlson. 1984. In *The Mycobacteria: A Sourcebook,* ed. G. P. Kubica and L. G. Waynu. New York: Marcel Dekker.
Thoen, C. O., A. G. Karlson, and A. F. Ranny. 1972a. Proc US Animal Health Assoc, pp. 423-26.
Thoen, C. O., W. D. Richards, and J. L. Jarnigin. 1972b. Proc US Animal Health Assoc, pp. 440-43.
Van Heelsbergen, T. 1929. *Handbuch der Geflugelkrankheiten und der Geflugelzucht.* Stuttgart: Ferdinand Enke.
Villemin. 1865. Gaz Hebdom.
Zschernitz, K. 1923. Inaug Diss U Leipzig, p. 16, Abstr ESR 51:782.

Omphalitis

Definition and Synonyms. Omphalitis is a wound infection of the navel or umbilicus of squabs caused by a variety of bacteria. It is characterized by an inflammation of the navel and unabsorbed yolk. Navel ill and mushy squab disease are synonyms for this disease.

Cause. Omphalitis is generally the result of a mixed infection. The usual bacteria include *Escherichia coli, Pseudomonas aeruginosa, Staphylococcus* sp., *Streptococcus* sp. in the dry form, and clostridial organisms in the anaerobic wet form. Extraneous fungi may also be involved.

Distribution

GEOGRAPHY AND HISTORY. Omphalitis is present wherever pigeons are raised.

INCIDENCE. Very few embryos are affected. If eggs become dirty or wet, bacterial penetration of the eggshell pores permits development of the infection prior to hatching. Posthatching infection of the navel can and does occur where the nest is filthy.

AGE, SEASON, SEX, BREED, AND SPECIES. The condition is evident within the first 10 days after hatching if the eggs hatch. No other factors appear to alter the incidence.

TRANSMISSION. The disease is not contagious from one egg to the other or from one breeding pair to another. The problem is essentially an individual egg infection that occurs whenever eggs are washed to remove the dirt. The protein bloom on the eggshell is also removed with the dirt. The bloom protects the egg from bacterial invasion. In addition, trouble occurs when eggs sweat or become wet and remain wet. Squabs do not transmit the infection to other squabs.

Epidemiology

DISTRIBUTION IN BODY. The organisms are initially largely confined to the navel and unabsorbed yolk, but septicemia with extension of infection to the entire body often occurs.

ENTRY. The infection may enter the porous eggshell or go through the wet unhealed navel of the newly hatched squab. The unhealed naval is essentially an open wound.

IMMUNITY. Most squabs will overcome the infection and live. No specific immunity is established.

Course. Omphalitis infections are usually over in a week, but walled-off egg yolk infections may persist in the body cavity of the squab for weeks.

Morbidity. A small percentage of squabs with the problem will continue to live but remain stunted.

Mortality. Maximum mortality occurs by the sixth day after hatching and losses usually cease by the tenth day. Total losses can reach 50% of affected squabs.

Signs. Mortality that occurs during the first 10 days after hatching is suggestive of omphalitis. Squabs will have pasty blood-tinged vents with scabs over the navel opening. The wet form is characterized by large, soft-bodied dead squabs that decompose rapidly with a bad odor. Squabs may have general weakness and lack of body tone. They lack activity and are stunted. Infection may precede hatching; therefore unhatched eggs must be considered as a sign of the problem.

Necropsy. Moist inflammation and hemorrhage of the navel and body cavity together with unabsorbed yolk are common findings. The liver and spleen are swollen and off-color.

The gallbladder is often distended.

In the wet form, the abdominal cavity is filled with putrid fluid and exudate from the rupture of the yolk sac. The swollen muscles are watery and soft.

Diagnosis. Specific causes for egg and squab infection, such as paratyphoid, must first be excluded. The presence of navel inflammation and abnormal unabsorbed yolk is sufficient for a diagnosis. It normally takes 3 days for the yolk to be absorbed and utilized by the squab after hatching. If the egg yolk sac is seared with a hot spatula and a loop culture is carefully taken at this point, a normal yolk will contain no bacteria. If bacteria are present, other eggs or squabs should be checked for duplicate results.

Differential Diagnosis. Salmonella infection must be considered.

Prevention and Control. Clean, dry nest boxes are essential, as is good ventilation in a loft to expel moisture. Eggs should not be washed.

Treatment. No treatment is indicated. Rectifying environmental conditions generally solves the problem.

Vibriosis

Definition and Synonyms. Vibriosis is an uncommon, usually acute infectious bacterial disease characterized by malaise, diarrhea, and death. It is also called Paracholera vibrio, campylobacter hepatitis, and infectious hepatitis.

Cause and Classification. The bacteria are listed as class Schizomycetes, tribe Spirilleae, order Eubacteriales, family Pseudomonadaceae, and genus *Campylobacter.*

Spanedda (1941) suggested *V. columbae* as the species in pigeons. *V. metschnikovii* was reported recovered from pigeons by Krause and Windrath (1919) and Kuzdas and Morse (1956) and from chickens by Peckham (1984) and Gamaleia (1888a). Mathey and Rissberger (1964) called the turkey isolate *V. meleagridis.* Abdullah and Winkenwerder (1966) found the sparrow strain to be similar to *V. fetus* I and II of sheep and cattle.

Nature. Bird vibrios are very similar morphologically and biochemically to cholera vibrio. They are short, thick, comma-shaped, gram-negative, bacterial rods with rounded ends. Coccoid types also occur. They are observed singly and in short-chain spirals. Old agar cultures form straight rods. They are aerobes and facultative anaerobes that lack spores and capsules. Corkscrewlike motility is produced by a single, short, polar flagellum. Vibrio colonies on solid media are small, round, slightly convex, glistening, faintly yellow colonies 1–2 mm in diameter (Kelser and Schoening 1948).

Distribution

GEOGRAPHY AND HISTORY. Gameleia (1888b) first described the disease in chickens and pigeons in Odessa, Russia. Krause and Windrath (1919) in Germany noted the infection in sunbirds. Csukas (1930) in Hungary subcutaneously infected and killed pigeons, mice, and guinea pigs but not rabbits. Force-fed geese were also infected. Spanedda (1941) studied the infection in pigeons, and because of the differences between *V. metschnikovii* and the organism in pigeons he suggested the name *V. columbae.* Kujumgiev (1957) in Bulgaria killed pigeons and guinea pigs by intramuscular injection. Chicks were infected orally. Smibert (1969) recovered vibrios from pigeons, blackbirds, starlings, and sparrows. In 1981 Weber et al. isolated *C. jejuni* from the droppings of 55% of 51 apparently healthy pigeons. Fenlon et al. (1982) in Scotland found *Campylobacter* sp. in 41% of 29 feces samples.

INCIDENCE. Vibriosis is not commonly seen in pigeons but is often observed in chickens.

SEASON. Chicken cases occur throughout the year, but most cases have been diagnosed in winter and spring months. There are too few pigeon cases to establish a pattern.

AGE, SEX, AND BREED. Chicks under 8 wk old are not commonly infected, and squab infection has not been observed. There is no indication that sex and breed have any relationship to the disease.

SPECIES. Tudor (1954) identified the disease condition in 1949 in chickens in New Jersey, where a pigeon case was also examined about the same time. Identification of the vibrio species was not made.

Avian species reported naturally infected with vibrio are pigeons (Spanedda 1941), chickens (Peckham 1962), turkeys (Mathey and Rissberger 1964), sparrows (Abdullah and Winkenwerder 1966), sunbirds (*Leiothrix luteus L.*) (Krause and Windrath 1919), and turkeys and pheasants (Kujumgiev 1957).

The following species are reported infected artificially by the methods indicated: intramuscular (IM), subcutaneously (SC), intraperitoneally (IP), parenterally (P), per mouth (PO), and respiratory tract (RT).

Pigeons–IM (Gamaleia 1888a), death in 24–48 hr; IM (Krause and Windrath 1919), death in 18–48 hr; IM (Kujumgiev 1957);

RT (Gamaleia 1888b); SC (Csukas 1930)
Chickens – RT (Gamaleia 1888b)
Chicks – PO (Gamaleia 1888a; Kujumgiev 1957)
Japanese Quail – IM (Mathey and Rissberger 1964)
Poults – IM (Mathey and Rissberger 1964)
Geese – PO (Csukas 1930)
Nightingales – IM (Krause and Windrath 1919), death in 18–48 hr
Rats – P (Krause and Windrath 1919)
Mice – P (Krause and Windrath 1919); SC (Csukas 1930)
Guinea Pigs – PO (Gamaleia 1888a), death in 48 hr; SC (Kujumgiev 1957); P (Krause and Windrath 1919); RT (Gamaleia 1888b); SC (Csukas 1930)
Rabbits – RT (Gamaleia 1888b)

Species reported refractory are as follows:

Chickens – P (Krause and Windrath 1919)
Gopher – method unknown (Gamaleia 1888a)
Guinea Pig – IP (Spanedda 1941)
Pigeon – PO (Gamaleia 1888a)
Rabbits – method unknown (Gamaleia 1888a); SC (Csukas 1930); P (Krause and Windrath 1919)

CARRIER. Inapparent carrier birds harbor the organism.

Transmission. Peckham (1962) produced a sinusitis and nasal discharge in 3-wk-old chickens. This sinus drainage is sufficient to contaminate water or feed and thus spread infection. Peterson et al. (1959) isolated vibrios from 80% of infertile turkey eggs. This offers another avenue of transmission. Contact transmission did not occur between infected animals, pigeons, chicks, and guinea pigs (Gamaleia 1888b).

Epidemiology

DISTRIBUTION IN BODY. The organism invades the respiratory or digestive tracts and migrates to the rest of the body. The liver, spleen, and sinuses are commonly infected.

ENTRY AND EXIT. Vibrio organisms enter by the mouth and respiratory tract and leave in the feces and in sinus discharges.

PATHOGENICITY. The disease may have an insidious onset but is usually an acute infection with sudden death.

IMMUNITY. It was observed (Gameleia 1888a) that pigeons resisted *V. metschnikovii* when previously vaccinated for Asiatic cholera and vice versa. In chickens the immunity appears short-lived because apparently recovered flocks experience further losses 10–12 mo later.

Incubation. Clinical signs develop in 2–5 days.

Course. Death may occur 2–3 days following onset of signs.

Morbidity. The infection can strike quickly and may involve a large percentage of the flock.

Mortality. Death in pigeons can occur quickly within 24–48 hr after IM inoculation, but when given orally no clinical disease develops (Gamaleia 1888a).

Signs. General inactivity; failure to fly; loss of appetite; greenish, loose droppings; and death are signs of infection.

Necropsy. Characteristic changes include congested pectoral muscles; intestinal inflammation; a mottled, dark, congested, friable liver with irregular hemorrhagic blotches and diffuse grayish-white stellate foci of necrosis; hemorrhagic viscera; pale small or mildly swollen spleen; and swollen kidneys and loss of weight.

Diagnosis. Dark-field microscope examination of fresh, warm feces aids in identifying the comma-shaped organisms with their characteristic motility. Smears stained with fluorescein-labeled specific antibody provide a presumptive diagnosis. Agglutination tests are essential in diagnosis.

Differential Diagnosis. Fowl cholera must be excluded as a cause of infection. *V. metschnikovii* can be differentiated from *V. cholera* of humans by guinea pig inoculation. Human cholera is never fatal for guinea pigs (Kelser and Schoening 1948).

Biological Properties

STRAINS. Strain differences are apparent within the two species affecting pigeons.

HEMATOLOGY. Strains vary in their ability to hemolyze blood.

TISSUE AFFINITY. The respiratory tract and liver are good sources for the recovery of the organism.

STAINING. Vibrios stain readily with ordinary aniline dyes.

STABILITY. The organisms are destroyed by drying, direct sunlight, and the usual disinfectants. *V. metschnikovii* is killed in 5 min at 50°C but not in 2 min (Gamaleia 1888a).

CULTURE MEDIA. Vibrio isolates vary in their growth and biochemical characteristics on artificial media. Peckham (1962) found 5% blood agar and 1% carbon dioxide sufficient for primary isolation of chicken strains. Kujumgiev (1957) used only nutrient agar for isolation. Spanedda (1941) reported that vibrios isolated from pigeons did not liquefy gelatin, hemolyze blood, produce indole or hydrogen sulfide, coagulate milk, or reduce nitrate. Gamaleia

(1888a), on the other hand, observed gelatin liquefaction, coagulation of milk, and indole production. Kuzdas and Morse (1956) gave the cultural characteristics for *V. metschnikovii* as no nitrate reduction, no urease activity, acid formation in litmus milk with an alkaline surface zone on 0.1% agar, and growth on semisolid broth containing 8% dextrose. Kujumgiev (1957) noted hemolysis, nitrate reduction, and catalase production in pigeon strains. Kelser and Schoening (1948), discussing *V. metschnikovii,* reported that gelatin is rapidly liquefied, litmus milk is not acidified or coagulated, nitrates are reduced, and it grows rapidly in bouillon.

Yolk sac inoculation of 6-day-old chick embryos is good for propagation of vibrios. Embryo death occurs in about 4 days (Siegmund 1979).

Prevention. Contaminated feed and water must be avoided. Feed should be given only in hoppers. Birds with diarrhea and or a nasal discharge should be isolated and the loft screened to prevent access by free-flying birds.

Control and Treatment. Since the organism causes liver damage, treatment to help liver recovery is indicated. Liver cord cells require 3–4 wk for regeneration. The following feed supplements have aided recovery: vitamin E increased 10%, 1 lb added methionine/t feed, 1–4 lb supplemental glucose/100 lb feed, vitamin A increased to 10,000 units/lb feed, 4 g vitamin K added/t feed, and increased cystine and choline in the diet. Prophylactic treatment with furazolidone at the rate 150 g/t for 5–7 days or chlortetracycline 100 g/t feed for 10 days may be helpful.

References

Abdullah, I. S., and W. Winkenwerder. 1966. Zentralbl Veterinaermed 13:338–44.

Csukas, Z. 1930. Allatory Lapok 53:173–76.

Fenlon, D. R., M. S. Reid, and I. A. Porter. 1982. Proc International Conference on Campylobacter, Aberdeen, Scotland, pp. 261–62.

Gamaleia, M. M. 1888a. Ann Inst Pasteur 2:482–88.

______. 1888b. Ann Inst Pasteur 2:552–57.

Kelser, R. A., and H. W. Schoening. 1948. *A Manual of Veterinary Bacteriology,* 5th ed. Baltimore: Williams & Wilkins.

Krause, W., and H. H. Windrath. 1919. Berl Muench Tieraerztl Wochenschr 35:468–69.

Kujumgiev, I. 1957. Vet Ital 8:1094–1102.

Kuzdas, C. D., and E. V. Morse. 1956. Am J Vet Res 17:331–36.

Mathey, W. J., and A. C. Rissberger. 1964. Poult Sci 43:1339.

Peckham, M. C. 1962. Annu Rep N Y State Vet Coll 1961–62, p. 60.

______. 1984. In *Diseases of Poultry,* 8th ed., ed. M. S. Hofstad et al., pp. 229–31. Ames: Iowa State University Press.

Peterson, E. H., R. D. Hendrix, and C. E. Worden. 1959. J Am Vet Med Assoc 135:219–22.

Siegmund, O. H. 1979. *The Merck Veterinary Manual.* 5th ed. Rahway, N.J.: Merck.

Smibert, R. M. 1969. Am J Vet Res 30:1437–42.

Spanedda, A. 1941. G Bacteriol Immunol 26:518–520.

Tudor, D. C. 1954. J Am Vet Med Assoc 125:219–20.

Weber, A., C. Lembke, and A. Kettner. 1981. Berl Muench Tieraerztl Wochenschr 94:449–51.

Staphylococcosis

Definition and Synonyms. Staphylococcosis is a common, acute or chronic, pus-producing bacterial infection of humans, animals, and birds. It causes arthritis, synovitis, bumblefoot, peritonitis, air sac infection, osteomyelitis, septicemia, endocarditis, meningitis, omphalitis, and wound infection in pigeons. It is also called *Micrococcus* or *Staphylococcus pyogenes* infection.

Cause and Classification. The organism is classified as class Schizomycetes; order Eubacteriales; family Micrococcaceae; genus and species *Staphylococcus aureus* (*S. pyogenes*) and *S. epidermidis* (*S. albus*). *S. citreus* is now classified with *S. albus* species. Devriese et al. (1983) reported *S. gallinarum* isolated from animals and chickens. According to Bergey (1948), the term *Staphylococcus* has been replaced with *Micrococcus* (Burrows 1954).

Nature. Staphylococci are round, gram-positive bacteria that form grapelike clusters and grow over a wide temperature range. They are nonmotile, non-spore-forming, noncapsulated, facultative anaerobes 0.7–1.2 μm in diameter that rarely form chains.

Dextrose, lactose, sucrose, maltose, glycerol, and mannitol are usually fermented with acid but no gas. Gelatin is usually liquefied. Aerobic cultures do not accumulate hydrogen peroxide because they produce catalase. *S. citreus* forms lemon-yellow to pink colonies on solid media whereas *S. albus* produces colorless to white colonies. On ovine, bovine, or human blood agar, pyogenic *S. aureus* forms golden-yellow pigmented colonies that are invariably surrounded by a wide clear zone of beta hemolysis caused by an elaborated hot-cold hemolysin. Alpha lysin may be observed on sheep, calf, and rabbit blood. Other hemolysins are also reported. In addition, *S. aureus* produces a coagulase enzyme that clots citrated or oxalated plasma. Many strains also produce enterotoxins that cause poisoning. Other pathogenic activity involves toxins that dissolve fibrin clots, kill leukocytes, and necrotize tissue (Burrows 1954; Davis et al. 1973).

Nearly all infectious strains produce the enzyme coagulase. *S. gallinarum* does not and is seldom associated with illness (Adegoke 1986). Since some non-pigment-producing strains form coagulase, all strains that produce coagulase are classified as *S. aureus*. *S. epidermidis* does not form coagulase and is usually not pathogenic.

Distribution

GEOGRAPHY. Staphylococcus organisms occur worldwide. They commonly reside on the skin and mucous membranes without causing infection. In 1878 Robert Koch first described the organism in human pus (Davis et al. 1973).

INCIDENCE. Pigeons are seldom clinically afflicted, but the infection is common in all animals and birds.

SEASON, AGE, SEX, AND BREED. These factors have little influence on the incidence of infection.

SPECIES. All animals and birds are probably infected at some time during their lives.

CARRIERS. Pigeons carry coagulase-positive staphylococci in their nasal passages but do not develop clinical infections (Oeding et al. 1970). Animals and birds carry the organisms on the skin and mucous membranes. *S. epidermidis* is a fairly constant part of normal skin flora.

Transmission. Contact with contaminated tissues and objects permits bacterial transfer. Airborne infection occurs, and people themselves may transfer organisms to their birds by and from their hands. Even chronic human sinus infections serve as a source of bacteria.

Epidemiology

DISTRIBUTION IN BODY. Coagulase and the deposition of fibrin help to wall off the foci of infection in various parts of the body, but the blood stream may carry viable organisms to other tissues. When systemic infection occurs, pure cultures may be obtained from most organs.

ENTRY AND EXIT. Infection often enters the skin or mucous membranes when the integrity of the tissues is altered by bruising, abrasions, cuts, or feather-pulling, but when the bird is under severe stress, body openings may themselves be invaded. This means that these bacteria may enter by the sinuses, eyes, or mouth. Eggshells may also permit pore penetration and thus squab infection.

The organism leaves in pus discharges from sites of infection.

PATHOGENICITY. The potential for bacterial invasion and infection largely resides with the species and strain of staphylococcus and in the degree of resistance that the host is capable of mobilizing. Stress and injuries lower the barriers to infection and help to overcome natural resistance. Lysins, enzymes, and toxins elaborated by the bacteria also play an important role in invasion. *S. epidermidis* is present on the skin and mucous membranes but is relatively nonpathogenic. *S. aureus,* which normally can be recovered from the skin, is pathogenic and penetrates body defenses where the integrity of the skin or mucous membrane is altered.

IMMUNITY. The body presents natural immunity to *S. epidermidis,* but acquired immunity to *S. pyogenes* infection is usually of low degree. Staph lysins are good antigens and appear to initiate significant acquired immunity (Burrows 1954).

Incubation Period. Clinical disease may be evident in only 5–7 hours.

Course. The period between clinical disease and recovery can be shortened by abscess drainage, proper medication, and adequate supportive care. Without essential treatment death can occur in 2 days.

Morbidity. Usually only individual birds in a loft become clinically infected. The disease condition does not spread to other birds.

Mortality. If the infection is walled off and lasts for some time or otherwise becomes systemic, birds die. This is usually an individual bird response.

Signs. Hot, localized collections of pus in the swollen footpad, the tibiotarsal leg joint, the marrow of the femoral neck at the hip, or the swollen sternal bursa suggest possible staphylococcus infection. Generalized septicemia is indicated by emaciation, dark injected blood vessels, dehydration, congested skin of the face and body, and dark nails and beaks.

Necropsy. Lanced localized abscesses yield semifluid, yellowish, turbid, stringy exudate. Acute systemic infections may develop following chronic localized foci. In both forms of infection a swollen dark liver and spleen and congested swollen kidneys are usually noted.

Diagnosis. Special care must be employed to avoid culturing contaminated lesions. The skin surface may be coated with alcohol and burned off and/or seared with a hot broad knife prior to culturing. Gram-positive cocci that are coagulase- and catalase-positive and ferment glucose are considered pathogens.

Biological Properties

STRAINS. Carnaghan (1966) and Smith et al.

(1961) bacteriophage-typed avian isolates and found that some strains belong to human phage types. This suggests rather serious implications for the exchange of organisms between birds and humans.

STAINING. The bacteria stain readily with the usual aniline dyes.

STABILITY. Staphylococci are quite resistant to drying. They are killed at 80°C dry heat for 30 min and at 60°C moist heat for 20 min. Phenol, saponified cresol compound, and formalin are effective at standard levels. Crystal violet at 1:300,000 inhibits growth (Kelser and Schoening 1948).

CULTURE MEDIA. Staphylococcus grows on most agar media. Gelatin is liquefied, and litmus milk is acidified and coagulated.

Prevention. People with diagnosed staphylococcus infections should not handle eggs or birds. Lofts should be routinely disinfected with a residual disinfectant such as Dowcide A (Dow Chemical Co.), at 1 lb/12 gal water, following thorough cleaning. Bird injuries should be cleaned and treated with the advice and assistance of a veterinarian. Clean, sanitary housing, avoidance of overcrowding and other diseases, and good nutrition are essential in overcoming stress, which often predisposes birds to staphylococcus infections. Because squabs may carry staphylococci, market-dressed squabs should be refrigerated and held at 4–6°C to prevent bacterial growth and the possible production of enterotoxin (Bergdoll 1972).

Treatment. Each case requires evaluation and individual attention. Many strains are drug-resistant, and antibiotic sensitivity tests may be required to determine the most effective treatment. In addition, abscesses require drainage. Thus it is wise to consult a veterinarian for advice and treatment.

References

Adegoke, G. O. 1986. Vet Microbiol 11:185–89.

Bergdoll, M. S. 1972. In *The Staphylococci*, ed. J. O. Cohes. New York: John Wiley, pp. 301–31.

Bergey, D. H. 1948. *Manual of Determinative Bacteriology.* 6th ed. Baltimore: Williams & Wilkins, p. 175.

Burrows, W. 1954. *Textbook of Microbiology.* Philadelphia: Saunders, pp. 306–16.

Carnaghan, R. B. A. 1966. J Comp Pathol Ther 76:9–14.

Davis, B. D., et al. 1973. *Microbiology.* 2d ed. Hagerstown, Md.: Harper & Rowe, pp. 728–39.

Devriese, L. A., et al. 1983. Int J Syst Bacteriol 33:480–86.

Kelser, R. A., and H. W. Schoening. 1948. *A Manual of Veterinary Bacteriology.* Baltimore: Williams & Wilkins, pp. 186–97.

Oeding, P., et al. 1970. Acta Pathol Microbiol Scan [B] 78:414–20.

Smith, W. J., et al. 1961. Am J Vet Res 22:388–90.

Streptococcosis

Definition and Synonyms. Streptococcosis is an uncommon, worldwide, acute or chronic bacterial disease that can occur in pigeons. Apoplectiform septicemia and endocarditis are synonyms for this infection.

Cause and Classification. The bacteria are classified as class Schizomycetes, order Eubacteriales, family Lactobacteriaceae, genus *Streptococcus.* Two important species are reported: *S. faecalis* in chickens by Povar and Brownstein (1947), Agrimi (1956), and Gross and Domermuth (1962), and in chicks by Huhtanen and Pensack (1965); and *S. zooepidemicus* (*S. gallinarum*) in chickens by Edwards and Hull (1937), Buxton (1952), Peckham (1966), and Gross (1984).

Nature. Individual streptococci are spherical. They divide in one plane and the united cells tend to form a chain of two or more cocci. They are gram-positive, nonmotile, non-spore-forming, catalase-negative, facultative anaerobes. The discrete colonies on agar media are tiny, translucent, convex, entire colonies.

Distribution

GEOGRAPHY AND HISTORY. Specific reports of the disease in pigeons have not been found, but due to the prevalence of streptococci, the disease must be considered to occur naturally in pigeons.

Nogaard and Mohler (1902) first reported infection of chickens in Virginia. In 1932 Volkmar observed a case in turkeys in Ohio. Other poultry reports include Damman and Manegold (1905) in Germany, Schmidt-Hoensdorf (1925) in Brazil, Buxton (1952) in England, Agrimi (1956) in Italy, and Sato et al. (1960) in Japan.

INCIDENCE. The disease is not common in pigeons.

SEASON, AGE, SEX, AND BREED. Birds of any age may be afflicted at any time of the year. Sex and breed are not relevant.

SPECIES. *S. zooepidemicus* infects chickens, horses, sheep, and goats. *S. faecalis* affects animals, humans, and birds, but birds are not considered a reservoir. Pigeons, turkeys, ducks, geese, mice, and rabbits have been experimentally infected with *S. zooepidemicus* (Gross 1984).

CARRIER. *S. zooepidemicus* carriers may harbor infection for several months and *S. faecalis* is common in the intestine of birds and animals (Gross 1984).

Transmission. Contaminated feed and water and infected birds transmit the organisms. Agrimi (1956) was able to transmit both *S. zooepidemicus* and *S. faecalis* by aerosol and also observed egg transmission.

Epidemiology

DISTRIBUTION IN BODY. Recovery of the organism may be made from the trachea, lungs, liver, spleen, heart blood, and/or joints.

ENTRY AND EXIT. The bacteria enter through body openings and breaks in the skin. Body discharges permit release of the organisms from an infected individual.

PATHOGENICITY. Beta-hemolytic streptococci such as *S. zooepidemicus* are serious pathogens; this is in part due to several toxic substances formed by the bacteria (Burrows 1954). Both *S. zooepidemicus* and *S. faecalis* are considered highly pathogenic but *S. faecium* and *S. durans,* which produce alpha hemolysis, are only mildly pathogenic (Gross 1984).

IMMUNITY. Attempts at immunizing experimental animals indicate that an antibody response can be stimulated by a specific polysaccharide. This suggests that a specific vaccine may have value in birds.

Incubation Period. The incubation period varies from 1 day to several weeks for *S. faecalis* in chickens (Gross 1984).

Course. In a flock the infection may persist for several months.

Morbidity. Generally only a few birds in a flock become infected at any one time.

Mortality. Usually only a few sporadic deaths occur, but Hudson (1933) in New Jersey reported an outbreak of hemolytic streptococcus in chickens with 50% mortality over a 4-mo period.

Signs. *S. zooepidemicus* may develop into a chronic flock infection. Infected birds may be pale and depressed and often have a high fever. The disease usually develops rapidly, resulting in loose droppings and death. *S. faecalis*-infected birds become emaciated and dehydrated and may have a subnormal temperature (Gross 1984).

Necropsy. In the acute disease livers are enlarged, brownish, and friable. Infarcts may also be present. Spleens and kidneys are congested and swollen. Breast and leg muscles are dark from hemolyzed blood. Bloody fluid may accumulate under the skin and in the peritoneal and pericardial cavities. Ova are congested and blood vessels are injected. Longer-lasting cases exhibit extensive necrosis and fibrinous coagulated exudate. Endocarditis often occurs, causing nodular heart valves. Joint infections may also be observed.

Micropathology. Cellular changes occur prominently in the lungs, liver, spleen, and kidney. Areas of cloudy swelling and diffuse necrosis of liver cord cells are present in lingering cases. Heart valves develop nodular lesions and degeneration of heart muscle is common.

Diagnosis. Streptococci may be recovered from yolk, heart, liver, spleen, or joint lesions. Blood agar and MacConkey's agar aid in the isolation and identification of the organism. MacConkey's agar permits the growth of *S. faecalis* and varieties of enterococci but not *S. zooepidemicus. S. zooepidemicus* and *S. faecalis* ferment sorbitol, but *S. durans* and *S. faecium* do not (Domermuth and Gross 1975). *S. faecium* ferments arabinose, but *S. durans* does not ferment arabinose or sorbitol. *S. zooepidemicus* in Lancefield antigenic group C is beta-hemolytic. It does not grow at pH 9.6 or in the presence of 6.5% sodium chloride, as does *S. faecalis.* The latter is a member of the fecal streptococci (enterococci) Lancefield group D and is not hemolytic. *S. faecalis* var. *zymogenes* is an exception and is beta hemolytic. *S. faecalis* ferments sorbitol, hydrolyzes gelatin, and reduces litmus milk, as does *S. faecalis* var. *liquefaciens* (Gross 1984).

Differential Diagnosis. Diseases with somewhat similar clinical signs that cause acute mortality include staphylococcus, paratyphoid, and swine erysipelas infections.

Biological Properties

HEMATOLOGY. Streptococci are distinguished as beta-hemolytic if they produce a clear zone of hemolysis about the colony on blood agar and as alpha-hemolytic if they produce a narrow, greenish zone in the media about the colony from the production of methemoglobin in 24–48 hr of incubation. They are nonhemolytic gamma organisms if no greenish color or hemolysis occurs.

TISSUE AFFINITY. All tissues are affected.

STAINING. Streptococci stain readily with the usual bacterial stains.

STABILITY. Common phenolic disinfectants are considered effective.

CULTURE. Blood agar permits rapid growth. Trypticase soy agar may be used, to which 2 ml defibrinated chicken blood has been added per petri dish. The agar must be cooled to 45–55°C following sterilization before the blood is added. In preparation of the media, sterile glass beads may be used to agitate and defibrinate 20 ml blood in 50 ml sterile saline.

Prevention and Control. Free-flying birds should not enter, and new birds should have a 30-day isolation period. Visibly sick birds should be isolated with flock treatment to prevent further infection. Water pans should be cleaned and disinfected daily with sodium hypochlorite, at 2 tbsp/pt water, rinsing the pans with fresh water afterward. The loft should be cleaned and disinfected, including washing the dust from the rafters.

Treatment. Sulfonamides and antibiotics are useful in treating infected birds. Drug resistance occurs from inadequate or indiscriminate use of chemotherapeutic agents. Thus it is wise to consult a veterinarian concerning drug therapy.

References

Agrimi, P. 1956. Zooprofilassi 11:491–501.
Burrows, W. 1954. *Textbook of Microbiology.* 16th ed. Philadelphia: Saunders, pp. 317–38.
Buxton, J. C. 1952. Vet Rec 64:221.
Damman, G., and O. Manegold. 1905. Dtsch Tieraeztl Wochenschr 13:577.
Domermuth, C. H., and W. B. Gross. 1975. In *Isolation and Identification of Avian Pathogens,* ed. S. B. Hitchner et al. Ithaca, N.Y.: Arnold.
Edwards, P. R., and F. E. Hull. 1937. J Am Vet Med Assoc 44:656.
Gross, W. B. 1984. In *Diseases of Poultry,* 8th ed., ed. M. S. Hofstad et al. Ames: Iowa State University Press.
Gross, W. B., and C. H. Domermuth. 1962. Am J Vet Res 23:320–29.
Hudson, C. B. 1933. J Am Vet Med Assoc 82:218.
Huhtanen, C. H., and J. M. Pensack. 1965. Poult Sci 44:830–34.
Nogaard, V. A., and J. R. Mohler. 1902. USDA BAI Bull 36.
Peckham, M. C. 1966. Avian Dis 10:413–21.
Povar, M. L., and Brownstein, B. 1947. Cornell Vet 37:49–54.
Sato, G., S. Miura, and J. Ushijima. 1960. Jpn J Vet Res 8:285.
Schmidt-Hoensdorf, F. 1925. Dtsch Tieraerztl Wochenschr 33:818.
Volkmar. F. 1932. Poult Sci 11:297.

Colibacillosis

Definition and Synonym. Colibacillosis is often an acute bacterial disease of birds, animals, and humans. It may initiate disease or be a secondary invader. Coliform infection is another name for colibacillosis.

Cause and Classification. The organism is classified as class Schizomycetes, order Eubacteriales, family Enterobacteriaceae, genus *Escherichia,* and species *coli* var. *communis* and *communior. E. coli* is grouped with the enteric bacteria and is the predominant organism in the large bowel of humans, animals, and birds.

Nature. The organism is a gram-negative, non-spore-forming aerobe and facultative anaerobe. Motility occurs by peritrichous flagella. Gelatin is not liquefied; litmus milk is acidified and coagulated and indole is produced but not hydrogen sulfide. Nitrites are formed from nitrates, and it is methyl red–positive and Voges-Proskauer-negative.

Initial cultures form smooth, round, shiny, convex, moist, translucent colonies 1–3 mm in diameter that, after continued subculturing, form irregular, rough, granular colonies on agar. Gelatin maple-leaf colonies appear opaque and grayish white, whereas nutrient agar colonies are grayish to yellowish brown on continued incubation. In broth it grows well, producing a turbid growth.

Gas fermentation occurs in dextrose, lactose, maltose, levulose, galactose, sorbitol, xylose, arabinose, dulcitol, rhamnose, and mannitol. Sucrose is variable, but polysaccharides starch, dextrin, and glycogen are not fermented. *E. coli* var. *communior* ferments sucrose, whereas *E. coli* var. *communis* ferments salicin but does not ferment sucrose (Burrows 1954). Some strains ferment lactose late, irregularly, or not at all (Davis et al. 1973).

Distribution

GEOGRAPHY AND HISTORY. In 1886 Escherich isolated the organism *Bacterium coli commune* from the feces of an infant. Since then this organism has been universally found in the intestinal tract of birds, animals, and humans. It is especially abundant in the colon, from which it derives its name. *E. coli* was first considered as a cause of septicemia in chickens by Legnieres in 1894. In the early fifties resistant pathogenic *E. coli* strains became increasingly more serious pathogens in chickens.

INCIDENCE. It is hard to determine the extent of *E. coli* infection in pigeons, because cultures are seldom taken in mild cases. Serious cases are not common in pigeons, and serology surveys have not been reported. Most references to the disease entity in this book apply to the infection in chickens and turkeys.

SEASON. Infection may occur at any time.

AGE, SEX, AND SPECIES. Younger birds appear to be more susceptible. Female adults are more often infected, following yolk rupture caused by rough handling or injury. Salpingitis has been observed in pigeons.

Birds, animals, and humans may be infected with *E. coli,* but according to Gross (1984), most of the serotypes isolated from birds produce disease only in birds. A few bird types may also be isolated from disease problems of animals.

CARRIERS. Pigeons are carriers of *E. coli* in their intestines, but serotypes have not been identified.

TRANSMISSION. Fecal organism penetration of eggshells is considered the major source of infection for embryos, particularly if the eggs have been washed. Fecal *E. coli* contamination also persists in dry dust in a loft and in rodent droppings.

Epidemiology

DISTRIBUTION IN BODY. The organism may be found throughout the body.

ENTRY AND EXIT. The bacteria gain entrance to the embryo through the eggshell. They enter the body in or on contaminated water or feed and in inhaled loft dust. They leave in the feces.

PATHOGENICITY. Ordinarily *E. coli* organisms have been considered nonpathogenic and confined to the intestine, but Harry and Hemsley (1965) found 10–15% of intestinal coliform serotypes to be pathogenic. These appear to migrate from the intestines and become established in other tissues, especially when resistance is compromised by stress or other disease. Several have shown *E. coli* to be a factor in chronic respiratory disease (Biddle and Cover 1957; Fabricant and Levine 1962).

Incubation Period. Stressed birds develop clinical disease more readily, which shortens the incubation period. Pathogenic strains also differ in virulence. The more virulent strains can produce clinical disease in 12 hr.

Course. Losses usually occur during the first week of infection.

Morbidity. Few pigeons in a flock will be affected at any one time.

Mortality. Losses of adult pigeons seldom occur from this organism.

Signs. *E. coli* infection may be suspected if early squab mortality is associated with omphalitis. General weakness and unthriftiness may be present, together with nasal or ocular discharges.

Necropsy. The organism stimulates air sac thickening and causes hazy to heavy purulent, yellowish exudate. When septicemia occurs, pericarditis and perihepatitis are often present. Livers become swollen and often turn greenish in color. Spleens are likewise enlarged, and enteritis usually is present. Eye and hock infection may occur.

Micropathology. Affected tissues present an infiltration of heterophiles and edema fluid followed by fibroblastic proliferation.

Diagnosis. Recovery of the organism from organ tissues other than the intestine, in conjunction with autopsy findings, is sufficient to establish a diagnosis. Serological testing enables identification of pathogenic strains.

Differential Diagnosis. Newcastle disease virus and paramyxovirus and mycoplasmal and psittacosis organisms may produce air sac lesions, which may accompany *E. coli* or be confused with coliform infection.

Biological Properties

STRAINS. *E. coli* is classified by O, K, and H antigens (Ewing et al. 1956). Fifteen O serotypes have been isolated from poultry. The somatic O antigen is the endotoxin released with the destruction of smooth bacterial colony organisms (Gross 1984). It is a polysaccharide-phospholipid with a boiling resistant protein factor (Harvey and Carne 1960). K antigens associated with virulence contain polymeric acid and 2% reducing sugars (Webster et al. 1952). H flagellar antigens are not related to virulence and are seldom used for identification (Gross 1984). Gross (1984) reported the serotypes isolated from poultry by various investigators throughout the world. The most common isolates were O-78: K80(B), O-2: K1(L), and O-1: K1(L). L and B designations refer to the degree of heat stability of the O antigen. Harry (1964) considered poultry house dust as a reservoir of septicemia-producing organisms. Those recovered were O-1, O-2, O-11, O-73, and O-78. Glantz et al. (1962) recorded the following O serotypes in Pennsylvania and New York chickens: 1ab, 2a, 13, 15, 17, 78, 88, 103, 109, 111ac, and 119. Ten other single isolates were noted.

TOXINS. Some strains excrete a potent, heat-labile enterotoxin that is active in the upper intestine but is not considered a factor in the disease of chickens (Gross 1984). In squabs there is some indication that it produces diarrhea.

TISSUE AFFINITY. *E. coli* has been isolated from air sacs, liver, spleen, kidney, heart, lungs, joints, and oviduct.

STAINING. The organism stains with the usual aniline dyes.

STABILITY. The bacteria are killed by exposure to 62°C for 10 min. The organism is somewhat susceptible to drying. Most disinfectants are effective at recommended levels (Kelser and Schoening 1948).

CULTURE MEDIA. The bacteria grow in the presence of bile; thus MacConkey's agar or broth is used as differential media. On Endo medium the colonies appear red with a metallic sheen, and on eosin methylene blue agar, colonies appear black or have black centers and also have a metallic sheen. Russell's double sugar medium becomes acidified and turns yellow when phenol-red indicator is used.

Prevention. Good sanitation is essential, with reduced loft dust and ammonia. Environmental stress should be avoided. Gross and Siegel (1982) have extensively studied the influence of socialization among birds and environmental stress as factors of infection. Crowding increases stress and must be avoided because it lowers resistance to disease. Indiscriminate medication and long-term low-level drug treatments are ineffective. Inactivated vaccines have been developed for specific poultry serotypes, but presently they are not needed for pigeons. Good nutrition is important; clean whole grains or heat-treated pellets are recommended.

Control. Improved sanitation and environmental conditions help to control the disease. *E.* coli survives in the dry state for sometime. In one test Harry (1964) wet the dust with water and reduced the *E. coli* population 84–97% in 7 days. Clean water and feed should be provided and sick birds segregated.

Treatment. Treatment is increasingly difficult. Highly resistant strains have developed that do not respond to previously used antibiotics and sulfonamides. A veterinarian can do sensitivity testing and give recommendations. Indiscriminate use of medications may further jeopardize the flock.

References

Biddle, E. S., and M. S. Cover. 1957. Am J Vet Res 18:405–58.

Burrows, W. 1954. *Textbook of Microbiology.* 16th ed. Philadelphia: Saunders, pp. 367–76.

Davis, D., et al. 1973. *Microbiology.* 2d ed. Hagerstown, Md.: Harper & Row, pp. 767–68.

Escherich. 1886. Die Darmbakterien des Sauglings. Stuttgart.

Ewing, W. H., et al. 1956. *Studies on the Serology of the* Escherichia coli *group.* Atlanta: Centers for Disease Control.

Fabricant, J., and P. P. Levine. 1962. Avian Dis 6:13–23.

Glantz, P. J., S. Narotsky, and G. Bubash. 1962. Avian Dis 6:322.

Gross, W. B. 1984. In *Diseases of Poultry,* 8th ed., ed. M. S. Hofstad et al. Ames: Iowa State University Press.

Gross, W. B., and P. B. Siegel. 1982. Am J Vet Res 43:2010–12.

Harry, E. G. 1964. Vet Rec 76:465–70.

Harry, E. G., and L. A. Hemsley. 1965. Vet Rec 77:35–40.

Harvey, D. G., and P. J. Carne. 1960. J Comp Pathol Ther 70:84–108.

Kelser, R. A., and H. W. Schoening. 1948. *A Manual of Veterinary Bacteriology.* 5th ed. Baltimore: Williams & Wilkins, p. 273.

Lignieres, J. M. 1894. C R Soc Biol 46:135–37.

Webster, M. E., M. Landy, and M. E. J. Freeman. 1952. J Immunol 69:135–42.

Listeriosis

Definition and Synonyms. Listeriosis is an acute bacterial disease of birds, animals, and humans that is characterized by septicemia, myocardial degeneration, and possible encephalitis in birds. It is uncommon in pigeons as a clinical disease.

Listerellosis and circling disease are synonyms for listeriosis.

Cause and Classification. The organism is placed in class Schizomycetes, order Eubacteriales, family Bacteriaceae, genus *Listeria,* and species *monocytogenes.*

Nature. The organism is a small, non acid-fast, gram-positive rod with a long terminal flagellum. It is aerobic or microaerophilic. It occurs singly, in parallel pairs, or in short chains and is non-spore- and non-capsule-forming. Gelatin is not liquefied and hydrogen sulfide and urease are not formed. Litmus milk produces no coagulation and only little acid. Nitrates are not reduced and indole are not formed. It is catalase- and methyl red–positive. Colonies are small and transparent under transmitted light. Acid but no gas fermentation occurs in dextrose, rhamnose, and salicin. Dextrin, sucrose, glycerol, and soluble starch produce acid slowly. Maltose and lactose are variable, slow acid formers. Arabinose, galactose, xylose, mannitol, dulcitol, inulin, and inositol are not fermented (Kelser and Schoening 1948).

Distribution

GEOGRAPHY AND HISTORY. Murray et al. (1926) first reported the isolation of the organism from rabbits and guinea pigs. Gray and Killinger (1966) reported infection in 17 bird species, including the pigeon, and on all continents except Africa and the Antartic. Pustovaya (1970) reported pigeons susceptible by artificial inoculation. Gray (1958) noted chicken isolations in 14 countries, including the United States. Lucas (1961) in France recovered the bacteria from pigeons, chickens, partridge, pheasants, ducks, and a canary.

INCIDENCE. Undiagnosed cases may occur in pigeons even though they are considered somewhat resistant. Packer (1975) indicated that chickens may have a 50% intestinal carrier status. This suggests that pigeons may also be heavily involved as carriers.

SEASON. Outbreaks in cattle and sheep usually occur during cold weather in the United States, but in Asia sheep are affected during dry hot weather (Siegmund 1979).

AGE, SEX, AND BREED. Younger birds appear more susceptible and waste away, but adult birds usually die suddenly. Sex and breed are not relevant.

SPECIES. The disease has a wide host range but is more common in sheep, cattle, and swine as a central nervous system infection. The organism, however, has been isolated from at least 42 domestic and wild mammals, including rodents and humans; 22 bird species; and fish, crustaceans, and insects (Siegmund

1979). Artificially inoculated mice, guinea pigs, rabbits, and rhesus monkeys are susceptible (Burrows 1954). Falcons and owls are not considered susceptible (Gross 1984).

CARRIERS. Apparently chickens may carry the organism for years and may serve as a source of infection for humans and animals. Brem and Eveland (1968) indicated that the L form of the bacteria may reside intracellularly and be transformed from this proposed carrier state to a pathogenic form.

Transmission. The widespread distribution of the organism increases the ease of transmission by direct or indirect contact. Thamm (1962) was the first to report the presence of *L. monocytogenes* in loamy soil where it may reside as a free-living organism. Kruger (1963) isolated types 1 and 4 on farms where the disease had been in sheep.

Epidemiology

DISTRIBUTION IN BODY. The organism may be isolated from the heart, liver, kidney, or spleen. If encephalitis is present, the organism may only be found in the brain.

ENTRY AND EXIT. *Listeria* may enter by the mouth, eye, or upper respiratory tract. Artificially infected chickens shed the bacteria in sinus discharges and feces but not in eggs (Gross 1984).

PATHOGENICITY. Strains vary in virulence, but pigeons are seldom clinically infected.

IMMUNITY. Graham et al. (1940) used living and killed suspensions of *Listeria* in an unsuccessful attempt to immunize chickens by subcutaneous inoculation. Later Pothmann (1944) failed to stimulate the production of agglutinins in chickens.

Incubation. Infection becomes established in about 3 days.

Course. The course in birds is usually short.

Morbidity. Very few birds will appear sick if the disease appears in a loft.

Mortality. Losses appear to vary within wide limits.

Signs. The clinical disease may be characterized by septicemia with depression that runs a rapid, fatal course. Other cases have a prolonged course lasting several days with loss of weight and diarrhea. When encephalitis is encountered, birds tend to walk in circles and have muscle tremors and torticollis.

Necropsy. Focal necrosis, congestion, and enlargement of the liver and spleen are usually observed. Peritonitis, enteritis, and salpingitis may also be seen, but extensive myocardial necrosis and pericarditis are perhaps the most important findings. Air sac infection and pulmonary edema may likewise be observed (Gray and Killinger 1966). Monocytes characterize the blood picture.

Diagnosis. Isolation of the bacteria establishes a diagnosis. Recovery of the organism is increased if tissue is ground and refrigerated for 1–4 wk and subcultured periodically (Packer 1975).

Differential Diagnosis. Paratyphoid, paramyxovirus, and coliform infections may be confused with listeriosis.

Biological Properties

STRAINS. Robbins and Griffin (1945) used flagellar antigens to identify 4 antigenic types.

HEMATOLOGY. A narrow ring of beta hemolysis occurs around colonies on blood agar.

TISSUE AFFINITY. *Listeria* is sometimes confined to the central nervous system, but it usually invades all bird tissues.

STAINING. It takes a Gram stain, but the flagellum stains with difficulty. Ordinary aniline dyes are effective stains.

STABILITY. Lehnert (1960) reported survival in soil, feces, and water for 1–2 yr and also resistance to freezing and thawing. *Listeria* can also grow in a 10% salt solution at a pH of 9.6 (Gross 1984). In addition, Lefen (1987) reported growth at a refrigerator temperature of 40°F.

CULTURE MEDIA. *Listeria* grows readily on ordinary media, but its growth is enhanced by enriched media such as blood. Chicken embryos may also serve as culture media.

Prevention and Control. Strict sanitation and rigid culling is essential in preventing the disease. Pigeons should be kept in separate facilities away from other animals. Acidification of premises to pH 3.5 or below as specified for paratyphoid will reduce soil and loft contamination.

Treatment. Levine (1965) reviewed antibiotic sensitivity in vitro and reported *Listeria* quite resistant to antibiotics such as penicillin (Foley et al. 1944); streptomycin, chloramphenicol, and bacitracin (Felsenfeld et al. 1950); and oxytetracycline (Zink et al. 1951).

It is sensitive to chlortetracycline (Gray et al. 1950); polymyxin D and neomycin (Felsenfeld 1950); and at high levels to tetracycline and erythromycin (Stenberg 1961) and oxytetracycline (Jasinska et al. 1962).

References

Brem, A. M., and W. C. Eveland. 1968. J. Infect Dis 118:181–87.

Burrows, W. 1954. *Textbook of Microbiology.* 16th ed. Philadelphia: Saunders, p. 477.

Felsenfeld, O., et al. 1950. J Lab Clin Med 35:428.

Foley, E. J., A. Epstein, and S. W. Lee. 1944. J Bacteriol 47:110.

Graham, R., C. C. Morrill, and N. D. Levine. 1940. Cornell Vet 30:291.

Gray, M. L. 1958. Avian Dis 2:296.

Gray, M. L., and A. H. Killinger. 1966. Bacteriol Rev 30:309–82.
Gray, M. L., F. Thorp, and S. L. Laine. 1950. Bacteriol Proc, p. 93.
Gross, W. B. 1984. In *Diseases of Poultry,* 5th ed., ed. M. S. Hofstad et al., pp. 261–63. Ames: Iowa State University Press.
Jasinska, S., T. Sobiech, and Z. Wachnik. 1962. Weter Wroclaw 12:155.
Kelser, R. A., and H. W. Schoening. 1948. *A Manual of Veterinary Bacteriology.* 5th ed. Baltimore: Williams & Wilkins, pp. 303–5.
Kruger, W. 1963. Arch Exp Veterinaermed 17:181–203.
Lefen, M. 1987. Poult Proc, June.
Lehnert, C. 1960. Zentralbl Bakteriol [Orig A] 180:350–56.
Levine, N. D. 1965. In *Diseases of Poultry,* 5th ed., ed. H. E. Biester and L. H. Schwarte, pp. 451–56. Ames: Iowa State University Press.
Lucas, A. 1961. Bull Off Int Epizoot 55:884.
Murray, E. G., R. A. Webb, and M. B. Swann. 1926. J Pathol Bacteriol 39:407–39.
Packer, R. A. 1975. In *Isolation and Identification of Avian Pathogens,* ed. S. B. Hitchner et al., pp. 80–83. College Station, Tex.: American Association of Avian Pathologists.
Pothmann, E. 1944. Dtsch Tieraerztl Wochenschr Rundsch 52/50:13,127.
Pustovaya, L. F. 1970. Veterinariia 8:60.
Robbins, M. L., and A. M. Griffin. 1945. J Immunol 50:237–45.
Siegmund, O. H. 1979. *The Merck Veterinary Manual.* 5th ed. Rahway, N.J.: Merck, pp. 357, 1060.
Stenberg, H. 1961. Zentralbl Bakteriol [Orig A] 182:485.
Thamm, H. 1962. Monatsh Veterinaermed 17:224–37.
Zink, A., G. D. deMello, and R. L. Burkhart. 1951. Am J Vet Res 12:194.

Erysipelas

Definition and Synonyms. Erysipelas is a sporadic, infectious, transmissible bacterial disease of animals and birds characterized by bacteremia and septicemia. It is particularly pathogenic for pigeons by injection but not common naturally.

Fish handlers' disease, diamond skin disease, and rotlauf are other names given this entity. Erysipeloid is the name of this disease in humans, not to be confused with human streptococcus erysipelas.

Cause and Classification. The organism is listed as class Schizomycetes, order Eubacteriales, family Corynebacteriaceae, genus *Erysipelothrix,* and species *rhusiopathiae* (*insidiosa*) (Davis et al. 1973).

Nature. *E. rhusiopathiae* is a gram-positive, small, slender rod 1–2 μm long. It is nonmotile, nonsporulating, and occurs singly, in short chains or in groups. Branching filaments have been described. It is a microaerophile but will grow under normal atmospheric conditions. Very tiny dewdrop-size, round, raised, smooth, clear colonies with a convex surface develop on solid media. They are about 0.1 mm in diameter. These smooth colonies are virulent; larger, rough colonies that develop later are less virulent.

Distribution

GEOGRAPHY AND HISTORY. In 1885 Loeffler discovered the bacillus causing swine erysipelas and called it *Rouget du Porc.* Jarosch (1905) was the first to report the infection in a turkey. Poels (1919) in Holland recognized the disease in 2 pigeons and 4 ducks. In 1945 de Mendonca Machado recognized the spontaneous disease in pigeons in Portugal. Evans and Narotsky (1954) first isolated the organism in chickens in the United States.

INCIDENCE. The disease is not common in pigeons.

SEASON. In swine it appears during the warmer months. Turkey outbreaks commonly occur September through November.

AGE AND SEX. Pigeons of any age are highly susceptible by injection. Either sex of pigeons can develop infection, but male turkeys are more prone to infection.

SPECIES. Rosenwald and Corstvet (1984) list naturally infected species as turkeys, chickens, pheasants, quail, peafowl, ducks, geese, wild mallards, white storks, herring gulls, golden eagles, mud hens, parrots, canaries, finches, parakeets, sparrows, eared grebes, thrushes, and blackbirds. Van Es and McGrath (1936), Drake and Hall (1947), Connell (1954), and Seibold and Neal (1956) listed humans, swine, cattle, sheep, horses, dogs, mink, mice, rats, chipmunks, porpoises, and several species of fish as naturally infected. Experimental hosts are pigeons, turkeys, mice, and rabbits. Guinea pigs and adult chickens are somewhat resistant.

CARRIERS. Infection is spread by infected animals or birds and by apparently healthy carrier animals such as swine, sheep, and rats. Pigeons do not have a persistent subclinical infection.

Transmission. Manure discharges from infected birds or animals are the chief source of infection. The author experienced a case in which chickens became infected after exposure to infected pig manure. Wellmann (1950) demonstrated mechanical transmission from sick mice to pigeons by stable flies, horseflies, mosquitoes, and other biting flies. Any break in the skin or mucous membrane will permit infection. Fish scale slime may also harbor the bacteria; thus fishmeal used as a protein source in pellet feed can carry the organism. Kubis (1942), however, indicated that the organism was killed in the small intestine of the pigeon after it was ingested.

Epidemiology

DISTRIBUTION IN BODY. The organism can be cultured from many parts of the body, but usually the liver, spleen, heart blood, and bone marrow are consistent sources unless the birds have been on antibiotics just prior to culturing. Corstvet et al. (1970) were able to recover the organism from the blood several weeks after inoculation of turkeys with the bacteria.

ENTRY AND EXIT. The bacteria can enter by bites, abrasions, and wounds of the skin or mucous membrane. They leave in the feces and discharges of the pigeon. Corstvet (1967) found the organism in feces in only a few turkeys up to 41 days after oral inoculation.

PATHOGENICITY. *E. rhusiopathiae* is highly infectious for pigeons by injection.

IMMUNITY. Immunization procedures that have been used on swine or turkeys may have value on pigeons. Mitrovic et al. (1961) used a formalin-adsorbed vaccine prepared with variant smooth strains of *E. insidiosa* to protect turkeys. Similar success was obtained by Adler and Spencer (1952) and Dickinson et al. (1953) using a formalin-killed aluminum hydroxide-adsorbed *E. rhusiopathiae* bacterin.

Incubation Period. Clinical disease and death can occur in 24 hr.

Course. Some pigeons may die in 24–48 hr. Others may live a week.

Morbidity and Mortality. In untreated pigeons illness and losses are similar.

Signs. Primary signs include weakness; inappetence; greenish diarrhea; ruffled feathers; cyanotic, reddish purple face and cere; and sudden death.

Necropsy. Enlargement and necrosis of the liver and spleen are the primary findings. The liver develops a brownish yellow engorged appearance, and the spleen becomes friable and hemorrhagic. Kidneys are swollen, and all blood vessels appear injected with blood. Diffuse hemorrhage may be found in all tissues and organs. Pigeons seldom reach the chronic stage when arthritis and endocarditis are evident.

Micropathology. Generalized engorgement of blood vessels, edema and hemorrhage, thrombosis of blood vessels, and scattered bacterial collections may be found. Degenerative changes in all body tissues may occur if the bird survives long enough. Inflammatory cell response is minimal because the pigeon dies too soon.

Diagnosis. A positive diagnosis can only be made by isolation of the organism on artificial media. The cultures are distinctive. Birds dead of the infection readily yield the organism from the liver, spleen, or bone marrow, but sick treated birds offer difficulty in culturing. Corstvet (1975) has outlined further diagnostic procedures.

Differential Diagnosis. Other bacterial diseases and poisoning must be considered.

Biological Properties

SEROLOGY. Agglutinins increase in the blood as a result of natural or artificial infection. These antibodies can be detected and measured.

TISSUE AFFINITY. The organism is found throughout the body.

STAINING. Macchiavello and Giemsa stains are reliable.

STABILITY. The bacteria are quite resistant. They can survive direct sunlight for 2 days. They remain viable in putrid material for 4 mo and survive in a dead hog carcass for 280 days. In strong brine the bacteria survive 26 days, but disinfectants readily destroy the organism (Kesler and Schoening 1948). Rowsell (1958) reported survival of the organism in swine stomach acidity of 4.3 pH but not in soil at 6.8–7.1 pH. In the author's experience the organism appears to overwinter in soil but not at pH 3.5 or below.

CULTURE MEDIA. Growth develops on plain trypticase soy agar in 24 hr at 37°C, but 5% serum or blood will enhance growth. "Test tube brush" growth develops along a nonliquefied, gelatin stab. Litmus milk may be partially acidified, but indole is not formed. Fermentation is variable; most strains ferment dextrose, lactose, galactose, and levulose with acid formation. Maximum growth occurs at 7.6 pH.

Prevention. Environmental stress factors such as overcrowding, drafty lofts, inclement weather, poor sanitation, and rapid temperature changes increase susceptibility and must be corrected. Marinelli (1928) has suggested that the diet of polished rice also predisposes pigeons to erysipelas; thus an adequate balanced diet must be provided as a means of prevention.

Keep pigeons isolated from other animals and birds and control the rodent population. A vaccine is presently not available for pigeons, but one may be developed.

Control and Treatment. The fancier must use caution in handling infected birds and litter to avoid personal infection. Rubber gloves should be used and disinfected in sodium hypochlorite (bleach) 2 tbsp/pt water. All birds should be confined until the outbreak is over, and dead birds should be burned or buried deeply and acidified with sodium acid sulfate.

Penicillin is the drug of choice in treating infected birds. It should be given at 200 g/t of feed for 10 days. Sick birds will not eat and must be injected subcutaneously under the skin on the upper side of the neck with 50,000 units of procaine penicillin morning and night. Prier and Alberts (1950) used 10,000 units of procaine penicillin at exposure and again in 24 hr in experimentally infected pigeons to reduce mortality. Van Es et al. (1945) also recommended penicillin for the infection in pigeons. In addition, Orlandella (1955) found erythromycin to be effective.

To prevent the further growth and spread of the bacteria the pH of the environment and soil must be lowered to 3.5 with sodium acid sulfate as suggested for paratyphoid.

References

Adler, H. E., and C. R. Spencer. 1952. Cornell Vet 42:238.

Connel, R. 1954. J Comp Med Vet Sci 18:22.

Corstvet, R. E. 1967. Poult Sci 46:1247.

———. 1975. In *Isolation and Identification of Avian Pathogens,* ed. S. B. Hitchner et al. Ithaca, N.Y.: Arnold, pp. 70–79.

Corstvet, R. E., C. A. Holmberg, and J. K. Riley. 1970. Commun Sci 3:149–58.

Davis, B. D., et al. 1973. *Microbiology.* 2d ed. Hagerstown, Md.: Harper & Row.

de Mendonca Machado, A. 1945. Repos Lab Pathol Vet Lisboa 6:63.

Dickinson, E. M., et al. 1953. Proc 90th Annual Meeting of the Am Vet Med Assoc, pp. 370–77.

Drake, C. H., and E. R. Hall. 1947. Am J Publ Health 37:846.

Evans, W. M., and S. Narotsky. 1954. Cornell Vet 44:32–35.

Jarosch, L. W. 1905. Oesterr Monatschr Tierheilkd 30:197.

Kelser, R. A., and H. W. Schoening. 1948. *A Manual of Veterinary Bacteriology.* 5th ed. Baltimore: Williams & Wilkins.

Kubis, L. 1942. Wein Tieraerztl Monatsschr 29:510.

Loeffler. 1885. Arbeiten aus dem Kaiserlichen Gesundheitsamte i, 46.

Marinelli, G. 1928. Folia Med 14:1478.

Mitrovic, M., P. H. Matisheck, and L. C. Lynch. 1961. Avian Dis 5:327.

Orlandella, V. 1955. Acta Med Vet Napoli 1:359.

Poels, J. 1919. Folia Microbiol 5:1.

Prier, J. E., and J. O. Alberts. 1950. J Bacteriol 60:139.

Rosenwald, A. S., and R. E. Corstvet. 1984. In *Diseases of Poultry,* 8th ed., ed. H. S. Hofstad et al., pp. 232–41. Ames: Iowa State University Press.

Rowsell, H. C. 1958. J Am Vet Med Assoc 132:357.

Seibold, H. R., and J. E. Neal. 1956. J Am Vet Med Assoc 128:537.

Van Es, L., and C. B. McGrath. 1936. Nebr Agric Exp Stn Res Bull 84.

Van Es, L., J. F. Olney, and I. C. Blore. 1945. Neb Agric Exp Stn Bull 141.

Wellmann, G. 1950. Zentralbl Bakteriol [Orig A] 155:109.

Fowl Cholera

Definition and Synonym. Fowl cholera is an uncommon, infectious, sporadic, bacterial disease of pigeons but a common, acute or chronic, septicemic disease of many other species of fowl. It results in persistent mortality in most species. Avian pasteurellosis is another term for the disease.

Cause and Classification. The organism is listed in class Schizomycetes, order Eubacteriales, family Parvobacteriaceae, genus *Pasteurella,* and species *multocida* (*avicida; aviseptica*).

Nature. *P. multocida* is a gram-negative, aerobic or facultative anaerobic, nonmotile, nonsporulating, elongated rod that measures 0.25–0.4 μm by 0.6–2.5 μm. Recent tissue or blood cultures are distinctly bipolar (Harshfield 1965). In addition, iridescent cultures form a capsular substance not observed in blue or gray cultures (Rhoades and Rimler 1984). The bacteria produce indole, oxidase, catalase, and ammonia and reduce nitrates. They ferment dextrose and sucrose with acid and no gas. Lactose and maltose are not usually fermented. Gelatin is not liquefied and milk is not coagulated. Colonies on primary isolation appear to have varying degrees of iridescence when viewed with a hand lens under obliquely transmitted light. Colonies are small, smooth, slightly convex, and circular. As they grow older or are subcultured, the larger colonies become rough and lose their iridescence and pathogenicity. Rhoades and Rimler (1984) reported that on primary isolation, blue colony organisms from birds mutated to form gray colonies composed of chains of avirulent unencapsulated bacteria, whereas iridescent or blue colony organisms occurred singly or in pairs.

Distribution

GEOGRAPHY AND HISTORY. Perroncito (1879), then Rivolta and Delprato (1880) in Italy described the organism. Pasteur (1880) isolated the organism in chicken broth. In the same year Salmon first studied the disease in the United States. Gray (1913) has credited Chabert with the first study of the disease, in

France in 1782. Maillet is cited by Manninger (1929) as the first to use the term *fowl cholera,* in 1836. Heddleston and Watko (1963) reported that pigeons died when exposed intranasally to *P. multocida.* Later Serdyuk and Tsimokh (1970) showed that pigeons, sparrows, and rats could be infected by contact with infected chickens. They also showed that pigeons and sparrows carried *P. multocida* without showing clinical evidence and that 10% of the rats had acute disease.

INCIDENCE. The disease is not common in pigeons, but it is widely distributed in other bird species.

SEASON. Chicken cholera cases are more prevalent during fall and winter months, but infection can persist year round in live birds, particularly in warm climates.

AGE AND SEX. All ages can be infected and there is no difference in incidence between sexes.

SPECIES. Intranasal exposure killed pigeons, rabbits, mice, and sparrows, according to Heddleston and Watko (1963). Chickens, turkeys, pheasants, grouse, geese, ducks, and starlings also serve as hosts. Faddoul et al. (1967) isolated *P. multocida* from a grackle, robins, a screech owl, an evening grosbeak, and a Baltimore oriole. Rhoades and Rimler (1984) listed horses, cattle, sheep, pigs, dogs, and cats as refractory to oral exposure but susceptible to intravenous inoculation.

CARRIERS. Live infected birds serve as chronic carriers. It is of interest that swine and possibly cats may also be carriers of organisms pathogenic for birds (Rhoades and Rimler 1984).

Transmission. Contact with newly purchased birds or with free-flying birds may serve to introduce infection. Mechanical carriers such as dogs, cats, and humans may track the infection on their feet, and mice and rats may shed the bacteria in their droppings.

Epidemiology

DISTRIBUTION IN BODY. The infection is basically a respiratory disease and often resides as a chronic sinus infection. Septicemic disease may develop, permitting recovery of the organism from liver, spleen, heart blood, and bone marrow.

ENTRY AND EXIT. The bacteria enter by the mouth and nostrils or sinuses and leave in the body discharges, including the droppings.

PATHOGENICITY. Blue colonies are usually isolated from low-grade chronic cases but have been isolated from chicken flocks with persistent mortality. Iridescent smooth colonies are generally considered to be highly pathogenic. This iridescence or fluorescence is associated with the capsule. Anderson et al. (1929) reported that smooth-colony organisms were 3–4 million times more virulent for pigeons than those from rough colonies.

IMMUNITY. Only low-grade resistance develops in most birds. Pigeons appear to have considerable natural resistance. Commercially produced killed bacterins have produced some immunity in chickens and turkey flocks but are not presently recommended for pigeons.

Incubation Period. Natural exposure to fowl cholera results in clinical disease in 4–9 days but in peracute disease losses may develop in 2 days.

Course. Apparent recovery may occur with treatment but low-level persistent infection may last the life of the bird.

Morbidity and Mortality. The majority of a flock may show fowl cholera infection at any one time. Losses may reach 50%.

Signs. Nasal and ocular discharge; swollen sinuses and eyes; diarrhea; wet, dirty hackle feathers; general inactivity; and death are suggestive of fowl cholera. Middle ear infection may cause loss of balance or neck twisting.

Necropsy. Dark blue congestion of the skin, breast, and viscera are common findings. Liver, spleen, and kidney swelling, together with necrosis, mucoid enteritis, inspissated discolored yolk, and pneumonia, may be evident.

Micropathology. Heterophilic infiltrations into the liver substance with coagulation necrosis is often noted, particularly in the longer-standing cases. Generalized congestion and edema is present throughout the body.

Diagnosis. Isolation of an organism from the nasal cleft, liver, spleen, or bone marrow that ferments dextrose and sucrose with acid but no gas and usually fails to ferment lactose and maltose but produces indole is considered *P. multocida* if clinical findings agree. Heddleston (1976) has listed the fermentation variations that further characterize the organism. Recovery of the organism should be made on trypticase soy agar by wire loop from a seared surface.

Differential Diagnosis. Several other diseases must be considered, such as paramyxovirus infection, herpes, erysipelas, paratyphoid streptococcus, and vibriosis.

Biological Properties

STRAINS. Brogden et al. (1978) report 16 different serotypes. Various methods have been employed by these and other workers to separate the serotypes.

HEMATOLOGY. No hemolysis of blood occurs.

TOXINS. Endotoxins are released within the bird by both virulent encapsulated and nonencapsulated strains (Rhoades and Rimler 1984).

TISSUE AFFINITY. The organism may be found throughout the body.

STAINING. *P. multocida* stains gram-negative and is readily stained with methylene blue, Giemsa, Wright's, and carbol-fuchsin stains.

STABILITY. Most standard disinfectants are effective following proper cleaning. The bacteria do not survive as saprophytes in the soil.

CULTURE MEDIA. Trypticase soy agar media is satisfactory for primary isolation. Beef extract broth and tryptose broth serve as suitable transfer media. Dunham's peptone water with 1% differential sugars and Andrade's indicator provide fermentation evaluation media.

Prevention. The infection is introduced with sick or carrier birds. All new birds must be isolated and young and old birds should be raised in separate lofts. Different species of birds and animals should be restricted. Free-flying birds and dogs and cats should be excluded. Feed and water containers should prevent entrance of droppings. Rodent control must be a part of any prevention program.

Control and Treatment. Any loft that is free of birds for 1 mo and is carefully cleaned and disinfected can be repopulated without fear of fowl cholera. No treatment is completely successful in eradicating subclinical carrier birds in chicken and turkey flocks. If an outbreak occurs in pigeons, however, the carrier state is less likely. This means that treatment can be more effective. Under these conditions treatment with Sulmet (American Cyanamid) may be given in water at 1 oz/gal for 5 days. This treatment should not be repeated in less than 5 days. Dry sulfaquinoxaline (Merck) is another effective drug. It may be mixed on vegetable oil–coated feed and given at 0.0125% continuously for a week. Mitrovic (1967) reported sulfadimethoxine (Albon, Roche) to be safe and effective against experimental chicken and turkey fowl cholera. A 2.5-lb bird should receive about 60 mg on day 1 followed by 30 mg thereafter for 5 days. If the temperature is above 80°F, water sulfa treatments must be restricted to half days, not full days. In addition to sulfa, various broad spectrum and gram-negative effective antibiotics have also been used with some success.

References

Anderson, L. A. P., M. G. Coombes, and S. M. K. Mallick. 1929. Indian J Med Res 29:611–22.

Brogden, K. A., R. Rhoades, and K. L. Heddleston. 1978. Avian Dis 22:185–90.

Faddoul, G. P., G. W. Fellows, and J. Baird. 1967. Avian Dis 11:413.

Gray, H. 1913. In *A System of Veterinary Medicine I,* ed. E. W. Hoare, p. 420. Chicago: Alexander Eger.

Harshfield, G. S. 1965. In *Diseases of Poultry,* 5th ed., ed. H. E. Biester and L. H. Schwarte, pp. 359–73. Ames: Iowa State University Press.

Heddleston, K. L. 1976. Am J Vet Res 37:745–47.

Heddleston, K. L., and L. P. Watko. 1963. Proc 67th Annual Meeting of the US Livestock Sanitary Assoc, pp. 247–51.

Manninger, R. 1929. In *Handbuch der Pathogenen Mikroorganismen,* ed. Kolle and Wasserman, pp. 529–62.

Mitrovic, M. 1967. Poult Sci 46:1153–58.

Pasteur, L. 1880. C R Acad Sci 90:239, 952, 1030.

Perroncito. 1879. Arch Tierheilkd 5:22.

Rhoades, K. R., and R. B. Rimler. 1984. In *Diseases of Poultry,* 8th ed., ed. M. S. Hofstad et al., pp. 141–56. Ames: Iowa State University Press.

Rivolta and Delprato. 1880. L'Ornitojatria, p. 546.

Salmon, D. E. 1880. Rep U S Comm Agric 401.

Serdyuk, H. G., and P. F. Tsimokh. 1970. Veterinariia 6:53–54.

Ulcerative Enteritis

Definition and Synonym. Ulcerative enteritis is an uncommon, acute, infectious bacterial disease of pigeons, quail, and other birds, characterized by sudden onset, inflammation and ulceration of the intestine, and death. It is also called quail disease.

Cause and Classification. The bacteria are listed as class Schizomycetes, order Eubacteriales, family Bacillaceae, genus *Clostridium,* and species *colinum* (Berkhoff et al. 1974b).

Nature. The anaerobe is a gram-positive bacillus that occurs singly and in pairs and appears as a straight or slightly curved rod with rounded ends. It is motile with peritrichous flagella and has oval subterminal spores (Peckham 1984). Good acid fermentation occurs in glucose, mannose, mannitol, raffinose, sucrose, and trehalose. Weak acid develops in fructose, maltose, salicin, and cellobiose. No reaction appears in arabinose, inositol, lactose, melibiose, sorbitol, xylose, gelatin, and milk (Berkhoff et al. 1974b). The organism fails to produce indole, hydrogen sulfide, catalase, urease, lipase, and lecithinase, but esculin is hydrolyzed and starch is usually hydrolyzed (Peckham 1984). Colonies on fresh, moist agar media tend to spread. They are white, round, semitranslucent, and convex, with a filamentous margin (Berkhoff et al. 1974a).

Distribution

GEOGRAPHY AND HISTORY. Morse (1907) in the USDA reported losses in quail and grouse in Great Britain and in the United States. In

1951, Glover in Canada and Peckham (1963) in New York observed infection in pigeons. Many other reports of the disease originated in the United States. Others have come from Germany (Schneider and Haass 1968), Great Britain (Harris 1961), and India (Shukla and Rajya 1968).

INCIDENCE. The disease is quite common in quail and grouse but much less so in pheasants, turkeys, and chickens. Pigeons seldom contract the infection.

AGE, SEASON, SEX, AND BREED. Young birds are more seriously affected. Season, sex, and breed appear to have no relationship to incidence.

SPECIES. Naturally infected species include pigeons (Glover 1951; Peckham 1963, 1984), quail and grouse (Gallagher 1924; Levine 1932; Pickens et al. 1932; Barger et al. 1934; Le Dune 1935), wild turkeys (Shillinger and Morley 1934), domestic poults (Bullis and Van Roekel 1944), chickens (Shillinger and Morley 1934; Witter 1952; Jungherr 1955), pheasants (Buss et al. 1958), and robins (Winterfield and Berkhoff 1977). Refractory species include guinea pigs (Peckham 1959; Berkhoff 1974a), rabbits, white mice (Shillinger and Morley 1934), and humans (Peckham 1984).

CARRIERS. The true carrier state does not appear to exist.

Transmission. Stress in the form of fasting or coccidiosis, according to Davis (1973), initiated infection in chickens exposed to the organism.

Epidemiology

DISTRIBUTION IN BODY. The organism may be recovered from the liver, spleen, and intestinal lesions.

ENTRY AND EXIT. Natural infection is established by mouth. The organism leaves in the droppings.

PATHOGENICITY. In quail 10^{-7} viable bacterial cells of a dialysis culture were sufficient to induce infection orally, but chickens given the same suspension orally failed to show clinical disease and toxins were not demonstrated (Berkhoff and Campbell 1974).

IMMUNITY. The degree of immunity in quail following exposure is sufficient to prevent reinfection, but pigeons appear to have inborn resistance. Shillinger and Morley (1934) were unable to infect pigeons with a suspension pathogenic for quail.

Incubation Period and Course. Death may occur in 1–3 days in experimentally infected quail, but in a flock the disease may last several weeks. Quail and chickens inoculated intravenously usually died in less than 20 hr (Berkhoff et al. 1974a). This indicates a short incubation.

Morbidity and Mortality. Only a few pigeons in a loft may possibly develop infection and losses are uncommon. In quail, however, losses may be 100%.

SIGNS. Pigeons become listless, depressed, and fluffed up and have persistent watery droppings. Birds appear to die from emaciation.

Necropsy. Findings include scattered intestinal crater ulcers and enteritis together with diffuse liver necrosis and splenic enlargement, often with hemorrhage. Frequently intestinal ulcers perforate with ensuing peritonitis.

Micropathology. In intestinal sections palisading rows of bacterial rods line ulcer depressions. Vascular engorgement, hemorrhage, coagulation necrosis, cellular degeneration, and edema involve the entire intestinal wall in deeper ulcers. Cellular infiltrations of lymphocytes and granulocytes occur in and about intestinal ulcers. Scattered liver foci are marked by cloudy swelling and liver cord necrosis.

Diagnosis. Gross intestinal autopsy lesions of ulcerative enteritis are diagnostic. Necrotic liver tissue crushed between two slides and stained with Gram stain may reveal the bacteria.

Differential Diagnosis. Coccidiosis and necrotic enteritis must be considered in arriving at a diagnosis.

Biological Properties

SEROLOGY. Berkhoff (1975) used an agar gel immunodiffusion test for diagnosis, but cross-reactions developed with other clostridial species. Morris (1948) developed an effective complement fixation test.

HEMATOLOGY. Strains vary in their ability to produce hemolysis (Berkhoff et al. 1974a).

TISSUE AFFINITY. Bacteria may be demonstrated in the liver, spleen, and intestine. Blood infection occurs only in the very late stages of the infection in quail.

STAINING. Brown and Brenn stain was used by Berkhoff and Campbell (1974) to demonstrate bacteria in tissue sections.

STABILITY. Freezing and lyophilizing preserves cultures (Berkhoff et al. 1974a). Peckham (1960) found cultures survived 70°C for 3 hr, 80°C for 1 hr, and 100°C for 3 min. Heat-resistant spores of the M-h strain are also resistant to octanol and chloroform (Ryter 1965; Balassa 1966). Berkhoff et al. (1974a) reported sensitivity in the laboratory to chlortetracycline, bacitracin, furacin, penicillin, and tetracycline but not streptomycin or polymyxin B at the levels used. Streptomycin, however, has been found effective in chick embryos at 10 mg/ml (Peckham 1959).

CULTURE MEDIA. The organism requires anaerobic conditions and an enriched medium for growth. Berkhoff and Campbell (1974) used 8% horse plasma with 0.2% glucose and 0.5% yeast in tryptose phosphate (Bacto) broth as a growth medium. Peckham (1959) used the chicken embryo yolk sac for isolation. Bass (1941) suggested thioglycollate medium with 5% added serum. Smith (1975) reported the use of tryptose-phosphate-glucose agar with 8% horse serum at a temperature of 37–40°C.

Prevention. Since spores in droppings remain infectious for a long time, daily thorough cleaning is essential in reducing the residual level of infection in a loft. Wire floors in exercise pens should be provided to prevent contact with spore-contaminated soil. In addition, fly control reduces mechanical transmission of spores on the feet of flies. Raising pigeons away from other birds also reduces the possibility of infection. Using clean waterpans and hoppers and avoiding floor feeding is essential in preventing the disease.

Control and Treatment. Control of the disease involves both sanitation and treatment. Feed or water treatments may be given for 7 days. Grain medications are calculated in grams per ton; streptomycin at 180 g, bacitracin at 200–400 g, or neomycin at 150 g is suggested. Water medication may be given in place of feed treatment. Water dosages are usually given in milligrams per gallon; bacitracin at 400 mg, neomycin at 200 mg, or streptomycin at 1 g is indicated. Treatments must be reevaluated at the end of 7 days before further treatments are given.

References

Balassa, G. 1966. Ann Inst Pasteur 110:175–91.
Barger, E. H., S. E. Park, and R. Graham. 1934. J Am Vet Med Assoc 84:776–83.
Bass, C. C. 1941. Conserv Rev, Summer, pp. 11–14.
Berkhoff, G. A. 1975. Am J Vet Res 36:583–85.
Berkhoff, G. A., and S. G. Campbell. 1974. Avian Dis 18:205–12.
Berkhoff, G. A., S. G. Campbell, and H. B. Naylor. 1974a. Avian Dis 18:186–94.
Berkhoff, G. A., et al. 1974b. Avian Dis 18:195–204.
Bullis, K. L., and H. Van Roekel. 1944. Cornell Vet 34:313–20.
Buss, I. O., R. D. Conrad, and J. R. Reilly. 1958. J Wildl Manage 22:446–49.
Davis, R. B. 1973. Poult Sci 52:1283–90.
Gallagher, B. A. 1924. Am Game Prod Assoc Bull, April, pp. 14–15.
Glover, J. S. 1951. Can J Comp Med Vet Sci 15:295–97.
Harris, A. H. 1961. Vet Rec 73:11-13.
Jungherr, E. 1955. East States Coop 31:6–8.
Le Dune, E. K. 1935. Vet Med 30:394–95.
Levine, P. P. 1932. Trans 19th Am Game Conf, pp. 437–41.
Morris, A. J. 1948. Am J Vet Res 9:102–3.
Morse, G. B. 1907. USDA BAI Circ 109.
Peckham, M. C. 1959. Avian Dis 3:471–77.
______. 1960. Avian Dis 4:449–56.
______. 1963. Annu Rept N Y State Vet Coll 1962–1963.
______. 1984. In *Diseases of Poultry,* 8th ed., ed. M. S. Hofstad et al., pp. 242–50. Ames: Iowa State University Press.
Pickens, E. N., H. M. DeVolt, and J. E. Shillinger. 1932. Md Conserv 9 (Spring):18–19.
Ryter, A. 1965. Ann Inst Pasteur 108:40–60.
Schneider, J. and K. Haass. 1968. Berl Muench Tieraerztl Wochenschr 81:466–68.
Shillinger, J. E., and L. C. Morley. 1934. J Am Vet Med Assoc 84:25–33.
Shukla, P. K., and B. S. Rajya. 1968. Indian Vet J 45:10–13.
Smith, L. D. S. 1975. In *Isolation and Identification of Avian Pathogens,* ed. S. B.Hitchner et al., pp. 91–92. College Station, Tex.: American Association of Avian Pathologists.
Winterfield, R. W., and G. A. Berkhoff. 1977. Avian Dis 21:328–30.
Witter, J. F. 1952. Proc 24th Northeast Conference on Avian Diseases, Maine.

Avian Pseudotuberculosis

Definition and Synonyms. Avian pseudotuberculosis (APTB) is an infectious bacterial disease found chiefly in birds, characterized by acute septicemia followed by the formation of chronic caseous nodules like those of avian tuberculosis. *Pasteurella* or *Corynebacterium* infection are terms applied to the infection.

Cause and Classification. The organism is listed as class Schizomycetes, order Eubacteriales, family Enterobacteriaceae, genus *Yersinia* (*Pasteurella*), and species *pseudotuberculosis.* (Buchanan et al. 1966).

Nature. The organism is a short, gram-negative rod with rounded ends that, according to Cook (1952), is slightly acid-fast. Coccoid and elongated filamentous forms may be found. It is an aerobe and facultative anaerobe. At 20–30°C peritrichous flagella may sometimes develop. At 22°C an envelope may be observed but neither spores nor capsules are formed. Catalase, urease, and ammonia are formed. Nitrates and methylene blue are reduced. Indole is not formed, blood is not hemolyzed, and hydrogen sulfide is usually not produced.

Plain agar colonies are smooth, round,

translucent, finely granular, grayish yellow and 0.5–1 mm in diameter. Blood agar colonies are larger, 2–3 mm in diameter. Gelatin is not liquefied, but uniform growth occurs.

Fermentation with acid but no gas occurs in arabinose, fructose, galactose, glucose, maltose, mannitol, mannose, melibiose, glycerol, rhamnose, and trehalose. Salicin and xylose are usually fermented, but no fermentation occurs in dulcitol, inositol, lactose, raffinose, sorbitol, and cellobiose. Sucrose and dextrin are variable (Kelser and Schoening 1948; Rhoades and Rimler 1984).

Distribution

GEOGRAPHY AND HISTORY. Malassez and Vignal (1883) recovered the agent from the forearm of a child and inoculated guinea pigs, the first reported isolation. Rieck (1889) reported a canary outbreak in Dresden, Germany. The organism was pathogenic for pigeons, mice, and sparrows by subcutaneous (SC) inoculation and for mice by feeding. Pfeiffer (1890) named the organism *Bacterium pseudotuberculosis rodentium.* In 1903 Wasielewski and Hoffmann in Berlin encountered the infection in buntings. Canaries, chaffinches, pigeons, mice, and rats were inoculated and infected. Pigeons died from 0.5 ml of an agar culture given intramuscularly. Pfaff (1905) in Prague lost canaries from the disease. Pigeons, sparrows, siskins, rabbits, and guinea pigs were laboratory infected. Pigeons died in 5 days after SC inoculation. Kinyoun in 1906 apparently was the first to observe the disease in birds in the United States. He examined canaries from Washington, D.C., in 1905. The recovered APTB bacteria were pathogenic for pigeons, rabbits, guinea pigs, white mice, sparrows, canaries, mockingbirds, thrushes, and parakeets, but not chickens. In 1916 in Germany, Dolfen isolated APTB from a 3-yr-old carrier pigeon and from others examined earlier. One chicken inoculated intravenously died. In Konigsberg, Krage and Weisgerber (1924) reported on an October outbreak in turkeys in which associated chickens, pigeons, and geese were not affected. Mice and guinea pigs were killed by the organism.

In 1925 Beck and Huck in Leipzig isolated APTB from turkeys on five occasions and from canaries and pigeons once. They indicated that this was not rare. In 1934 Lesbouyries in France observed the disease in 30 pigeons in which 20 were sick and 10 died. Beaudette (1940) isolated the organism from a blackbird in New Jersey. In 1953 Clapham isolated APTB from pigeons in Norfolk, England, and from other bird species. The following year Marthedal and Velling (1954) in Denmark noted seven pigeon outbreaks. In Maryland, Clark and Locke (1962) observed a persistent widespread outbreak in a winter roost of grackles (*Quiscalus quiscula*). Such an episode in migratory birds could provide an important means of dissemination of the agent.

INCIDENCE. The disease is apparently common in European countries, particularly Germany, France, and England. Several reports originating in the United States involve turkeys (Rosenwald and Dickinson 1944; Karlson 1945; Mathey and Siddle 1954; Wise and Uppal 1972).

SEASON. APTB may occur at any time, but cold, wet weather increases the mortality.

AGE, SEX, AND BREED. Birds of any age are susceptible, but younger birds are considered more easily infected. Sex and breed have little influence.

SPECIES. Naturally infected species include the previously mentioned pigeons, canaries, buntings, blackbirds, and turkeys. Others are finches and titmice (Bryner 1906); hares (Lerche 1927); ducklings and pheasants (Truche and Bauche 1930, 1933); a swan (Truche 1935); a cat, an ox, a horse, and a human (Boquet 1937); chickens (Truche and Isnard 1937); toucans (Urbain and Nouvel 1937); doves, a wood pigeon, a lark, a rabbit, a gray partridge, a bobwhite quail, pheasants, a jackdaw, a rook, and a hare (Clapham 1953); canaries, snow buntings, and waxwings (Marthedal and Velling 1954); a partridge and a long-eared owl (Borg and Thal 1961), and a grackle (Clark and Locke 1962).

Artificially inoculated species include the previously mentioned pigeons, mice, sparrows, chaffinches, mice, rats, siskins, rabbits, guinea pigs, mocking birds, thrushes, parakeets, and chickens. Others are monkeys (Truche and Bauche 1933) and baboons (Rhoades and Rimler 1984). Refractory species include white rats and hamsters (Rhoades and Rimler 1984).

CARRIERS. The reservoir of infection resides in sick birds, but a true carrier state is questionable.

Transmission. Sick pigeons or other birds may introduce infection into a loft. Contaminated clothing, shoes, and equipment may also carry the disease. Dolfen (1916) indicated that the disease appeared in a loft after sick pigeons were introduced.

Epidemiology

DISTRIBUTION IN BODY. The organism is commonly recovered from liver, spleen, or lung nodules.

ENTRY AND EXIT. Ingestion appears to be the chief means of establishing APTB infection, but breaks in the skin or mucous membrane undoubtedly also permit infection. Liver infection naturally involves the gallbladder. The bacteria can thus exit in the droppings. Other body discharges may also permit release of the agent.

IMMUNITY. Live, nonvirulent strains have protected against experimental challenges with atoxic strains and subtoxic levels of toxic strains (Thal 1954).

Incubation Period. The period for APTB varies from 3 to 6 days.

Course. Pfaff (1905) reported death of pigeons in 5 days following SC infection. Lesbouyries (1934) observed losses in 3–8 days in naturally infected pigeons.

Morbidity and Mortality. Over half the flock may show evidence of disease during an outbreak, and losses can reach 30% or more (Lesbouyries 1934).

Signs. Weakness, loss of weight, lameness, loss of appetite, and death are signs of infection.

Necropsy. Tubercular-like nodules develop in the liver, spleen, lungs, and sometimes in the duodenum.

Micropathology. The nodules are typical of caseation necrosis. Local foci are bounded by lymphocytes and granulocytes.

Diagnosis. Diagnosis is based on necropsy findings together with isolation and identification of the organism. Initial isolation of APTB may be made on trypticase soy agar incubated for 24 hr at 37°C.

Differential Diagnosis. Tuberculosis and salmonellosis must be considered together with other *Yersinia* species.

Biological Properties

STRAINS. Thal (1954) studied 186 strains of pseudotuberculosis and recovered 33 from 8 species of birds.

SEROLOGY. Strains are defined on the basis of O and H antigens. There are 5 serotype groups identified by the O heat-stable antigen. Three of these are further divided into A and B groups by absorption agglutination tests. There are also 5 heat-labile H antigens that are formed as a part of the flagella (Thal 1954).

TOXINS. Most strains in group III produce heat-labile exotoxins that can form toxoids (Rhoades and Rimler 1984).

STAINING. Aniline dyes stain irregularly, but bipolar staining has been observed (Kelser and Schoening 1948).

STABILITY. A temperature of 60°C kills the organism in 15 min (Kelser and Schoening 1948). It is easily destroyed by heat, sunlight, drying, and the usual disinfectants, but lyophilization preserves the agent for years (Rhoades and Rimler 1984).

CULTURE MEDIA. Plain trypticase soy agar may be used for isolation. MacConkey's agar has also given good growth at 37°C. The bacteria do not grow readily unless glutamic acid, thiamine, pantothenate, cystine, and nicotinamide are added (Burrows and Gillett 1966).

Prevention. The loft must be isolated and screened to exclude free-flying birds. Visitors should be restricted. Routine cleaning and disinfection practices are important.

Control and Treatment. If losses occur, a reliable diagnosis is necessary. Sick birds must be destroyed since humans are susceptible. Hot lye at the rate of 1 lb/5 gal water should be applied with a broom to the floor and lower walls; rubber gloves, boots, and eye protection must be used. A veterinarian may have suggestions. Vaccines are not available and specific treatment is not advised.

References

Beaudette, F. R. 1940. Am Vet Med Assoc 97:151–57.

Beck, A., and W. Huck. 1925. Centralbl Bakteriol 95:330–39.

Boquet, P. 1937. Ann Inst Pasteur 54:341–81.

Borg, K, and E. Thal. 1961. Svenska Lakartidin 58:1923.

Bryner, A. 1906. Inaugural diss., University of Zurich, pp. 5–53.

Buchanan, R. E., J. G. Holt, and E. F. Lessel. 1966. Index Bergeyana. Baltimore: Williams & Wilkins.

Burrows, T. W., and W. A. Gillett. 1966. J Gen Microbiol 45:333–45.

Clapham, P. A. 1953. Nature 172:353.

Clark, M. C., and L. N. Locke. 1962. Avian Dis 6:506–10.

Cook, R. 1952. J Pathol Bacteriol 64:228–29.

Dolfen, H. 1916. Inaug diss, Hannover, pp. 5–46.

Karlson, K. 1945. Skand Vet Tidskr 35:673.

Kelser, R. A., and H. W. Schoening. 1948. *A Manual of Veterinary Bacteriology,* 5th ed. Baltimore, Md.: Williams & Wilkins, pp. 213–14.

Kinyoun, J. J. 1906. Science 23:217.

Krage and Weisgerber. 1924. Tieraerztl Rundsch 20 (Jahrgang).

Lerche. 1927. Centralbl Bakteriol 104:493–502.

Lesbouyries, G. 1934. Bull Acad Vet France 7:103–7.

Malassez, L., and W. Vignal. 1883. *Archives de Physiologie Normale et Pathologique,* Series 3, pp. 369–412.

Marthedal, H. E., and G. Velling. 1954. Nord Vet Med 6:651–65.

Mathey, W. J., and P. J. Siddle. 1954. J Am Vet Med Assoc 125:482.

Pfaff, F. 1905. Centralbl Bakteriol 38:275–81.

Pfeiffer, A. 1890. Zentralbl Bakteriol [Orig A] 7:219.

Rhoades, K. R., and R. B. Rimler. 1984. In *Diseases of Poultry,* 8th ed., ed. M. S. Hofstad et al., pp.

157–60. Ames: Iowa State University Press.
Rieck, M. 1889. Dtsch Z Tiermed Vgl Pathol 15:68–80.
Rosenwald, A. S., and E. M. Dickinson. 1944. Am J Vet Res 5:246.
Thal, E. 1954. Berlingska Boktryckeriet, Lund, Sweden.
Truche, C. 1935. Bull Acad Vet France 8:278.
Truche, C. and J. Bauche. 1930. Bull Acad Vet France 3:391.
______. 1933. Bull Acad Vet France 6:43–46.
Truche, C., and S. Isnard. 1937. Bull Acad Vet France 10:38.
Urbain, A., and J. Nouvel. 1937. Bull Acad Vet France 10:188.
Wasielewski, V., and W. Hoffmann. 1903. Arch Hyg 47: 44–56.
Wise, D. R., and P. K. Uppal. 1972. J Med Microbiol 5:128–30.

Pseudomonas Infection

Definition and Synonym. *Pseudomonas aeruginosa* is frequently found in water, soil, and organic matter and is the only bacterium causing disease in plants; mammals, including humans; birds; reptiles; and insects. Pyocyanea has been used as a synonym.

Cause and Classification. The bacteria are listed as class Schizomycetes, order Eubacteriales, family Enterobacteriaceae, genus *Pseudomonas,* and species *aeruginosa.*

Nature. The gram-negative, non-acid-fast organisms usually appear as small, slender rods 1.5–3 μm long and 0.5 μm broad. They frequently form pairs and short chains. Motility results from 1–3 polar flagella. Capsules have been described by Cetin et al. (1965), but no spores are formed. Most strains produce smooth, moist colonies with irregular, wavelike, spreading edges. It grows readily on most media and causes hemolysis on blood agar. A metallic sheen has been observed in about half of the cultures (Wahba 1964). Some strains ferment D dextrose with acid. The organism is indole-negative, hydrogen sulfide–positive; and catalase- and oxidase-positive. Almost all strains liquefy gelatin, and aniline dyes are effective stains (Lusis and Soltys 1971).

Distribution

HISTORY. Poultry infection was recorded by Kaupp and Dearstyne (1926). In 1971 Lusis and Soltys provided an excellent review of the infection.

INCIDENCE. The bacterium is found worldwide. It is seldom recovered from pigeons, but becomes established in traumatized tissue by continued fighting and picking in crowded quarters.

Transmission. The organism is often present in small numbers in normal intestinal flora of humans and animals (Lusis and Soltys 1971). This suggests intestinal migration when stress occurs. Certainly *Pseudomonas* bacteria are often recovered following the rupture of ova from rough handling or trauma.

Epidemiology

DISTRIBUTION IN THE BODY. In systemic infections the organism may be isolated from the liver, spleen, and heart blood. It is often associated with enteritis, hepatic necrosis, purulent conjunctivitis, genitourinary problems, otitis externa, and pneumonia.

ENTRY AND EXIT. *Pseudomonas* invades when the integrity of the skin or mucous membranes is altered, particularly after burns. When abscesses rupture the organism is released in the pus. Kidney, nasal, and fecal discharges also carry organisms.

PATHOGENICITY. For some time *P. aeruginosa* has been regarded as a harmless saprophyte. It is now perceived to be involved in many pus-producing abscesses and a number of extracellular products enhance its pathogenicity. Its activity is perhaps also improved by combination with other organisms such as streptococcus and staphylococcus.

Pigeons and mice are less susceptible than guinea pigs or rabbits (Burrows 1954). Gross (1984) also reports isolations from chickens, turkeys, and pheasants.

IMMUNITY. Immunity can be produced by small, nonlethal filtrates of cultures (Alexander et al. 1966). Formalized bacterins have been successful in goats (Toidze 1954).

Course. The course is usually under 2 wk since medication is generally not indicated.

Mortality. Individual birds may die from infection. It is not considered a flock problem.

Signs. The presence of blue-green pus in an abscess and the isolation of the organism from an infection or from unhatched eggs prompts a diagnosis. Edema of the head has been noted in some cases.

Biological Properties

STABILITY. *Pseudomonas* organisms withstand high levels of radiation and can live in water with minimal nutrients, as in a boiling water reactor.

Prevention. Avoid washing hatching eggs without proper disinfectants.

Treatment. In treating infection, sulfadiazine, gentamycin, and streptomycin are drugs of choice (Lusis and Soltys 1971).

References

Alexander, J. W., et al. 1966. Surg Gynecol Obstet 123:965–77.

Burrows, W. 1954. *Textbook of Microbiology.* 16th ed. London: Saunders, p. 470.
Cetin, E. T., K. Toreci, and O. Ang. 1965. *J Bacteriol* 89:1432–33.
Gross, W. B. 1984. In *Diseases of Poultry,* 8th ed., ed. M. S. Hofstad et al. Ames: Iowa State University Press.
Kaupp, B. F., and R. S. Dearstyne. 1926. J Am Vet Med Assoc 65:484–88.
Lusis, P. I., and M. A. Soltys. 1971. Vet Bull 41:169–77.
Toidze, A. I. 1954. Veterinariya (Moscow) 7:24–29.
Wahba, A. H. 1964. Nature (London) 204:502.

Avian Arizonosis

Definition and Synonyms. Arizonosis is a bacterial infection caused by Arizona enteric bacteria, which are pathogenic for cold-blooded animals, fowl, and humans. The disease is very similar to salmonellosis, and it occurs in pigeons. This disease is also called paracolon infection.

Cause and Classification. The bacteria are listed as class Schizomycetes, order Eubacteriales, tribe Salmonelleae, family Enterobacteriaceae, genus *Arizona.*

The *Arizona* enteric organisms were listed within the paracolon group (Fields et al. 1967). Members are closely related serologically to salmonellae, which they resemble. They are coded in Arabic numbers (O antigen : H antigen).

Nature. The facultative anaerobe is a gram-negative non-spore-forming rod that is usually motile by peritrichous flagella. Acid with gas occurs in dextrose, mannitol, maltose, and sorbitol. Most cultures promptly ferment lactose but some take 1–2 wk. No fermentation occurs in dulcitol, salicin, adonitol, and inositol, and sucrose usually does not ferment. Gelatin slowly liquefies, but urea is not hydrolyzed. Methyl red, sodium malonate, and hydrogen sulfide are positive. Simmons' citrate is used, nitrates are reduced, but Voges-Proskauer and Jordan's tartrate are negative. Potassium cyanide and indole are usually negative. Betagalactosidase and the decarboxylases lysine and ornithine are positive, but arginine is usually delayed. Phenylalanine deaminase is negative (Ewing and Edwards 1960; Edwards and Ewing 1972).

Distribution

GEOGRAPHY AND HISTORY. Caldwell and Ryerson (1939) observed fatal infections in reptiles and described the organism now known as *Arizona* O-1,2:H-1,2,5. Lewis and Hitchner (1936) isolated a slow lactose fermenter; this is considered to be the first report of the disease in chicks. In 1941 Kauffmann studied the organism and classified it as *S. Arizona.* In 1945 he reclassified *Arizona* as a genus separate from *Salmonella* (Edwards et al. 1947).

INCIDENCE. The disease is not common in pigeons, but a percentage of apparent clinical uncultured *Salmonella* cases are in fact *Arizona.*

SEASON. The disease may occur at any time of the year.

AGE, SEX, AND BREED. Birds of any age may contract the infection but young squabs are more susceptible. Sex and breed do not appear to affect the incidence.

SPECIES. The *Arizona* organisms have a wide natural host range. Hinshaw and McNeil (1947) recovered the bacteria from reptiles (snakes). Edwards et al. (1947, 1956, 1959) reported infection in humans, turkeys, chickens, ducks, a pheasant, canaries, parrots, a macaw, dogs, a cat, sheep, swine, reptiles, a Gila monster, a capybara, monkeys, a mink, opossums, and guinea pigs. Goetz and Quortrup (1953) noted gopher infection. Sharma et al. (1970) isolated the agent from snakes, rats, millipeds, and humans. Windingstad et al. (1977) found infection in a sandhill crane. In 1981 Winsor et al. in Texas identified *Arizona* in turkey vultures. Sambyal and Sharma in India in 1972 observed infection in doves and a cow.

CARRIER. Infected although apparently healthy intestinal carriers occur frequently and serve to spread the organism. This occurs in pigeons and other species of animals and birds.

Transmission. *Arizona* is transmitted in much the same way as salmonella. Eggs and feed have been important methods of spreading the infection in turkeys and chickens. Meat scrap and bonemeal incorporated into pellets can carry the bacteria. Goetz (1962) reported a 90% infection rate in rats and 50% in mice on one turkey premise. This means that rodent droppings can serve to infect pigeon feed. Hinshaw and McNeil (1944, 1947) considered reptiles as a source of infection for poultry. This source of infection is unlikely as a cause of pigeon disease.

Epidemiology

DISTRIBUTION IN BODY. The organism may be recovered from the liver, spleen, heart blood, lungs, kidney, and unabsorbed yolk. Unhatched eggs should also be cultured.

ENTRY AND EXIT. The bacteria enter by mouth or by breaks in the skin or mucous membranes. They leave the body in droppings and excretions.

PATHOGENICITY. *Arizona* organisms are serious pathogens and have killed squabs.

Incubation. The incubation period is 3–5 days.

Course. Losses will usually occur within 7–10 days following the onset of the disease.

Morbidity and Mortality. Many squabs will be affected during the outbreak. Losses are greater under 3 wk of age and can reach 50% but are usually under 10%.

Signs. Symptoms are not specific. Birds may appear listless, become weak, and have diarrhea and pasted vents. Shivering, trembling, and sudden death may be noted, but adult birds may not show any evidence of infection.

Necropsy. Squabs may have unabsorbed yolk, navel infection, an enlarged spleen, liver swelling and necrosis, and intestinal inflammation and congestion.

Micropathology. Cloudy swelling and fatty infiltrations of the liver precede generalized liver necrosis. Other degenerative cellular changes may be found in involved tissues.

Diagnosis. A diagnosis depends on the isolation and recovery of the bacteria. Those cultures that conform to the typical reactions of *Arizona* strains may be further studied serologically and typed antigenically.

Differential Diagnosis. The culture may be confused with salmonella and nonpathogenic *Citrobacter* and *Escherichia coli* organisms. Salmonellae do not utilize sodium malonate, whereas *Arizona* cultures do. *Arizona,* on the other hand, fails to ferment dulcitol but slowly liquefies gelatin.

Biological Properties

STRAINS. Williams (1984) reported over 400 serotypes and McWhorter et al. (1977) listed identified cultures. They are presently divided into 34 O agglutinin (Ag) groups and 43 H antigens with 332 serotypes. Some of the O and H Ags cross-react with salmonellae according to Davis (1973). Edwards et al. (1956, 1959) reported *A. hinshawii* 7:1,7,8 to be the most frequent serotype isolated in poultry, animals, and humans in the United States.

STAINING. The gram-negative organisms stain readily with most basic aniline dyes. Crystal violet, methylene blue, or basic fuchsin may be used.

STABILITY. *Arizona* survives in water, soil, and feed for prolonged periods. It is killed by a pH of 3.5 or below or by very alkaline conditions. The usual disinfectants are effective.

CULTURE MEDIA. Initial liver, spleen, or egg cultures may be made on trypticase soy agar plates, but tetrathionate brilliant green enrichment broth and brilliant green sulfapyridine agar media are recommended. Bismuth sulfite agar plates are also suggested (Williams 1975). Tetrathionate broth at 35°C or selenite F broth at 43°C and brilliant green agar plates at 35°C were used by Greenfield and Bigland (1971) in their turkey egg studies. *Arizona* colonies are circular, slightly raised, smooth, and lustrous black or brown, often with a metallic appearance on bismuth sulfite agar. On brilliant green sulfapyridine agar, colonies appear transparent pink on light red media. On triple-sugar-iron agar slants *Arizona* organisms develop an alkaline red surface and an acid yellow butt with gas bubbles (Williams 1975).

Prevention, Control, and Treatment. Procedures outlined for paratyphoid infection are effective for this disease.

References

Caldwell, M. E., and D. L. Ryerson. 1939. J Infect Dis 65:242.

Davis, B. D., et al. 1973. *Microbiology.* 2d ed. Hagerstown, Md.: Harper & Row, p. 776.

Edwards, P. R., and W. H. Ewing. 1972. *Identification of Enterobacteriaceae.* Minneapolis: Burgess.

Edwards, P. R., M. G. West, and D. W. Bruner. 1947. Ky Agric Exp Stn Bull 499.

Edwards, P. R., A. C. McWhorter, and M. A. Fife. 1956. Bull World Health Org 14:511.

Edwards, P. R., M. A. Fife, and C. H. Ramsey. 1959. Bacteriol Rev 23:155.

Ewing, W. H., and P. R. Edwards. 1960. Int Bull Bacteriol Nomencl Taxon 10:1.

Fields, B. N., et al. 1967. Am J Med 42:89–106.

Goetz, M. E. 1962. Avian Dis 6:93–99.

Goetz, M. E., and E. R. Quortrup. 1953. Vet Med 48:58–60.

Greenfield, J., and C. H. Bigland. 1971. Avian Dis 15:254–61.

Hinshaw, W. R., and E. McNeil. 1944. Cornell Vet 34:248–54.

______. 1947. J Bacteriol 53:715–18.

Kauffmann, F. 1941. Acta Pathol Microbiol Scand 18:351.

Lewis, K. H., and E. R. Hitchner. 1936. J Infect Dis 59:225–35.

McWhorter, A. C., et al. 1977. HEW Publ 78-8363, Centers for Disease Control, Atlanta.

Sambyal, D. S., and V. K. Sharma. 1972. Br Vet J 128:50–55.

Sharma, V. K., Y. K. Kaura, and I. P. Singh. 1970. Indian J Med Res 58:409–12.

Williams, J. E. 1975. In *Isolation and Identification of Avian Pathogens,* ed S. B. Hitchner et al. Ithaca, N.Y.: Arnold.

______. 1984. In *Diseases of Poultry,* 8th ed., ed. M. S. Hofstad, pp. 130–40. Ames: Iowa State University Press.

Windingstad, R. W., D. O. Trainer, and R. Duncan. 1977. Avian Dis 21:704–7.

Winsor, D. K., A. P. Bloebaum, and J. J. Mathewson. 1981. Appl Environ Microbiol 42:1123–24.

Mycoplasmosis

Definition and Synonyms. Mycoplasmosis is a slowly developing, chronic, infectious disease caused by tiny pleomorphic organisms that share some of the characteristics of bacteria in that they grow on artificial media but lack cell walls. In pigeons the clinical disease is characterized by respiratory signs, a nasal discharge and air sac clouding. A persistent subclinical respiratory infection is common.

In pigeons the infection is known as air sac infection and pleuropneumonia infection.

Cause and Classification. The organisms are assigned to class Schizomycetes, order Eubacteriales, family Mycoplasmataceae, and genera *Mycoplasma* (which requires sterols for growth) and *Acholeplasma* (which grows without sterols). Mycoplasma strains also have the ability to grow aerobically but are unable to utilize urea.

The infection in pigeons is caused by three species of *Mycoplasma: M. columbinum* (N). *M. columborale* (BO), and *M. columbinasale* (NA). *M. columbinum*-type strain MNPI was isolated from the trachea of a healthy pigeon in 1973 and *M. columborale*-type strain MMP4 was also recovered from the propharynx of a healthy pigeon in the same year by Shimizu et al. (1978). *M. columbinasale*-type strain 694 was formerly designated as L serotype by Yoder and Hofstad (1964), Dierks et al. (1967), Barber and Fabricant (1971), Stepkovitis and El-Ebeedy (1977), and Jordan et al. (1982). These three L-type strains were officially adopted by the International Organization for Mycoplasmology in 1980.

Nature

FORM. Mycoplasmas, in general, are the smallest known free-living organisms. They pass filters with an average pore size of 150 nm. They vary in shape, are pleomorphic, and because of this particle plasticity, are difficult to measure. Electron photomicrographs reveal a diameter of 150–300 nm and a limiting membrane of 75–100 Å (Davis et al. 1973). With pigeon mycoplasmas Shimizu et al. (1978) noted many pleomorphic *M.* (N) and *M.* (BO) organisms, including coccoid, ring, and conglomerate forms. These organisms were surrounded by a triple-layered membrane.

GROWTH CHARACTERISTICS. On special solid media, *Mycoplasma* spp. form tiny, translucent round colonies that require a magnifying lens for observation. Uncovered petri dishes permit low-power light microscope examination of colonies. The *M.* (N) colonies develop a "fried egg" appearance with a thicker central zone of growth on solid media, and they form "film and spots" on a medium with 20% horse serum. *M.* (BO) colonies do not form film and spots on either media (Shimizu et al. 1978). Yoder and Hofstad (1964) also noted the dense central elevated zone in L serotype colonies, which were smooth, entire, and 0.4 mm in diameter. According to Jordan et al. (1982), *M.* (NA), like *M.* (N), produces film and spots. Another characteristic noted by Jordan et al. (1981) and Howse and Jordon (1983) was that *M.* (BO) outgrows *M.* (N) in vitro. This may account for the prevalence of one species over another.

REPRODUCTION. Cell division of mycoplasmas differ from that of typical bacteria, which have a rigid cell wall and form a well-defined septum as they divide and multiply. The mechanism of reproduction is debated, but it appears that separation of sections of filamentous mother cells containing DNA components permits the development of new organisms. Domermuth et al. (1964), studying *M. gallisepticum* of chickens, noted internal dense bodies that appear to form elementary bodies located within a budding protrusion of the cell membrane.

Distribution

HISTORY. Dodd (1905) reported an epizootic pneumoenteritis in turkeys. In 1926 Tyzzer called the respiratory problem in turkeys sinusitis. Nelson (1933, 1936), working at Rockefeller Institute in New Jersey, noted coccobacilliform bodies in coryza of chickens. Dickinson and Hinshaw (1938) termed the condition in turkeys infectious sinusitis or swell head. Delaplane and Stuart (1943) isolated the organism from chickens and later from turkeys with sinusitis. They cultivated the organism in chicken embryos and designated the condition in chickens as chronic respiratory disease. Smith et al. (1948) studied the Nelson coryza agent and grew it in 30% horse serum infusion broth. In 1952 Markham and Wong placed the agent in the pleuropneumonia group. Van Roekel and Olesiuk (1953) cultivated the organism from chickens and from turkeys and noted the similarity. Grumbles et al. (1953) described some cultural and biochemical characteristics of an agent that was presumably *M. gallisepticum.* Winterfield (1953) first demonstrated mycoplasma in pigeons by chick embryo inoculation. Adler and Yamamoto (1956) noted a relationship between *Hemophilus gallinarum* and Nelson's coccibacilliform bodies in long-term coryza. Freundt (1957) in *Bergey's Manual* gave the name *Mycoplasma* to the avian pleuropneumonialike organism causing chronic respiratory disease (CRD). Edwards and Freundt (1967) proposed Mollicutes as the name of a new class that listed *M. gallinarum* as a nonpathogenic avian species. In 1956 Mathey et al. were the first in the United States to isolate pleuropneumonialike organisms from pigeons. Later Schrag et al. (1974) in West Germany was the first in Europe to report mycoplasmas in pigeons. Yoder and Hofstad (1964) isolated L serotype mycoplasma from pigeons and Gianforte et al. (1955) isolated mycoplasma in fluid media from a Connecticut pigeon with air sac infection.

GEOGRAPHICAL DISTRIBUTION. Mycoplasmas have been recovered from pigeons in England (Sinclair 1980; Jordan et al. 1981; MacOwan et al. 1981; Howse and Jordan 1983), Japan (Shimizu et al. 1978; Shimizu 1982), West Germany (Schrag et al. 1974), Germany (Gerlach 1977, 1978; Gerlach and Gylstorff 1979; Stepkovits and El-Ebeedy 1977), and the United States (Winterfield 1953; Gianforte et al. 1955; Mathey et al. 1956; Yoder and Hofstad 1964).

INCIDENCE. The prevalence of mycoplasmas in pigeons has only recently been assessed, and virtually all flocks and birds have experienced infection. Shimizu (1982) isolated mycoplasmas from 68.7% of 262 pigeons and their nestlings. In several lots of pigeons from different cities in Japan, isolations ranged from 54% to 100%. Schrag (1974) considered virtually all pigeons to be infected.

SEASON. The infection may occur at any time of the year. Subclinical infection may develop into a clinical disease under conditions of stress, such as inclement weather.

AGE, SEX, AND BREED. Birds may be infected at any age. Both sexes and all breeds may become infected.

SPECIES. Only pigeons become naturally infected with *M.* (N, BO, NA). It is of interest that both *M.* (N, BO) have been recovered from the same pigeon (Jordan et al. 1981). MacOwan et al. (1981) demonstrated the air sac infection of 3-wk-old specific pathogen–free chickens by *M.* (BO) inoculated into the left abdominal air sac.

CARRIERS. Pigeons may remain symptom-free carriers for life. Howse and Jordan (1983) demonstrated the carrier state in pigeons.

Transmission. Direct and indirect contact serves to transmit the agent. Contaminated premises and equipment may thus be the source. Transovarian egg passage was established by Gerlach and Gylstorff (1979), and Shimizu (1982) reported transmission by crop milk.

Epidemiology

DISTRIBUTION IN BODY. The agent has largely been recovered from the oropharynx, sinus turbinates, and tracheal exudate, but lung, brain, and air sac isolates have also been reported. *M.* (N, BO) organisms can often be recovered from the oropharynx, esophagus, and trachea. They are less commonly found in the lungs and brain (Jordan et al. 1981). *M.* (N, BO, NA) have been isolated from the oropharynx and nasal sinuses, *M.* (N, NA) from the brain, *M.* (N) from the lungs, and *M.* (BO) from the air sacs (Keymer et al. 1984). Cloacal swabs were not successful in the isolation of mycoplasmas (Jordan et al. 1981). Shimuzu et al. (1978) isolated mycoplasmas from 35.2% of 54 tracheas and from 66.7% of 21 oropharynx cultures of pigeons. No isolations were made from reproductive tracts. Shimuzu (1982) also noted the predominance of *M.* (BO) in 64.5% of 262 oral isolations. *M.* (N) was next with 24.8%. Jordan et al.(1981) were also unable to isolate mycoplasmas from the cloaca.

ENTRY AND EXIT. Mycoplasmas may gain entrance to the body by the mouth, nares, and eyes. Discharges from the mouth, nasal openings, eyes, and vent spread the agent.

PATHOGENICITY. The agent spreads rather slowly and is not nearly as contagious as a paramyxovirus. Gerlach (1977) considered all types to have negligible effect on pigeons.

IMMUNITY. Gerlach (1978) found the occurrence of mycoplasmal "humoral antibodies" to be exceedingly rare. Serological results, however, indicate that antibodies to *M.* (N) exist in some groups of pigeons infected with this organism (Keymer et al. 1984).

CONTRIBUTORY FACTORS. Mycoplasmal infection in chickens is enhanced by intercurrent factors and diseases. If pigeon stress factors such as *E. coli,* crowding, *Hemoproteus* spp., poor hygiene, ammonia, inclement weather, and long races are reduced, the chances of clinical infection are negligible.

Incubation Period. The incubation period is not easily identified if clinical disease fails to develop.

Course. The course is indefinite. Birds remain infected for a prolonged time.

Morbidity. During the course of infection it is difficult from casual observation to identify clinically infected birds. Keymer et al. (1984) reported 10% of pigeons in five lofts with clinical signs of respiratory infection but failed to find clear evidence that *Mycoplasma* caused the problem. *Mycoplasma* was isolated from 28% of 58 pigeons in these lofts.

Mortality. The number of deaths varies considerably within individual flocks. Keymer et al. (1984) noted that dead pigeons with *Mycoplasma* also had other infections. This fact is supported by the lack of documentary evidence to show that *Mycoplasma* alone can cause the disease.

Signs. Mild respiratory signs, including mucus clicks, coughing, slight nasal discharge, and respiratory rales, may be noted. A few birds may present a watery, ocular discharge from one or both eyes. On the other hand, healthy pigeons showing no signs may yield *Mycoplasma* (Gerlach 1978; Shimizu et al. 1978; Jordan et al. 1981).

Necropsy. Examination of dead pigeons may reveal varying amounts of mucus and some air sac clouding with yellowish material lining the air sac membranes, but gross lesions may be entirely absent in serologically positive birds.

Diagnosis. Diagnosis is based on the recovery of the organism. Sterile swabs of the nasopalatine cleft, trachea, lung, or air sac exudate may be transferred to mycoplasma broth media. Transfers to solid media enables further study.

The tube or plate agglutination tests, the growth inhibition test, and the immunofluorescence technique may be used in diagnosis. Keymer et al. (1984) used the metabolism inhibition test in classifying pigeon mycoplasmas as described by Leach (1973). Rosendal and Black (1972) used both direct and indirect immunofluorescence of nonfixed and fixed mycoplasmal colonies in their studies. Kelton and Van Roekel (1963) classified 19 serotypes using the tube agglutination test and growth inhibition test and included strain 694 as the L serotype.

Differential Diagnosis. Other diseases that give similar findings include one-eye cold, vitamin A deficiency, *Chlamydia psittaci* infection, *Escherichia coli* infection, paramyxovirus infection, and Newcastle disease.

Biological Properties

FILTRATION. Broth cultures may be diluted with phosphate-buffered saline at pH 7.2 and be filtered using filters (Millipore Corp.) with an average pore size of 450–220 nm.

HEMATOLOGY. The three pigeon species do not cause hemagglutination of turkey or chicken erythrocytes (Yoder 1984).

STAINING. With Giemsa stain *Mycoplasma* appear as tiny pleomorphic cocci or short rods and they stain weakly gram-negative. Diene's stain applied to a coverslip and permitted to dry is ideal for staining colonies. A small block of agar with growth, placed on a slide, may be covered with the coverslip stain side down.

STABILITY. It is assumed that most commonly used disinfectants are effective against *Mycoplasma.* Lye at the rate of 1 lb/5 gal hot water is an effective disinfectant. Safety precautions including rubber gloves and boots and eye goggles must be employed in its application. It must not be sprayed because of the danger of burns.

Cultures are quite susceptible to ordinary refrigeration temperatures. Quick freezing in special low-temperature refrigeration units (Revco) at −30°C preserves the organisms. *Mycoplasma* remain infective for long periods suspended in tryptose phosphate broth and held at this temperature. Freeze-drying also preserves the organism.

CULTURE MEDIA. *Mycoplasma* sp. from avian sources usually require enrichment media with serum and yeast components for growth. Penicillin and thallium acetate are employed to retard growth of bacteria. Swab cultures may be transferred from broth to agar after incubation for 1–3 days at 37°C. Hayflick's modified broth and agar are suggested by (Keymer et al. 1984). Shimizu et al. (1978) used the following medium for primary pigeon *Mycoplasma* isolations: "*Mycoplasma* broth base (Pfizer Diagnostic Division) supplemented with 15% heat inactivated swine serum, 0.025% thallium acetate, and 1,000 units of penicillin per ml, beta-diphosphopyridine nucleotide (Sigma Chemical Co.) and L-cysteine (200 μg/ml each) and Eagle vitamin solution (Difco Lab.; ×100, 1%)." They found that all strains of *M.* (N) and *M.* (BO) grew anaerobically, as well as aerobically, without the addition of diphosphopyridine nucleotide and both species required serum for growth. They further designated these as new *Mycoplasma* species and *M.* (N) as an arginine-positive fermenting species and *M.* (BO) as a glucose fermenting species. Jordan et al. (1982) and Yoder (1984) reported *M.* (NA) as an arginine fermenting species. Each species utilizes one or the other but not both. Triphenyl tetrazolium chloride was aerobically reduced in broth cultures by *M.* (NA) (Yoder and Hofstad 1964). *M.* (N) and *M.* (BO) both reduced triphenyl tetrazolium anaerobically but not aerobically and reduced tellurite anaerobically and aerobically (Shimizu et al. 1978).

The author has grown *M. gallisepticum* on broth media similar to the Shimizu pigeon *Mycoplasma* media, using the following procedure. Mix 22.5 g *Mycoplasma* broth base (Pfizer 340A), 3.0 g dextrose; 0.025 g phenol red, 0.2 g thallium acetate, 500,000 units penicillin potassium, 970 ml deionized water, and 150 ml swine serum heat inactivated at 56°C for 30 min. After allowing the mix to stand 10 min, add 0.1 g beta DPN (diphosphopyridine nucleotide); 0.1 g L cysteine hydrochloride in 30 ml deionized water. The final solution is adjusted to pH 7.7 with 0.5 Normal sodium hydroxide and sterilized by filtration. For agar plates, use 3% Nobel agar with the preceding broth base. The filter-sterilized enrichments should be added after the agar base cools to 45°C. All equipment and filters must be washed with deionized water. The phenol red indicator serves to signal an acid pH change with a glucose fermenter, *M.* (BO). When the pH changes, the culture may be lost unless the culture is transferred. Plates should be incubated aerobically at 37°C 2–5 days.

CULTURE IN EMBRYOS AND BIRDS. Tracheal exudate or scrapings may be suspended in broth and treated with 10,000 units of penicillin/ml. This is permitted to stand at room temperature before it is inoculated into the yolk sac of 7-day-incubated specific-pathogen-free chicken eggs. Streptomycin must not be used for bacterial control in the inoculum since mycoplasmas are sensitive to it. After 5–6 days of further incubation the infected embryos upon harvest will present evidence of respiratory involvement.

Disease-free squabs or chicks may be used for test inoculations.

Prevention. The very nature of the racing pigeon situation exposes live noninfected birds to infected birds. Strict isolation as a means of prevention is not practical. The breeding of resistant birds has merit, but this alone will not prevent the disease. Avoidance of ammonia fumes in a loft is essential because this may trigger the clinical disease. Immunization with active and inactivated vaccine has met with varied success in chickens. Similar results may be expected with pigeons. Antibiotic egg dipping, egg inoculation, and egg heating have been effective in overcoming the problem in chickens. The use of lincomycin-spectinomycin in combination as an injection into embryonated eggs may be successful. The author injected streptomycin into 9000 chicken hatching eggs from a positive flock with the eradication of the agent.

Control. Serologic testing may be used to identify positive birds and flocks. Medication of positive flocks may be helpful in reducing the shedding and thus the transmission level within a loft.

Treatment. Lincomycin, tiamulin hydrogen fumarate (Dynamutilin, Squibb), and spiramycin were used without success (Keymer et al. 1984). Howse and Jordan (1983) previously used tiamulin hydrogen fumarate without success. The drug suppressed *M.* (BO) and *M.* (N) fairly rapidly but did not eradicate the organisms. Domermuth and Johnson (1955) tested several antibacterial agents in situ on an avian strain of *Mycoplasma* and reported drugs to be active in the following order of diminishing activity: (1) magnamycin; (2) terramycin, streptomycin, furazolidone; (3) chlorotetracycline chloromycetin, (4) neomycin, and (5) penicillin. The administration of Tylosin (Eli Lilly) to chickens at 2 g/gal drinking water for 5–7 days or injected subcutaneously in the neck region at 3–5 mg/lb body weight has been reasonably effective in reducing the clinical disease but not in eradicating it.

References and Selected Readings

Adler, H. E., and R. Yamamoto. 1956. Cornell Vet 46:337–43.

Barber, T. L., and J. Fabricant. 1971. Avian Dis 15:125–38.

Davis, B. D., et al. 1973. *Microbiology.* 2d ed. Hagerstown, Md.: Harper & Rowe, pp. 930–44.

Delaplane, J. P., and H. O. Stuart. 1943. Am J Vet Res 4:325–32.

Dickinson, E. M., and W. R. Hinshaw. 1938. J Am Vet Med Assoc 93:151–56.

Dierks, R. E., J. A. Newman, and B. S. Pomeroy. 1967. Ann N Y Acad Sci 143:170–89.

Dodd, S. 1905. J Comp Pathol Ther 18:239–45.

Domermuth, C. H., and E. P. Johnson. 1955. Poult Sci 34:1395–99.

Domermuth, C. H., et al. 1964. J Bacteriol 88:727.

Edward, D. G., and E. A. Freundt. 1967. Int J Syst Bacteriol 17:267–68.

Freundt, E. A. 1957. In *Bergey's Manual of Determinative Bacteriology,* 7th ed., ed. R. S. Breed et al., pp. 914–26. Baltimore: Williams & Wilkins.

Gerlach, H. 1977. Berl Muench Tieraerztl Wochenschr 90:140–43.

______. 1978. Inaug Diss, Poultry Faculty in Animal Medicine, Ludwig-Maximilians Universitat, Muenchen, p. 193.

Gerlach, H., and I. Gylstorff. 1979. Berl Muench Tieraerztl Wochensch 92:256–62.

Gianforte, E. M., E. L. Jungherr, and R. E. Jacobs. 1955. Poult Sci 34:662–69.

Grumbles, L. C., et al. 1953. Southwest Vet 6:166–68.

Howse, J. N., and F. T. Jordan. 1983. Vet Rec 112:324–26.

International Order of Microplasmology. 1980. Rep Consultations, Sept., Blue Bell Lodge, Custer, S. Dak.

Jordan, F. T., et al. 1981. Vet Rec 109:450.

Jordan, F. T., et al. 1982. Int J Syst Bacteriol 32:108–15.

Kelton, W. H., and H. Van Roekel. 1963. Avian Dis 7:272–86.

Keymer, I. F., et al. 1984. Avian Pathol 13:65–74.

Leach, R. H. 1973. J Gen Microbiol 75:135–53.

MacOwan, K. J., et al. 1981. Vet Rec 109:562.

Markham, F. S., and S. C. Wong. 1952. P Sci 31:902–4.

Mathey, W. J., H. E. Adler, and P. J. Siddle. 1956. Am J Vet Res 17:521–22.

Nelson, J. B. 1933. J Exp Med 58:297–304.

______. 1936. J Exp Med 63:515.

Rosendal, S., and F. T. Black. 1972. Acta Pathol Microbiol Scand 80:615.
Schrag, L., H. Enz, and H. Klette. 1974. *Healthy Pigeons: Recognition, Prevention, and Treatment of Major Pigeon Diseases.* Hengersburg, West Germany: Verlag Schober.
Shimizu, T. 1982. Rev Inf Dis 4:242.
Shimizu, T., H. Ern, and H. Nagatomo. 1978. Int J Syst Bacteriol 28:538.
Sinclair, D. V. 1980. Vet Rec 106:466.
Smith, W. E., J. Hillier, and S. Mudd. 1948. J Bacteriol 56:589-601.
Stepkovits, L., and A. A. El-Ebeedy. 1977. Zentralbl Veterinaermed [B]:218-230.
Tyzzer, E. E. 1926. Cornell Vet 16:221-24.
Van Roekel, H., and O. M. Olesiuk. 1953. Proc 90th Annual Meeting of the American Veterinary Medicine Assoc, pp. 289-303.
Winterfield, R. W. 1953. Vet Med 48:124-25.
Yoder, H. W. 1984. In *Diseases of Poultry,* 8th ed., ed. M. S. Hofstad et al., p. 187. Ames: Iowa State University Press.
Yoder, H. W., and M. S. Hofstad. 1964. Avian Dis 8:481-512.

5

Mycotic Diseases

Mycotic organisms are fungi that grow in part by mycelial strands. They have no chlorophyll and reproduce asexually.

Aspergillosis

Definition and Synonyms. Aspergillosis is an infectious fungus disease of birds, animals, and humans caused by several *Apergillus* species. It is usually characterized in the pigeon as a chronic infection of the lungs and air sacs. Pneumomycosis is another name for this disease.

Cause and Classification. The organism is placed in group Thallphyta, division Mycota, class Ascomycetes, order Eurotiales; family Eurotiaceae, and genus *Aspergillus. A. fumigatus* is considered the most common and also the most pathogenic species (Rivolta 1887; Weidman 1929; Mastrofrancisco and Raimo 1940; Swierstra et al. 1959). Other species have also been reported in pigeons (Bonizzi 1876). In particular, *A. glaucus,* a group of fungi, has been reported to cause skin diseases producing yellow scaly spots and broken feathers. *A. nidulans* resulted in necrotic nodular lesions of the liver and lungs (Russo and Graziosi 1950), *A. fumisalordes* var. *roseus* n. sp. produced damage to the feathers (Sartory 1942), and *A. flavescens* affected the serous membranes of the oviduct (Schieblich 1915, 1921).

Nature

DEVELOPMENT. The fungi (from *fungus*–mushroom) have been regarded as "plant life." The organisms grow by branching, twiglike tubular filaments called hyphae and by extensions or elongating mycelial structures. *Aspergillus* fungi grow as multicellular, fluffy mold colonies. They may be free-living in the soil and on vegetation or they may be parasitic living in or on birds, animals, and humans.

REPRODUCTION. Conidiophores, which produce spores, form on fruiting stalks, and these serve as a means of reproduction.

Distribution

GEOGRAPHY. The fungi are found worldwide (Barden et al. 1971).

INCIDENCE. Only occasional fungus infections are recognized in pigeons even though *Aspergillus* fungus spores are widely distributed.

AGE, SEX, AND BREED. A young squab is more likely to develop infection because of close skin contact with floor or nest box. Spore contamination is usually concentrated on the floor or in the litter. Both sexes and all breeds are affected. Infection can occur in all animals and birds.

CARRIERS. Pigeons may serve as carriers, but a respiratory tract–infected bird is virtually never a source of infection for other birds. A skin infection, however, may be the reservoir for spores.

Transmission. Transmission occurs by ingestion and inhalation of spores, which are released by the fruiting bodies of the fungus. The spores are easily inhaled when airborne and grow readily on mucous membranes of the lungs and air sacs. Skin and feathers that remain wet also serve as media for growth. *A. fumigatus* given by mouth produces a chronic disease, but parenterally it causes a rapidly fatal infection (Battelli 1944). Others (Anon. 1935; Henrici 1939) also established infection by forced feeding. In 1960 Ciurea et al. infected pigeons intravenously and intraperitoneally with acute infection within 12–30 hr. Others (Myczkowskyi 1934; Chute 1984) have established infection by various means. Transmission has also been observed by Eggert and Barnhart (1953) to be eggborne in a chick hatchery where the fungus penetrated the eggshell. This is unlikely in pigeon eggs if the nests are dry.

Epidemiology

DISTRIBUTION IN BODY. The organism has been found throughout the body.

ENTRY AND EXIT. Fungus spores gain entrance to the body by the nostrils, mouth, or eyes. Skin infections may develop following prolonged surface contamination. Spores very seldom leave the body cavity by way of the respiratory tract because they become trapped by body moisture.

PATHOGENICITY. Heavily spore-contaminated air may overcome body defenses. Many inhaled spores are trapped by the nasal and tracheal membranes and are expelled with mucus discharges. If one or more spores enter the lungs or air sacs, however, they have a good chance of survival. Chute et al. (1956) observed *A. fumigatus* colonies inside broiler chicks, but they were not always pathogenic. Each fungus species varies in its pathogenicity.

Incubation. *Aspergillus* spore colonies develop within 12–24 hr. The course is prolonged.

Morbidity and Mortality. *Aspergillus* infection seldom occurs as a flock condition. Occasionally individual birds will develop infection. Losses result largely from culling, but Taylor (1966) reported a 78% loss in one pigeon flock.

Signs. Depression, ruffled feathers, inactivity, respiratory difficulty, rapid breathing, and diarrhea often occur in late stages of internal infection. Lahaye in 1928 noted yellow, scaly skin infection in pigeons that was caused by *A. glaucus.* Skin infection may result in constant preening, feather-pulling, and self skin destruction.

Necropsy. Lung infection may cause generalized grayish white solidification of lung tissue much like tubercles. Other whitish lesions often occur at the bifurcation of the bronchi and on the syrinx, larynx, tongue, and hard palate. Air sac lesions are flattened, saucerlike, yellowish to grayish green caseous masses. Eye infection may involve the interior or exterior of the eye and produce cloudiness of the chambers and cornea.

Diagnosis. Necropsy and culture procedures may reveal the presence of fungi. Tissue sections may also demonstrate mycelial strands and fruiting bodies.

Biological Properties

TISSUE AFFINITY. Infection has been found in the lungs, air sacs, skin, eyes, intestine, and brain. Pinoy (1936) also observed spleen infection following experimental infection.

STAINING. Clark and Hench (1962) used acridine orange stain to demonstrate fungi.

STABILITY. The spores are quite resistant to destruction by ordinary means. Steam sterilization at 15-lb pressure for 30 min is effective but is limited to laboratory or kitchen pressure-cooking operations.

CULTURE MEDIA. Sabouraud's agar and Czapek's solution agar are useful in growing the fungus at or about room temperature, 37°C. The fungus produces cottony, velvety, surface colonies that are blue-green and darken with age. The fungus has septate, walled-off, mycelial strands that produce spores on the ends of upright fruiting stalks.

MYCOTOXINS. *Aspergillus* toxin is discussed in Chap. 8.

Prevention. Fungus infections must be prevented or treated. Control is not the aim. Damp litter or building materials may harbor fungi. Where possible lofts should be constructed with nonabsorbent building materials that shed water. Lofts should also be constructed to prevent leaks and provide adequate ventilation to reduce moisture condensation. Shed roofs without a front overhang permit free movement of air and moisture away from the loft through louvered front rafter spaces. If an overhang is present, the warm moist air is trapped as in an A roof. Roof cooling condenses the moisture in the air, and it then falls or precipitates to the floor below.

Moldy litter and feed should be avoided, and feed and water containers should be constructed of impervious material and be kept clean.

Treatment. No treatment is recommended for *Aspergillus* respiratory infections in pigeons. Bird disposal is indicated. Skin infections can be successfully treated. The author has used STA (salicylic acid 3 g, tannic acid 3 g in ethyl alcohol – q. s. 100 ml) or copper sulfate (1:2000 dilution) sponged biweekly on small areas of the skin for 2–3 mo (Tudor 1983). Lahaye (1928) used a solution of mercuric chloride diluted 1:500 sponged on pigeon skin infections. Following treatment, the birds were rinsed with lukewarm water.

References

Anon. 1935. Mouvem Sanit 12:625.

Barden, E. S., et al. 1971. *A Bibliography of Avian Mycosis.* 3d ed. Orono: University of Maine.
Battelli, C. 1944. Boll Soc Ital Med Ig Trop, Asmara 4:4-6.
Bonizzi, P. 1876. (cited by Newmann as reviewed [1908]). J Comp Pathol Ther 21:260-64.
Chute, H. L. 1984. In *Diseases of Poultry,* 8th ed., ed. M. S. Hofstad, pp. 309-15. Ames: Iowa State University Press.
Chute, H. L., et al. 1956. Am J Vet Res 17:763-65.
Ciurea, V. et al. 1960. Lucr Stint Inst Agron Bucur, pp. 635-49.
Clark, R. F., and M. E. Hench. 1962. Am J Clin Pathol 37:237-38.
Eggert, M. J., and J. V. Barnhart. 1953. J Am Vet Med Assoc 122:225.
Henrici, A. T. 1939. J Immunol 36:319-38.
Lahaye, J. 1928. Imprimerie Steinmetz-Haenen, Remouchamps, p. 393.
Mastrofrancisco, N., and H. F. Raimo. 1940. Rev Ind Anim 3, 4:71-101.
Myczkowskyi, M. 1934. Experimental Aspergillosis in Pigeons. Thesis, Bucharest.
Pinoy, P. E. 1936. Rev Pathol Comp 36:367-69.
Rivolta. 1887. Gior di Anat Fisiol Pathol Anim, p. 121.
Russo, G., and F. Graziosi. 1950. Rend Inst Super Sanit, Rome, 13:46-56.
Sartory, A. 1942. Acad Sci Colon Paris C R 214:565.
Schieblich. 1915. Berlin tieraerztl Wochenschr 31:3.
______. 1921. Berlin tieraerztl Wochenschr 37:76.
Swierstra, D., and J. Jansen. 1959. Tijdschr Diergeneeskd 84:892-900.
Taylor, P. A. 1966. Can Vet J 7:262-63.
Tudor, D. C. 1983. Vet Med Small Anim Clin 78:249-53.
Weidman, F. D. 1929. Arch Pathol 8:1019-20.

Thrush

Definition and Synonyms. Thrush is a common acute or chronic opportunistic fungus infection of the digestive tract caused by a yeastlike organism which has a mycelial phase. Mycosis, oidiomycosis, soor, muguet, sour crop, *Candida* infection, and *Monilia* infection are other names for this mycotic disease.

Cause and Classification. The fungus is placed in class Deuteromycetes, order Moniliales, family Cryptococcaceae (Lennette et al. 1980), genus *Candida,* and species *albicans.* In 1939 the Third International Microbiological Congress replaced the generic name, *Monilia,* with *Candida* (Skinner 1947).

Nature. On the surface of enriched agar media, *C. albicans* is usually found in the budding unicellular yeast stage. Mold-type mycelia are formed only under semianaerobic conditions, as may be found deeper in media or tissues. Both forms may be found in infected tissues and in most cultures. The septate mycelia or pseudohyphae formed by elongated yeast cells occasionally form swollen, spherical, thick-walled chlamydospores.

Distribution

GEOGRAPHY. Langenbeck (1838) first described the organism in the digestive tract of a person. It has since been found in many species worldwide. Cassamagnaghi (1949) reported the disease in pigeons in Uruguay, as did Wells (1955) in Malaya. Numerous reports (Jungherr 1933; and Levi 1941; Chute 1984) have identified the disease in the United States in pigeons.

INCIDENCE. In pigeons the digestive tract infection occurs quite frequently, although most cases are unreported. Mayeda (1966) reviewed poultry diagnostic cases received in California and noted a higher frequency following the use of furazolidone and tetracyclines in feed.

SEASON. The infection is most often seen in late spring and early summer, but it occurs year-round.

SPECIES. Thrush has been reported in pigeons, chickens, turkeys, geese, pheasants, quail, ruffed grouse, peafowl, and parakeets (Chute 1984).

CARRIERS. Free-flying birds may introduce thrush infection. Debilitated birds may harbor the organism in the digestive tract and spread it when they take a drink.

Transmission. The yeastlike organism may be transferred on or in food and water. Unclean containers are often sources of the fungus.

Epidemiology

DISTRIBUTION IN BODY. The yeast is largely confined to the digestive tract but can become systemic, usually in combination with other disease conditions. In the tissues it may develop the mycelial phase.

ENTRY AND EXIT. The yeast organism enters by the mouth and leaves in the oral discharges and feces.

PATHOGENICITY. The yeast is commonly present in the digestive tract but is not pathogenic under normal conditions of health. It is not highly virulent. Other predisposing factors or debilitating conditions favor its growth.

Incubation. The incubation period is indefinite and the course is prolonged.

Morbidity and Mortality. Many birds in a loft may have thrush at one time, but it is usually the weaker birds that are seriously infected. Young birds are often the ones that die.

Signs. Birds develop a sour, fetid, water-filled crop. They appear listless, stunted, and have rough feathering and unsatisfactory growth in squabs. Pigeons often vomit, go off feed, and have no desire to move or fly. One chicken report in Israel by Kuttin et al. (1976) describes only dermatitis and feather-pulling as an indication of thrush infection.

Necropsy. Thickening of the crop lining, giving it a grayish white turkish towel appearance with circular raised ulcers and slimy mucus patches of necrotic tissue may cover a relatively noninflamed mucous membrane. The esophagus and proventriculus may be involved with hemorrhagic diphtheroid lesions.

Micropathology. Crop linings may show destruction of the stratified epithelium, often with walled-off ulcers or diphtheritic membranes present. Jungherr (1933) also reported periportal focal necrosis in the liver, suggesting a toxic reaction.

Diagnosis. Gross turkish towel growth on the crop mucosa is sufficient to establish a necropsy diagnosis of thrush. This can be confirmed by the recovery of the organism with heavy growth on first isolation on artificial Pagano Levin media (Squibb). The microscopic examination of a scraping of a washed lesion saves time. A plastic coverslip centered over the tissue on the slide and pressed firmly provides a thin layer for the high dry objective. A blue filter on the lamp aids identification of hyphae and spores.

Differential Diagnosis. Lahaye (1928) noted the similarity between pox and thrush in pigeons. He demonstrated poxvirus in cases suspected of being thrush. In the author's experience, pox often precipitates a heavy growth of *C. albicans* on the crop mucosa. Both agents can be present at the same time.

Biological Properties

TISSUE AFFINITY. The crop is the most commonly affected organ, but other parts of the digestive tract may be infected. Systemic *Candida* infections may establish the organism anywhere in the body.

CULTURE. The rabbit is highly susceptible to ear vein infection, using a 1% suspension of a fresh culture. Mice may also be used. In addition, young turkeys can be readily infected per os.

In cultures on artificial media, the oval, yeastlike organisms form bacterialike, creamy white, highly convex colonies in 24–48 hr at 37°C. As the colonies grow older they appear furrowed and rough. In gelatin stab cultures the yeast grows on the surface and radiating mycelial strands extend into the media (Burrows 1954). Commonly used media include Sabouraud's agar, cornmeal agar, and Pagano Levin media.

In Dunham's peptone 1% sugar water with 1% Andrades indicator, acid and gas develops with the addition of dextrose, levulose, maltose, or mannose. Some acid forms in galactose and sucrose but not in lactosc, raffinose, or inulin.

Prevention and Control. To prevent or control thrush, sanitary conditions should be improved. Debilitated birds should be removed. Crowding should be avoided and adequate nutrition provided. Excessive use of antibiotics may initiate a problem; therefore no antibiotics should be given at any time unless clearly indicated.

Treatment. Nystatin is perhaps the best known of a number of polyene antifungal drugs that can be incorporated in pellets to help overcome mycosis in chickens. Nystatin (Mycostatin 20, Squibb), has been used with some success in chicken feed at 100 g/t of feed or 50 mg/lb feed for 4–6 wk. Wind and Yacowitz (1960) also used nystatin in the drinking water at 62.5–250 mg/L with sodium lauryl sulfate (7.8–25 mg/L) for 5 days for chicken mycosis. Other antibiotics must not be used during an outbreak of thrush.

Copper sulfate has been used at 8 oz/gal water in preparing a stock solution for chickens. Two tbsp stock is then added to 1 gal drinking water for 7–10 days in nonmetal containers. Underwood et al. (1956) reported, however, that copper sulfate was ineffective in treating or preventing the disease in chicks and poults.

References

Burrows, W. 1954. *Textbook of Microbiology.* 16th ed. Philadelphia: Saunders, pp. 613–16.
Cassamagnaghi, A. 1949. Rev Med Vet Montevideo 24:925–28.
Chute, H. L. 1984. In *Diseases of Poultry,* 8th ed., ed. M. S. Hofstad, pp. 316–19. Ames: Iowa State University Press.
Jungherr, E. 1933. J Agric Res 46:169–78.
Kuttin, E. S., A. M. Beemer, and M. Meroz. 1976. Avian Dis 20:216–18.
Lahaye, J. 1928. Imprimerie Steinmetz-Haenen, Remouchamps, p. 393.
Langenbeck, B. 1838. Arch Med Exp Anat Pathol 12:145–47.
Lennette, E. H. et al. 1980. *Manual of Clinical Microbiology.* 3d. ed. Washington, D.C.: American Society of Clinical Microbiology, p. 572.
Levi, W. M. 1941. *The Pigeon.* Columbia, S.C.: Bryan.
Mayeda, B. 1966. Annu Meeting of the Georgia Veterinary Assoc, June 17–19.
Skinner, C. E. 1947. Bacteriol Rev 11:227–74.
Underwood, P. C., et al. 1956. Poult Sci 35:599–605.
Wells, C. W. 1955. Fed Malaya, Rep Vet Dep 1954, Kuala Lumpur Government Press.
Wind, S., and H. Yacowitz. 1960. Poult Sci 39:904–5.

Favus

Definition. Favus is an infectious fungus disease of birds, animals, and humans characterized by a chronic dermatomycosis resulting in a dry, grayish white, scaly appearance of exposed areas. Infection is pigeons has not been recognized in literature.

Cause and Classification. The classification is the same as for aspergillosis except for the genus and species. *Trichophyton megnini* (*Achorion gallinae*) is the agent.

Nature. The slow-growing fungus forms small, round, velvety colonies with central cups and radial grooves. Twisted, branched mycelial hyphae are irregularly septate. Spores appear in clusters.

Distribution

GEOGRAPHY AND INCIDENCE. The fungus is found worldwide. The disease may occur in pigeons, but isolations have not been reported.

CARRIERS. Birds, animals, and humans have been infected. Pigeons are unlikely to carry infection subclinically but it can occur.

Transmission. Direct and indirect contact serves to transmit the spores from bird to bird and from contaminated equipment, but this is rare in the United States.

Epidemiology

DISTRIBUTION IN BODY. The organism resides on the skin or in the subcutaneous tissue.

ENTRY AND EXIT. Spores probably enter the skin by the hair or feather follicles. Exfoliation of infected skin, scales, and feathers carry spores from the body.

PATHOGENICITY. Favus infection is more likely in birds under stress and when high spore levels are encountered. The infection spreads slowly, but once infected, few birds actually recover completely.

IMMUNITY. Very few birds become infected with favus, suggesting a degree of individual innate immunity. Others argue that this represents only a low exposure rate.

Incubation. Several days or weeks may be required to establish clinical infection. The course of favus is prolonged.

Morbidity and Mortality. The number of birds with favus is usually low. Unless treated, affected birds will die after a long course.

Signs. Young birds are most likely to be affected on the legs or face. As the disease progresses, the grayish white scaly deposits become thicker, forming a rough crust. The skin becomes thickened, especially in and about adjacent feather follicles. A moldy odor may be noted. Persistent infection results in emaciation, weakness, and anemia.

Necropsy. In favus infection, lesions have been observed in the bronchi and lungs, and crop infection may occur.

Micropathology. In the absence of positive cultures, skin sections may reveal subcutaneous mycelial hyphae.

Diagnosis. Gross lesions are usually sufficient for diagnosis of favus, but cultures and histological sections may be necessary.

Biological Properties

TISSUE AFFINITY. The fungus prefers unfeathered areas of the body.

STAINING. Stains employed for other fungi are effective. Lactophenol cotton blue and methylene blue may be employed.

STABILITY. Spores are very stable.

CULTURE. The favus fungus can be cultivated on Sabouraud's glucose agar. The media becomes reddish as growth proceeds.

Prevention. Visibly favus-infected birds should not be introduced.

Control and Treatment. Birds badly infected with favus should be destroyed and the other birds treated. Infected houses must be thoroughly cleaned and disinfected. An iodine-base disinfectant is suggested.

Individual birds may be treated with tincture of iodine and glycerine. 1 part to 6 parts (Van Heelsbergen 1929). Riedel (1950) reported success with a 2% mixture of equal parts of alkyl-dimethyl benzyl ammonium chloride and alkyl-dimethyl dichlor benzyl ammonium chloride applied to the lesions.

References

Riedel, B. B. 1950. Poult Sci 29:741.

Van Heelsbergen, T. 1929. *Handbuch der Geflugelkrankheiten und Gefluegelzucht.* Stuttgart: Ferdinand Enke, pp. 322–27.

Cryptococcosis

Definition and Synonyms. *Cryptococcus* infection is not a clinical disease of pigeons. The yeastlike fungus occurs sporadically in humans and animals, producing a fatal meningitis in humans. The fungus grows in accumulated pigeon manure and is thus a public health hazard. *Torula histolytica* or *Debaryomyces* infection, Busse-Buschke's disease, and European blastomycosis are other names for the disease.

Cause and Classification. The organism is placed in group Thallphyta, division Mycota, class Basidiomycetes (Davis et al. 1973), family Endomycetaceae (Burrows 1954), and genus and species *Cryptococcus neoformans.* Only *C. neoformans* is pathogenic for humans (Davis et al. 1973).

Nature. Cryptococci ordinarily appear in both infected tissue and cultures as round to oval yeast cells 5–6 μm in diameter with mucilaginous capsules. Under certain growth conditions hyphae form with clamp connections. Sexual forms have not been observed. Reproduction is by budding. Colonies are smooth, white to light tan, and produce urease (Davis et al. 1973).

Distribution

OCCURRENCE. The fungus lives as a free-living, nonparasitic saprophyte in the soil (Emmons 1950, 1951; Schwabe 1964; Walter and Atchison 1966); in pigeon manure (Emmons 1955; Fragner 1962; Hull 1963; Schneidan 1964), and old pigeon manure may provide a preferential medium for the saprophytic growth of cryptococci (Emmons 1955). The feral pigeon roost may serve as a reservoir for the organism (Littman 1959; Muchmore et al. 1963; Yamamoto 1957).

GEOGRAPHY AND HISTORY. In 1893 Busse and Buschke reported the first human case of yeast infection which they called systemic (European) blastomycosis (Burrows 1954). In 1938 Bisbocci isolated *C. neoformans* from a pheasant. It was not until 1955 that Emmons in Bethesda, Md., reported the recovery of *C. neoformans* from pigeon droppings in the Washington, D.C., area. In 1957 Kao and Schwarz recovered the fungus from pigeon nests in Cincinnati. Bishop (1960), working in Morgantown, W.V., obtained the organism from 6 of 13 pigeon nests. One isolation was *C. neoformans* var *innocuous.* In 1963 Bergman in south Sweden diagnosed *C. neoformans* in the excreta of 3 of 450 pigeons.

INCIDENCE. The fungus is found throughout the world. It is very common in old pigeon droppings, but it does not cause clinical disease in pigeons. A few pigeons may have the yeast in the digestive tract as an incidental agent.

SEASON, AGE, SEX, AND BREED. These factors do not appear to alter the incidence of cryptococcosis.

SPECIES. Littman et al. (1965a) demonstrated the pathogenicity of *C. neoformans* for feral and loft pigeons by intracranial inoculation. They also showed that pigeons can harbor the organism for extended periods without signs of illness (Littman et al. 1965b). In 1962 Fragner isolated *C. neoformans, C. albidus, C. diffluens,* and *C. laurentii* from the feces of 48 pigeons, 13 fowl, 7 pheasants, 10 house martins, 4 jackdaws, and 3 chaffinches. Infection has also been reported in dogs, cats, cattle, horses, sheep, goats, wild animals, and humans (Siegmund 1979) and in marmosets, monkeys, and cheetahs (Chute 1984). *C. neoformans* is pathogenic for mice by intracerebral inoculation. An LD_{50} is produced by only 1000 cells. Other laboratory animals are also susceptible (Davis et al. 1973).

CARRIERS. Littman and Borok (1968) found the pigeon to be a mechanical carrier of *C. neoformans* on its beak and feet. This was demonstrated in pigeons with *C. neoformans*–negative excreta. Sethi and Randhawa (1968) recovered *C. neoformans* from 18 of 26 pigeons fed a virulent strain, but neither death nor illness was evident in these birds.

Transmission. Pigeon manure dust transmission of *C. neoformans* was studied by Staib and Bethauser (1968).

Epidemiology

DISTRIBUTION IN BODY. The organism may reside in the digestive tract of the pigeon, but Littman et al. (1965a) have shown by intracerebral inoculation that it can grow in other tissues.

ENTRY AND EXIT. The organism enters by the mouth and leaves in the excreta.

IMMUNITY. Pigeons and birds in general are highly resistant to natural infection (Davis et al. 1973).

Incubation. Without a disease there is no incubation period. Only by inoculation can a clinical infection of cryptococcosis be established.

Morbidity and Mortality. Pigeons do not develop the clinical disease naturally and do not die from the presence of the organism in the digestive tract.

Signs and Necropsy. There is no clinical evidence of infection in pigeons. At necropsy a mild enteritis may be observed.

Diagnosis. Fink et al. (1968) demonstrated cryptococcal antibodies by the indirect fluorescent and precipitating techniques. Sabouraud's dextrose media (Littman 1959) provided a distinctive blue-gray color as growth developed. This was recommended by Walter and Coffee (1968). Davis et al. (1973) suggested serological testing to identify the soluble capsular polysaccharide in serum or urine of animals and humans.

Public Health Aspects. Emmons in 1955 isolated *C. neoformans* from 16 of 19 nesting and roosting sites and from 63 of 111 pigeon dropping samples but not from tissues or digestive tracts of 20 pigeons. This demonstrated that the organism is an adventitious saprophyte that grows preferentially in pigeon droppings but not in chicken droppings (Walter and Yee 1968). Littman and Schneierson (1959) compared 20 human and 72 pigeon dropping strains and found them morphologically and physiologically identical. Following this, McDonough et al. (1966) isolated *C. neoformans* from pavement sweepings below feral pigeon roosts and Littmann and Borok (1968) showed that pigeon excreta enabled the rapid multiplication of the organism. About this time, Walter and Atchison (1966) used the complement fixation fluorescent antibody test to identify the prevalence of the infection in pigeon fanciers. Of 134 fanciers 22% were positive compared with 3% of 36 controls with no pigeon contact. Only types A and B were identified. This suggests that subclinical human infection is more common than is realized. People who are constantly exposed to pigeon-contaminated environments have a greater risk of acquiring cryptococcal infections than those with less contact (Walter and Atchison 1966). There have been documented human cases of *Cryptococcus* infections following the cleaning of pigeon roosts. All of this work serves to emphasize the public health hazard presented by accumulated old pigeon droppings.

Biological Properties

STRAINS. There are three strains, A, B, and C, which are differentiated on the basis of the polysaccharide capsular antigen.

TISSUE AFFINITY. Cryptococci may pass through the digestive tract, but other tissues are not normally invaded.

STAINING. Mucicarmine is the stain of choice for spore and capsule identification (Chute 1984). India ink is also useful.

STABILITY. The fungus remains viable in dried material for months (Davis et al. 1973).

CULTURE MEDIA. Mouse inoculation is frequently used in diagnosing *cryptococcus.* In 1951 Brueck and Buddingh grew the organism in the yolk sac of chick embryos. Later Staib (1962) tested bird manure as media for the growth of the organism. Fresh canary and pigeon manure supported growth for 6 days only, but old, dry, hard manure served well. Bird urine, uric acid, zanthine, and guanine were used by *Cryptococcus* spp., but creatinine was assimilated by only *C. neoformans.* Littman and Borok (1968) studied the heat tolerance level of *C. neoformans* human and pigeon strains. With increasing temperatures fewer strains grew at 39–43°C and survived at 40–45°C. None grew at 44°C. The rectal temperature of pigeons varied from 41.5 to 43.3°C. This explains why pigeons seldom carry the organism in their excreta even though it grows readily in pigeon droppings.

Prevention. Routine cleaning and removal of pigeon droppings is essential. It is advisable to wear a dust respirator whenever a loft or feral pigeon roost is cleaned. The dust should be settled with a water spray prior to cleaning.

Control. The application of aqueous alkaline hydrated lime at 1 lb/3 gal water fortified with 18 g sodium hydroxide was effective in eliminating *C. neoformans* in pigeon lofts (Walter and Coffee 1967).

Treatment. Amphotericin B has been quite effective even in very ill persons (Davis et al. 1973).

References

Bergman, F. 1963. Acta Med Scand 174:651–55.
Bisbocci, G. 1938. Nuova Ercolani 43:290–314.
Bishop, R. H., R. K. Hamilton, and J. M. Slack. 1960. Abstr West Virginia Bull 26:31–32.
Brueck, J. W., and G. J. Buddingh. 1951. Soc Exp Biol Med Proc 76:258–61.
Burrows, W. 1954. *Textbook of Microbiology.* 16th ed. Philadelphia: Saunders, pp. 607–10.
Chute, H. L. 1984. In *Diseases of Poultry,* 8th ed., ed., M. S. Hofstad, p. 321. Ames: Iowa State University Press.
Davis, B. D., et al. 1973. *Microbiology.* 2d ed. Hagerstown, Md.: Harper & Row, pp. 984–86.
Emmons, C. W. 1950. Proc 7th International Botanical Congress, Uppsala, pp. 416–21.
______. 1951. J Bacteriol 62:685–90.
______. 1955. Am J Hyg 62:227–32.
Fink, J. N., J. J. Barboriak, and L. Kaufman. 1968. J Allerg 41:297–301.
Fragner, P. 1962. Czech Epidemiol Mikrobiol Immunol 11:135–39.
Hull, T. G. 1963. *Diseases Transmitted from Animals to Man.* 5th ed. Springfield, Ill. Charles C Thomas.
Kao, C. J., and J. Schwarz. 1957. Am J Clin Pathol 27:652–63.
Littman, M. L. 1959. Am J Med 27:976–98.
Littman, M. L., and R. Borok. 1968. Mycopathol Mycol Appl 35:329–45.
Littman, M. L., and S. S. Schneierson. 1959. Am J Hyg 69:49–59.
Littman, M. L., R. Borok, and T. J. Dalton. 1965a. Am J Epidemiol 82:197–207.
______. 1965b. Bacteriol Proc 18.
McDonough, E. E., A. L. Lewis, and L. A. Penn. 1966. Publ Health Rep 81:1119–23.
Muchmore, H. G., 1963. N Engl J Med 268:1112–14.
Schneidan, J. D. 1964. Science 143:525.
Schwabe, C. 1964. *Veterinary Medicine and Human Health.* Baltimore: Williams & Wilkins.
Sethi, K. K., and H. S. Randhawa. 1968. J Infect Dis 118:135–38.
Siegmund, O. H., ed. 1979. *The Merck Veterinary Manual.* Rahway, N.J.: Merck, pp. 449–50.
Staib, F. 1962. Zentralbl Bakteriol [Orig A] 186:233–47.
Staib, F., and G. Bethauser. 1968. Mykosen 11:619–24.
Walter, J. E., and R. W. Atchison. 1966. J Bacteriol 92:82–87.
Walter, J. E., and E. G. Coffee. 1967. Am J Epidemiol 87:173–78.
______. 1968. Sabouraudia 6:165–67.

Walter, J. E., and R. B. Yee. 1968. Am J Epidemiol 88:445–50.
Yamamoto, S. K., K. Ishida, and A. Sato. 1957. Jpn J Vet Sci 19:179–94.

Histoplasmosis

Definition and Synonym. In pigeons, histoplasmosis is uncommon. It occurs sporadically as an acute, fatal or a benign, chronic, noncontagious, systemic fungus disease. The fungus has yeastlike and mycelial phases that produce lung and air sac infection. It is also called Darling's disease.

Cause and Classification. The organism is placed in group Thallphyta, division Mycota, class Deuteromycetes (Fungi Imperfecti), order Moniliales, family Moniliaceae, and genus and species *Histoplasma capsulatum.*

Nature. In tissues the fungus forms small, oval, encapsulated, yeastlike cells 3–5 μm in diameter. On Sabouraud's media, slow-growing, fine mycelial, moldlike growth develops that is white at first and becomes tan in about two wk. By enriching the media hyphae convert to yeast cells at the edge of the colony and form smooth, creamy, round colonies (Davis et al. 1973). Double-walled chlamydospores are of two types: small spherical, spiney spores 3–4 μm in diameter and spherical or rarely club-shaped spores 8–12 μm in diameter with fingerlike projections (Chute 1984). The fungus grows readily in soil in the presence of droppings from chickens, bats, and birds.

Distribution

GEOGRAPHY AND HISTORY. The fungus was first discovered by Darling in 1906 during human autopsies in Panama. The first human case in the United States was reported the same year (Burrows 1954). The fungus is prevalent in areas along the Ohio, Missouri, and Mississippi rivers. It has also been reported from Mexico, Panama, Trinidad, Brazil, French Guinea, Peru, Venezuela, Congo, South Africa, and Malaya (Ajello 1964) and in Europe, Russia, and Java (Chute 1984). In 1953 Wildervanck studied two human cases near Paramibo, Surinam, and found insects, fowl, and pigeons infected. Murdock et al. (1962) recovered *H. capsulatum* from a starling-infested cemetery in Memphis, Tenn. Later Dodge et al. (1965) recovered the same organism from a starling roost at a Milan schoolyard in Italy.

INCIDENCE. Histoplasmosis is not a common problem in pigeons or birds, probably because the organism does not thrive at the high body temperature of pigeons (105°F, 40.5°C) (Davis et al. 1973).

SEASON, AGE, AND SEX. There appears to be no relationship between these factors and the development of histoplasmosis.

SPECIES. Histoplasmosis has been recognized in humans and domestic and wild animals. Experimental infection of rats, guinea pigs, and rabbits has been produced by Hansmann and Schenken (1934). Chute (1984) reported the disease in dogs, cats, rats, mice, woodchucks, skunks, opossums, foxes, raccoons, and humans. Menges (1954) used histoplasmin in a skin test survey in Kansas and Iowa and found evidence of infection in dogs, cattle, horses, sheep, pigs, fowl, and turkeys. Histoplasmin may cross-react with blastomycosis; thus results of the test may be questioned (Chute 1984).

CARRIERS. Apparently, active subclinical bird or animal carriers do not occur.

Transmission. Histoplasmosis is acquired from the soil reservoir by inhalation of fungus spores. It is rarely contracted by the oral route. Both yeastlike and mycelial forms are infectious experimentally (Burrows 1954).

Epidemiology

DISTRIBUTION IN BODY. The yeastlike *Histoplasma* organisms may be found widely distributed throughout the body, but in most cases nodules are restricted to one or more organs, including spleen and bone marrow (Davis et al. 1973).

ENTRY AND EXIT. The spores enter the body by inhalation. They leave the body in discharges and feathers.

PATHOGENICITY. Histoplasmosis is not contagious from one bird to another. Schwarz et al. (1957) inoculated pigeons intravenously with *H. capsulatum* and recovered the organism from the liver and spleen of 19 birds 24 days later.

IMMUNITY. Pigeons and other birds have an innate resistance to histoplasmosis. Exposure to overwhelming numbers of spores may induce disease.

Incubation. It is difficult to determine the onset of histoplasmosis in most birds, but some develop an acute infection within 2 wk. The course is usually prolonged.

Morbidity and Mortality. Only individual birds within a loft will contract histoplasmosis. Recovery is not expected.

Signs. Initial lung infection may pass unnoticed or appear as a self-limiting respiratory disease. Progressive, widely disseminated infection results in fever, poor appetite, anemia, emaciation, and death.

Necropsy. Granulomatous lesions may be

found in the lungs. Old lesions become calcified. In humans, skin and intestinal ulceration have been observed (Burrows 1954).

Micropathology. There is extensive proliferation of reticuloendothelial cells, and many may contain the yeastlike organisms.

Diagnosis. Culture and identification of the fungus on Sabouraud's media and the presence of the yeastlike (1–5 μm) organisms in tissue sections provides a positive diagnosis of histoplasmosis. Blood serum used for complement fixation, immunodiffusion, immunofluorescence, or latex agglutination tests aid the diagnosis (Siegmund 1979.

Differential Diagnosis. Other granulomatous diseases, such as toxoplasmosis, blastomycosis, coccidioidomycosis, nocardiosis, tuberculosis, and tumors may cause confusion.

Biological Properties

STRAINS. *H. capsulatum* var. *duboisii* with large spores has been reported in Africa (Davis et al. 1973).

HEMATOLOGY. Blood smears may reveal the organism in macrophages.

TISSUE AFFINITY. In infected individuals, macrophages engulf the oval yeast cells in the blood. Guidry and Spence (1967) found *H. capsulatum* deposited in feathers of inoculated chick embryos and Tewari and Campbell (1965) isolated the organism from chicken feathers of intravenously inoculated birds.

STAINING. Wright's, Giemsa's, or periodic acid–Schiff stain may be used to stain tissue sections (Davis et al. 1973). Bauer's or Gridley's stain is also helpful (Chute 1984). Berliner and Reca (1966) used aqueous Janus Green 0.01% to stain dead cells blue. Living cells remained colorless.

STABILITY. Yeast cells are less stable than spores and are readily killed by drying, freezing, or heating (Davis et al. 1973).

CULTURE. *H. capsulatum* colonies develop readily on Sabouraud's media at room temperature. The yeastlike phase has special growth requirements, which include a temperature of 37°C (98.6°F), high humidity, carbon dioxide, and blood or yeast extract enrichment media (Vanbreuseghem 1958). Brueck and Buddingh (1951) grew the fungus in the yolk sac of embryonated eggs.

Prevention. Pigeons should have restricted access to soil in exercise yards in geographical areas where *H. capsulatum* is prevalent. No loft should be constructed with a dirt floor. Impervious floors or 1-in.-mesh plastic-coated wire floors should be used. In prevalent areas attendants should wear masks. Sick birds should be destroyed. If a positive diagnosis is made, the loft should be thoroughly cleaned after the birds are removed. The entire loft, including the ceiling, walls, and floor, should be soaked with water prior to cleaning to settle the dust. Removable debris should be buried below 2 ft of soil. If burnable litter is used in the loft, it should be sprayed with fuel oil prior to removal and should then be burned. The loft and the contents should be sprayed with a 5% phenolic disinfectant solution.

Treatment. Amphotericin B (0.1%) solution given intravenously is helpful, but the drug is toxic (Siegmund 1979). In recognized cases treatment has not been fully successful.

References

Ajello, L. 1964. Publ Health Rep 79:266–70.

Berliner, M. D., and M. E. Reca. 1966. Sabouraudia 5:22–29.

Brueck, J. W., and G. J. Buddingh. 1951. Proc Soc Exp Biol Med 86:258–61.

Burrows, W. 1954. *Textbook of Microbiology.* 16th ed. Philadelphia: Saunders, pp. 621–23.

Chute, H. L. 1984. In *Diseases of Poultry,* 8th ed., ed. M. S. Hofstad, p. 321. Ames: Iowa State University Press.

Darling, S. T. 1906. J Am Med Assoc 46:1283–85.

Davis et al. 1973. *Microbiology.* 2d ed. Hagerstown, Md.: Harper & Row, pp. 988–90.

Dodge, H. J. 1965. Am J Publ Health 8:1203–1211.

Guidry, D. J., and H. A. Spence. 1967. Sabouraudia 5:278–82.

Hansmann and Schenken. 1934. Am J Pathol 10:731.

Menges, R. W. 1954. Cornell Vet 44:21–31.

Murdock, W. T., et al. 1962. J Am Med Assoc 179:73–75.

Schwarz, J., et al. 1957. Mycopathol Mycol Appl 8:189–93.

Siegmund, O. H., ed. 1979. *The Merck Veterinary Manual.* Rahway, N.J.: Merck, pp. 446–47.

Tewari, R. P., and C. C. Campbell. 1965. Sabouraudia 4:17–22.

Vanbreuseghem, R. 1958. *Mycoses of Man and Animals.* Springfield, Ill.: Charles C Thomas.

Wildervanck, A., W. A. Collier, and W. E. Winckel. 1953. Doc Ned Geogr Trop 5:108–15.

III

NONINFECTIOUS DISEASES

6

Nutritional Diseases

Pigeon nutrition deals with pigeon nutritional requirements, the feed ingredients and their procurement, and the aspects of ingestion, absorption, nutrient utilization, and the interrelationships of nutrients and their metabolites. Unfortunately little has been published on the actual nutritional requirements or the metabolism of nutrients of pigeons; thus most of the information in this chapter had to be derived from research on other species. The emphasis is on causes, effects, and prevention of nutritional problems.

A ration must be adequate in all essential ingredients. When dietary deficiencies or excesses occur, specific signs often appear that are characteristic of the problem. Good nutrition of the pigeon depends on providing a sufficient quantity of feed and a balance of good-quality nutrients essential to meet the requirements of the bird. These nutritional requirements change as the bird grows older, as seasons change, and as activity and energy demands are altered. As the young squab matures the food is utilized more for maintenance and repair than for growth. When winter approaches, more energy is required to keep the body warm, and as exercise is increased, food intake must be adjusted to prevent depletion of body reserves.

To understand how these needs are satisfied, each nutrient is considered separately: proteins and amino acids; carbohydrates, starches, and sugars; lipids; vitamins; minerals; energy; and water.

Protein Requirements and Imbalances

The regular intake of protein is needed for growth, maintenance and replacement of damaged tissues, disease resistance, nursing squabs, and egg production. One-fifth to one-fourth of the fat-free body of birds consists of proteins (Griminger and Gamarsk 1972), and 20–30% of body protein is found in the feathers, which constitute 5–10% of the body weight (Griminger and Scanes 1986). Proteins are found in important structural parts of body tissues, such as in muscle and connective tissue proteins and keratin proteins found in the skin, nails, and the horny part of the beak. Proteins form body enzymes and hormones, blood cells and serum proteins, hydrolysis products, and other nitrogen-bearing complexes. They are directly involved in disease resistance in the formation of such elements as white blood cells, hemoglobin, fibrinogen, blood globulin, antibodies, complement, and precipitins. Without adequate dietary protein, birds fail to thwart infection and have prolonged recovery and increased mortality. During the period of rapid growth, protein demands are highest. In young growing birds, if proteins are definitely lacking or if a deficiency of a single essential amino acid occurs, growth stops (Scott et al. 1969).

Proteins are classified as simple proteins (those yielding amino acids on hydrolysis), conjugated proteins (those simple proteins combined with a nonprotein radical), and derived proteins (those compounds that represent altered and degraded products of natural proteins) (Maynard and Loosli 1962). Protein molecules are complex nitrogen-bearing compounds composed of hundreds of different amino acids. A single protein, however, may contain many of the same kind of amino acid. They are formed by carbon, hydrogen, oxygen, nitrogen, and in three instances, sulfur.

Amino Acid Requirements. There are 20 naturally occurring, nutritionally important amino acids in protein, termed either *essential* (indispensable) or *nonessential* (dispensable). Essential amino acids are those that *must* be present in the diet because they cannot be synthesized or cannot be made at a rate sufficient to promote maximal growth. Nonessential amino acids can be synthesized from other amino acids and do not have to be present in amounts needed to meet the full requirement. If the diet contains an adequate level of nonessential amino acids, the essential amino acids need not be converted to missing nonessential amino acids; thus a sparing effect occurs. The essential amino acids must be present in the diet, plus a sufficient nonspecific source of nitrogen to synthesize missing nonessential amino acids, nucleic acids, and other nitrogen-bearing compounds. If the essential amino acids are not available in the diet, healthy body protein is broken down to supply the required amino acids. It is obviously more economical and efficient to provide all of the amino acids in the diet.

Birds have specific requirements for certain amino acids and less specific ones for others. Those nine amino acids found to be essential for all bird species and ages are histidine, isoleucine, leucine, lysine, methionine, phenylalanine, threonine, tryptophan, and valine. Arginine is essential for growth in some species. In chicks proline and glycine or serine must be included as essential. Cystine and tyrosine may be synthesized from methionine and phenylalanine; thus cystine, tyrosine, glycine, and proline may be required under certain conditions (Griminger 1986). Only levorotatory (L) forms of amino acids occur in natural proteins, and these are the only forms utilized by animals (Schaible 1970). Synthetic sources of methionine and lysine are composed of D and L isomers; D forms must be converted to L forms in the body before the cells can synthesize them into proteins.

NUTRITIONAL VALUE OF AMINO ACIDS AND PROTEINS. Feed ingredients themselves contain different kinds and levels of protein, and each protein varies in its amino acid content. The quality or nutritional value of a protein thus depends, in part, on the number and kind of amino acids and on the extent to which the amino acid content of the protein agrees with the dietary needs of the bird for certain amino acids. Nutritional value is also influenced by the total nitrogen content (TNC), which averages 14–18% in most proteins (Schaible 1970). For pigeons, which are largely seed eaters, the TNC of a grain diet is essentially only grain protein nitrogen; thus the TNC percentage is a fairly stable value (Maynard and Loosli 1962). Protein metabolism and nitrogen excretion are interrelated with energy utilization and production. Griminger (1986) reported pigeon endogenous N excretion to be 18.8 mg N/Kg/hr. This figure corresponds to 2.89 mg N/Kcal, which compares with 2.3–2.9 mg N/Kcal reported for all homeotherms.

Animal and plant proteins differ in the balance of amino acids that they contain. Animal protein such as high-quality fish meal has a complete amino acid composition similar to the body proteins of animals and birds. The greater the similarity of dietary proteins to the makeup of body proteins, the lower the overall protein requirement. Particular attention must therefore be given in formulating a ration to include a balance of amino acids that approximate body needs.

Cereal grain or plant proteins, on the other hand, are quite different. They are generally weak in lysine and methionine, but this deficiency can be overcome by blending plant and/or animal protein sources or by the addition to pellet feed of specific pure synthetic amino acids. Pure amino acid supplementation is limited, however, because amino acids tend to lower palatability and thus feed intake. When combining several protein sources, one protein compensates for the amino acid deficiency of the other. This practice utilizes a minimum of costly protein. The use of excess protein in an unbalanced diet is wasteful because the excess is not stored but lost as heat (Scott et al. 1969). During protein metabolism, the nitrogen is removed and the molecule is converted to keto acid and carbohydrate and may be used for energy or converted to fat. Excess nitrogen is excreted as uric acid.

The specific requirements of the pigeon for each amino acid is not known, but it is assumed that pigeons require the same essential amino acids as do chickens and turkeys. The most critical amino acids are lysine, arginine, tryptophan, methionine, and cystine, and each must be considered individually (Schaible 1970).

Tables 6.1 and 6.2 list the protein sources

Table 6.1 Average composition of pigeons feedstuffs (air-dry base)

	Protein, %	Protein, Digestible, %[a]	Metalizable Energy, K/cal/lb	Fat, %	Crude Fiber, %	Calcium, %	Phosphorus, Total, %	Sodium, %	Potassium, %	Chloride, %	Manganese, ppm	Zinc, ppm
Alfalfa meal, dehydrated	17.0		720	3.0	24.0	1.30	0.24	0.18	2.50	0.37	44	32
Barley, irrigated	11.5	10.0	1250	1.9	6.0	0.08	0.42	0.02	0.60	0.15	16	15
Buckwheat	11.0	7.4	1200	2.5	11.0	0.11	0.33	0.05	0.45	0.04	34	9
Corn												
Dent #2 yellow	8.6	6.6	1530	3.9	2.0	0.02	0.30	0.01	0.38	0.04	5	10
Flint[a]	9.8	7.5		4.3	1.9		0.33					
Pop[a]	11.5	8.9		5.0	1.9		0.29					
Fishmeal, menhaden	61.0		1400	9.0	1.0	5.50	2.80	0.30	0.70	1.20	36	150
Hempseed	31.0	25.1		6.2	23.8	0.25	0.43					
Meat meal, 55%	55.0	45.8	910	8.0	2.0	8.00	4.00	0.70	0.65	1.00	10	105
Oatmeal, feed	16.0	14.4	1420	5.8	4.0	0.08	0.40	0.05	0.40	0.05	44	140
Peas, field dry	22.0	20.1	1000	1.1	6.0	0.15	0.30	0.04	0.40	0.06		30
Peanut meal[b]	51.0	46.9	1250	1.6	7.0	0.10	0.50	0.08	1.10	0.03		20
Rapeseed meal	35.0	28.5	625	7.0	12.0	0.90	1.20				60	
Rice polishings	12.0	5.7	1300	12.0	3.0	0.04	1.40	0.07	1.10	0.07		
Sesame meal	45.0	38.9	870	5.0	5.0	2.00	1.50	0.04	1.20	0.06	48	100
Sorhum, milo	10.0	8.8	1480	2.8	2.0	0.03	0.30	0.01	.35	0.08	13	17
Soybeans[c]	38.0		1600	18.0	5.0	0.25	0.60	0.02	1.50	0.03	30	20
Soybean meal[d]	48.5	40.7	1150	0.8	3.0	0.26	0.62	0.01	2.00	0.02	40	45
Sunflower meal	37.0		1050	5.0	15.8	0.30	1.20	0.01	1.10	0.19	23	100
Tallow			3200	100.0								
Wheat, hard red northern	14.0	11.3	1500	2.2	2.5	0.04	0.40	0.06	0.40	0.08	20	14
Wheat, soft	10.0		1500	2.0	2.5	0.05	0.40	0.06	0.40	0.08	20	14
Vetch seed[a]	29.6	23.7		0.8	5.7							
Oyster shell						37–39.00						
Dicalcium phosphate						24–28.00	18–21.00					

Source: Adapted from Scott et al. (1982).
[a]Adapted from Morrison (1951).
[b]Solvent, dehulled.
[c]Unextracted, properly processed.
[d]Dehulled.

Table 6.2. Average amino acid composition of pigeon feedstuffs, %

	Arginine	Cystine	Glycine	Histidine	Isoleucine	Leucine	Lycine	Methionine	Phenyl-alanine	Threonine	Tryptophan	Tyrosine	Valine
Alfalfa meal, dehydrated	0.80	0.32	0.90	0.31	0.80	1.20	0.65	0.28	0.75	0.65	0.23	0.55	0.80
Barley, irrigated	0.53	0.18	0.36	0.27	0.53	0.80	0.40	0.18	0.62	0.36	0.18	0.36	0.62
Buckwheat	1.00			0.26	0.35	0.53	0.62	0.18	0.44	0.44	0.18		0.53
Corn, dent #2 yellow	0.50	0.18	0.40	0.20	0.40	1.10	0.20	0.18	0.50	0.40	0.10	0.41	0.40
Fishmeal, menhaden	3.80	0.94	4.40	1.40	3.60	5.00	5.00	1.80	2.70	2.60	0.80	2.00	3.40
Meat meal, 55%	3.90	0.70	8.10	1.00	1.90	3.50	2.80	0.70	2.00	1.80	0.36	0.84	2.80
Oatmeal, feed	0.90	0.24	0.65	0.30	0.55	1.00	0.54	0.20	0.65	0.48	0.20	0.70	0.75
Peas, field dry	1.40	0.17	1.10	0.72	1.10	1.80	1.60	0.31	1.30	0.94	0.24		1.30
Peanut meal[a]	5.20	0.80	2.60	1.10	2.20	3.20	1.80	0.50	2.50	1.40	0.50	1.90	2.20
Rape seed meal	2.00		2.00	0.77	1.30	2.20	1.90	0.66	1.40	1.50	0.45	0.90	1.90
Rice polishings	1.30	0.37	0.74	0.52	0.56	1.10	0.71	0.27	0.71	0.57	0.09	0.63	0.84
Sesame meal	4.80	0.60	4.20	1.10	2.10	3.40	1.30	1.40	2.20	1.60	0.78	2.00	2.40
Sorghum, milo	0.36	0.15	0.40	0.19	0.46	1.40	0.20	0.13	0.47	0.36	0.12	0.70	0.53
Soybeans[b]	2.80	0.64	2.00	0.89	2.00	2.80	2.40	0.51	1.80	1.50	0.55	1.20	1.80
Soybean meal[c]	3.80	0.80	2.30	1.20	2.60	3.80	3.20	0.73	2.70	2.00	0.65	2.00	2.70
Sunflower meal	3.20	0.33	2.40	1.00	2.10	2.50	1.30	0.50	2.10	1.50	0.60	0.70	1.80
Tallow													
Wheat, hard red northern	0.92	0.25	0.70	0.63	0.60	1.00	0.39	0.25	0.66	0.42	0.18	0.31	0.60
Wheat, soft	0.66	0.20	0.50	0.45	0.43	0.80	0.28	0.23	0.47	0.31	0.12	0.22	0.45

Source: Adapted from Scott et al. (1982).
[a]Solvent, dehulled.
[b]Unextracted, properly processed.
[c]Dehulled.

included in most pigeon feed and indicate the average percentage composition of any nutritional ingredients, including the percentage of certain amino acids in the feedstuff. Note that peas and soybeans, which are good legume protein sources, are high in arginine and lysine and thus balance cereal grains, such as corn, which are low in these amino acids.

AMINO ACID FUNCTIONS. Aside from the general body requirement for amino acids, the following specific functions observed in other bird species may also have clinical manifestations in pigeons.

Arginine deficiency in chicks causes wing feathers to curl upward (Austic and Scott 1984). It is also essential for rapid feathering (Schaible 1970). Chicks cannot synthesize arginine at all (Almquist 1957). Dietary proteins must therefore supply arginine for protein synthesis in chickens, and it is from the breakdown of arginine that some urea appears in chicken urine, which is predominately uric acid (Scott et al. 1969).

Isolucine is needed for growth because it is not synthesized (Phillis 1976).

Leucine is also needed for growth. A deficiency of leucine, isoleucine, and phenylalanine is reported to cause a curled tongue deformity in turkey poults (Schaible 1970).

Lysine deficiency is evident in abnormal melanin pigmentation in black or reddish feathers of bronze poults (Fritz et al. 1946). Lysine also maintains the proper function of the enzyme tyrosinase (Harrow 1946). The deficiency is also reported by Hegsted et al. (1941) to cause twisted feathers in chicks.

Methionine deficiency may increase the effects of choline and cobalamine deficiency because of its role as a methyl group donor (Austic and Scott 1984). Methionine requirements can only be met by methionine, but cystine requirements can be met by either acid, according to the National Academy of Sciences–National Research Council (NAS-NRC 1977). Methionine and cystine are sulfur-bearing and are required for egg production and growth (Schaible 1970). Feed proteins are notably deficient in growth-promoting methionine, but when added at high levels, methionine is growth-depressing (Scott et al. 1969). Methionine is available as dry D, L-methionine and as a liquid sodium salt.

Phenylalanine can be transformed to tyrosine in the liver, but the requirement for phenylalanine can only be met by phenylalanine (NAS-NRC 1977). The enzyme tyrosinase converts tyrosine into melanin, a brown or black pigment of skin or feathers (Harrow 1946).

Tryptophan is essential in hemoglobin and niacin formation (Almquist 1957). The conversion of tryptophan to niacin, however, is incomplete with a deficiency of vitamin B complex (Schaible 1970).

Glycine is essential for rapid feathering (Schaible 1970). Either glycine or serine can be used in a diet, and if the other amino acid requirements are adequate, usually glycine or serine levels are also satisfactory (NAS-NRC 1977). Adult chickens do not require glycine, but chicks cannot synthesize it rapidly enough for growth and other needs (Maynard and Loosli 1962).

Proline can be transformed into hydroxyproline or to ornithinic as well as glutamic acid (Harrow 1946), but chicks have difficulty in synthesizing proline, glycine, and glutamic acid fast enough to meet their needs (Almquist 1957). Proline, along with serine and glycine, are most abundant in feathers (Harrap and Wood 1967).

Cystine is mainly found in the keratin of the skin and feathers and is synthesized from serine with sulfur from methionine (Phillis 1976).

The amino acid taurine is found in fish and meat and is essential in the prevention of retinal atrophy, heart dilation, and degeneration. A deficiency of the aminosulfonic acid results from heat processing of food, oxidation, and desulfurizing. Heat-processed pellets, overheated grains, or old grain could be deficient in taurine; however, pigeons reportedly synthesize taurine in sufficient amounts.

Total Protein Requirements

PROTEIN SOURCES. Aside from the minimum requirements for essential amino acids, there is an overall requirement for total protein. Following are the protein sources generally included in pigeon feed, together with comments concerning their value.

High-quality fishmeal may be used in pellets. It is considered a complete amino acid source (Schaible 1970). It is only limited by its cost and availability.

Good-quality meat scrap used in pellets tends to provide a surplus of lysine, glycine, and arginine. It is deficient in methionine, cystine, and tryptophan (Almquist 1957; NAS-NRC 1977).

Field peas, which are well utilized up to 10% of the ration (Ewing 1963), are a fairly good source of arginine and lysine, but only about half the amount is supplied by soybeans (NAS-NRC 1977).

Peanuts (hulled) are high in arginine and tryptophan but deficient in lysine, threonine, methionine, and cystine (Almquist 1957). They contain some harmful heat-resistant trypsin inhibitors found chiefly in the skin (Scott et al. 1969). *Aspergillus* toxins may be present in the peanuts and are often present in the meal. Such toxins are harmful and are unaffected by heat.

Soybeans (cooked) are high in arginine, lysine, and tryptophan but low in sulfur-bearing amino acids, methionine, and cystine (Almquist 1957; NAS-NRC 1977). According to Scott et al. (1969), "With the exception of the sulfur amino acids, all essential amino acids are supplied by soybean meal in excess of the NRC pattern." These researchers also indicated that uncooked soybeans contain harmful protein factors that are detrimental to chicks. They inhibit trypsin, increase the excretion of bile acids, lower the absorption of fat, inhibit growth, lower metabolizable energy, enlarge the pancreas, and affect methionine oxidation. Fortunately heat treatment of the meal destroys the factors.

Wheat is quite variable in protein content and compared with the other grains is low in most amino acids, with the exception of cystine.

PROTEIN LEVELS. The Bulletin 684 from the U.S. Department of Agriculture (1960) indicates the total protein level to be 13.5–15% for squab growth. Wolter et al. (1970) tested protein levels from 12–26% using ad lib cereal grains and peas and obtained best results at 18%. Griminger (1981) also indicated that 18% protein was optimum for the hatch and growth of offspring. He further reported that the percentage dietary requirements for protein paralleled the amino acid level in crop milk. Vogel (1980) reported that dried crop milk contained approximately 56–59% protein. The protein demands of pigeons used as breeders are therefore increased.

DIGESTIBLE PROTEIN. Table 6.1 provides the tabulated percentage of total protein and digestible protein in the ingredients of pigeon rations, giving an indication of the actual feeding value of each feedstuff. For example, field peas have a total protein of 22%, whereas the digestible protein, that percentage utilized by the bird, is about 20.1%. Shelled peanuts, another protein source, have a total protein of 51% and a digestible protein level of approximately 46.9%. The nutritive value, along with availability and cost factors, largely determines which protein sources are selected for a ration.

Carbohydrate Requirements and Imbalances

Carbohydrates are composed of carbon, hydrogen, and oxygen, and the latter two elements are always present in the same proportion as in water. They are classified according to the number of carbohydrate groups they contain: mono-, di-, tri-, and polysaccharides. Monosaccharides are the simple sugars, such as the six carbon sugars (e.g., glucose, fructose, and galactose) and the five carbon sugars (e.g., arabinose and xylose). The disaccharides normally split to give two molecules of simple sugars. Sucrose (table sugar) forms glucose and fructose, lactose (milk sugar) forms galactose and glucose, and maltose (malt sugar) forms two molecules of glucose. Trisaccharides include raffinose from the cotton seed. The polysaccharides include starch and cellulose. Grouped with starch are dextrin, inulin, and glycogen. Cellulose is included with gums, pectin, and agar.

Monosaccharides are the building blocks of all carbohydrates. Multiple sugars and polysaccharides must be broken down to simple sugars or monosaccharides to gain entrance into the bloodstream. The simple sugars D-galactose, D-glucose, D-xylose, and D-fructose are preferentially adsorbed (Griminger 1986). Grains are the chief source of carbohydrates in pigeon rations, the carbohydrates being found mostly in the form of starch. Grains such as corn provide a higher percentage of starch than do field peas or soybeans. Corn is thus fed primarily for its starch or carbohydrate content, but like proteins, carbohydrates are not totally utilized in the body. The digestive enzymes of the chicken appear to break down starch and dextrin efficiently but are unable to derive any benefit from cellulose (Norris and Scott 1965). The hull of most grains is composed of indigestible fiber or cellulose, which reduces the digestibility of the carbohydrate content.

In the chemical analysis of a ration, the carbohydrate content is separated into crude indigestible fiber and nitrogen-free extract. The latter value approximates the digestible portion or the sugars and starch together with minor quantities of indigestible pentosans and lignins (Norris and Scott 1965).

Lipid Requirements and Imbalances

Lipids are composed of carbon, hydrogen, and oxygen and include all ether soluble materials in foodstuffs. They are classified as simple, compound, or derived lipids. Simple lipids are fats, oils, and waxes. Fats (solids and semisolids) and oils (liquids) are esters of glycerol and fatty acids. Waxes are fatty acid esters combined with a nonglycerol alcohol. Compound lipids are glycerol esters of fatty acids linked with phosphoric acid and an added chemical group, which yield on hydrolysis fatty acids, alcohol, and other substances. Included in the group are the phospholipids lechithin, cephalin, and sphingomyelin, which are present in nerve tissue. Lecithins, for example, all contain choline. Those that contain linoleic and linolenic acids occur in soybeans and are utilized in the plasma membranes of the brain. Derived lipids are formed by hydrolysis of simple and compound lipids and include the sterols, such as ergosterol, cholesterol, and sitosterol (Scott et al. 1969). The lipids of primary interest include fatty acids, glycerides, phospholipids, cerebrosides, and cholesterol and other alcohols, which include the fat-soluble vitamins (Phillis 1976).

The simple fats are triglycerides, the basic molecule of which consists of three fatty acid molecules attached to one glycerol molecule. Practically all the fatty acids found in nature contain an even number of carbon atoms. Differences in fats depend on the length of the carbon chain and the degree of saturation. In pigeons, stored triglycerides are largely energy reserves predominantly containing 16–18 carbon chain fatty acids (Griminger 1986).

The unsaturated fatty acids are characterized by the presence of one or more double chemical bonds in their molecules. The degree of saturation is directly related to the total number of double bonds present. Saturated fats lack double bonds and their physical properties are related to their molecular weight. Fatty acids that contain ten or fewer carbon atoms, such as oleic acid, are liquids at room temperature. The remainder are solids. Fatty acids with fewer than four carbon atoms are miscible in water and solubility decreases as the chain lengthens.

Many different triglyceride mixtures form the lipids involved in pigeon nutrition. Thus the nature of the fat or oil and its digestibility largely depends on the fatty acids incorporated into the glyceride molecule (Best and Taylor 1945). Fatty acids with less than 12–16 carbon atoms in the chain are preferred because they are assimilated better (Jones 1985).

Some of the common saturated fatty acids that pigeons may receive include myristic (milk and vegetable fats), palmitic and stearic (most vegetable and animal fats), and arachidic (peanut oil). Included in the unsaturated list are oleic (most fats and oils), linoleic (cotton seed oil), linolenic (linseed oil), and arachidonic. Each of these unsaturated fatty acids may be found in fish meal.

Arachidonic acid is a metabolic product of linoleic acid. The latter cannot be synthesized by poultry. Both are important parts of cell membranes of organelles and fatty tissue and serve as forerunners of prostaglandins (Austic and Scott 1984). Some also include linolenic as essential, but this is questioned (Griminger 1986). Earlier Russell et al. (1942) showed that fat is important in the absorption of fat-soluble vitamins. Although present in only small amounts, lipids are very important in nutrition and they serve as concentrated potent sources of energy. Fatty acids also are essential in the diet of chicks for growth and for the prevention of respiratory diseases and fatty livers (Ross and Adamson 1961; Hopkins and Nesheim 1967).

Vitamin Requirements and Imbalances

Vitamins are chemically unrelated organic dietary compounds that are essential in small but definite amounts for health, growth, and maintenance (Table 6.3). Vitamins A, D, E, and K and the essential fatty acids are fat-soluble. The water soluble vitamins are thiamin (B_1), riboflavin (B_2 or G), pantothenic acid, niacin, pyridoxine (B_6), folic acid, cobalamin (B_{12}), and choline. All recognized vitamins except ascorbic acid or vitamin C and inositol are dietary essentials for poultry (Austic and Scott 1984).

Most vitamins occur naturally in foodstuffs, but quantities vary. In preparing cost-efficient rations it is often necessary to use synthetic vitamin supplements to ensure against deficiencies. Most of the vitamins are integral parts of the vital enzymes or mechanisms that participate in the metabolism of nutrients after absorption by the intestine. These cellular chemical reactions are concerned with the synthesis, transformation, and assimilation of nutrients needed by the body; their breakdown; and the release of waste products and energy. The omission or malabsorption of one vitamin can alter cellular metabolism and result in a deficiency disease.

Vitamin A

DEFINITION. Vitamin A_1 alcohol is a pale yellow, crystalline, fat-soluble, organic compound containing a beta-ionone ring and is found in

Table 6.3. Average vitamin composition of pigeon feedstuffs

	Riboflavin mg/lb	Niacin mg/lb	Pantothenic acid mg/lb	Choline g/lb	Cobalamin µg/lb	Pyridoxine mg/lb	Biotin, mg/lb	Folic acid mg/lb	Alpha Tocopherol IU/lb	Linoleic acid %	Xanthophylls, mg/lb	Provitamin A IU/g
Alfalfa meal, dehydrated	7.0	25.0	15.0	0.73		5.00	0.16	1.80	64.00	0.52	140	530
Barley, irrigated	0.9	26.0	3.0	0.50		1.30	0.07	0.23	16.00	0.85		
Buckwheat	5.0	8.2	5.9									
Corn, dent #2 yellow	0.6	10.0	2.6	0.28		3.20	0.03	0.16	10.0	1.90	11	8
Fish meal, menhaden	2.2	25.0	4.0	1.60	40.0	1.60	0.12	0.09	4.00	0.12		
Meat meal, 55%	2.4	26.0	2.2	1.0	20.0	1.40	0.12	0.02	0.45	0.29		
Oatmeal, feed	0.7	4.6	1.5	0.20						0.70		
Peas, field dry	0.8	17.0	4.6			0.45	0.08	0.16				
Peanut meal[a]	5.0	77.0	24.0	0.90		4.50	0.18	0.16	1.40	0.38		
Rape seed meal	1.7	70.0	4.2	3.00					9.00	1.00		
Rice polishings	0.8	240.0	26.0	0.59			0.28		42.00	3.60		
Sesame meal	1.5	14.0	2.7	0.68		5.70				1.90		
Sorghum, milo	0.5	18.0	5.0	0.31		1.80	0.08	0.11	5.50	1.10		
Soybeans[b]	1.2	10.0	7.0	1.30		4.90	0.17	1.00	23.00	9.00		
Soybean meal[c]	1.4	10.0	6.5	1.20		3.60	0.15	1.60	1.50	0.40		
Sunflower meal	3.3	48.0	18.0	1.47		7.30			9.00	2.90		
Tallow												
Wheat, hard red northern	0.5	27.0	6.0	0.45		1.80	0.05	0.18	5.00	0.60		
Wheat, soft	0.5	27.0	6.0	0.45		1.80	0.05	0.14	5.00	0.60		

Source: Adapted from Scott et al. (1982).

[a]Solvent.

[b]Unextracted, properly processed.

[c]Dehulled.

nature only in animal tissues. The deficiency is largely characterized by a deterioration of epithelial health. The problem is observed in pigeons.

SOURCE. Animals and birds derive much of their vitamin A from plant sources, which contain yellow carotene and xanthophyll pigments but no true vitamin A. These provitamins are converted to vitamin A by the bird. Beta-carotene is the most important relatively abundant provitamin. For chickens vitamin A activity is possessed by vitamin A_1 alcohol (retinol; all-trans retinyl acetate), vitamin A aldehyde (retinene; retinal) and vitamin A acid (retinoic acid) (Scott et al. 1982). Fish liver oils, alfalfa leaf meal, and yellow corn are natural vitamin A sources for pigeons. Synthetic vitamin A is manufactured as all-trans retinyl acetate or palmitate (Scott et al. 1969).

STANDARD. One USP unit or one international unit (IU) equals the activity of 0.3 μg retinol or 0.6 μg pure beta-carotene or 0.25 mg pure vitamin A or 0.344 μg vitamin A acetate (Morrison 1951; Scott et al. 1969; Siegmund 1979; Stecher 1960).

FUNCTION. Vitamin A is essential for life, growth, and health and is perhaps better known as the anti-night blindness factor. It also promotes appetite and digestion and is necessary in maintaining normal epithelial structure, and function. It is required for egg production, fertility, and hatchability, and it increases resistance to infection and to some parasites.

UTILIZATION. Beta-carotene in the ration is converted in the intestinal wall to vitamin A alcohol (retinol), retinal, and retinyl ester (Scott et al. 1969). Vitamin A esters are hydrolyzed by pancreatic lipase to vitamin A_1 alcohol, absorbed as the alcohol, and partially re-esterified in the intestinal wall. Vitamin A_1 alcohol and esters are transported in lymph to the liver and other tissues. In the liver it is stored predominately in the ester form. During transportation in blood, carotenoids and probably vitamin A are bound to plasma proteins (Ganguly et al. 1952).

SIGNS. Vitamin A-deficiency signs are related to the degree of deprivation and are primarily found associated with epithelial mucous membrane linings. An absence of the vitamin results in white cheesy exudate under the eyelids, pasty eyelids, lachrymation, semimucoid exudate in the nostrils, ruffled feathers, incoordination, weakness, and lack of yellow pigment in the skin. Fertility and hatchability are reduced, and in squabs lack of vitality and cessation of growth may be noted. Death may occur if the condition persists.

NECROPSY. Necropsy findings in vitamin A deficiency include creamy pustulelike lesions of the mucous glands of the mouth and/or esophagus (Fig. 6.1), enlarged pale kidneys, urate-distended ureters and gout, gelatinous infiltrations around the heart and over the breast muscle, cankerous growths about the palatine cleft, enlargement of the proventriculus and gallbladder, and pus formation in the ears and sinuses.

MICROPATHOLOGY. In vitamin A deficiency, there is atrophy and degeneration of the respiratory tract mucous membrane. The ciliated columnar epithelial cells are replaced by stratified squamous cells. There is degeneration but not inflammation of the central and peripheral nerves. Distended mucous glands of the palate and esophagus become filled with stratified keratinized epithelial cells (Seifried 1930).

TOXIC MANIFESTATIONS. Retinol and its esters are toxic for chickens at about 1,000,000–1,500,000 IU/kg diet. This is about 500 times the minimum normal required level. Retinal is somewhat more toxic. Toxicity of retenoic acid is encountered at 50–100 times the required

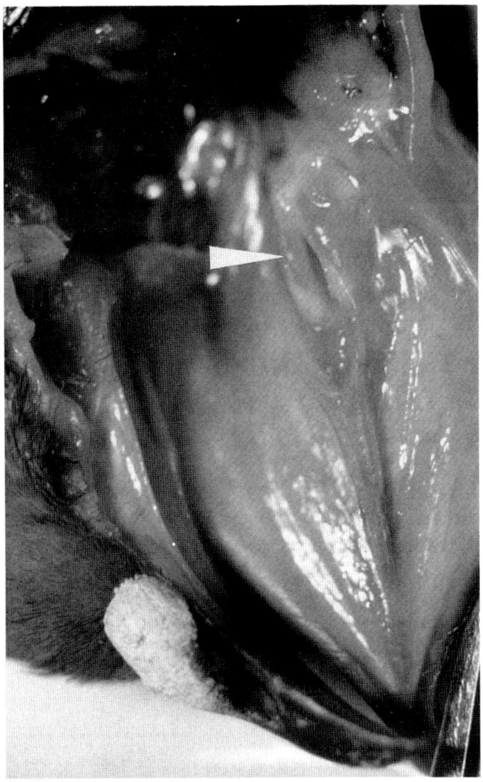

Fig. 6.1. Localized pustule in the larynx, caused by vitamin A deficiency.

level (Scott et al. 1969). Continued excessive use of the vitamin produces decreased feed intake, loss of weight, rarefaction of the bones, eyelid swelling, and inflammation.

STABILITY. Vitamin A and carotene are reduced in potency by oxygen and ultraviolet light. Minerals such as manganese, cobalt, copper, iron, and iodine catalyze the oxidative destruction of vitamin A and carotene. Minerals such as ground limestone, bonemeal, sulfur, and charcoal are destructive when directly mixed together with the vitamin. In addition, heat, rancid fat, dried milk, dried whey, and meat scrap are also destructive. Antioxidants such as ethoxyquin are thus essential in the protection of vitamin A.

DIFFERENTIAL DIAGNOSIS. Early vitamin A-deficiency eye signs are confused with pox, mycoplasmal, and aspergillosis infections and irritations caused by ammonia and lime dust.

PREVENTION. Factors that lower vitamin A potency should be avoided and rations should contain an adequate level. Birds should receive 1400–2200 USP units/lb feed. The NAS-NRC (1977) level is 3000–5000 IU/kg feed for poultry.

TREATMENT. If a positive diagnosis of vitamin A deficiency is made, a water-miscible stabilized form of vitamin A at the rate of 10,000 USP units/lb feed for 7–10 days only is suggested. A light coating of corn oil on the grain permits adherence of a dry vitamin powder. This should be followed by the preventative level. A stable feed form of vitamin A is just as acceptable as a water form.

Vitamin D

DEFINITION. Vitamin D is the antirachitic fat-soluble vitamin concerned with the proper absorption and metabolism of calcium and phosphorus. The deficiency occurs in pigeons on occasion.

SOURCES. The plant form of vitamin D, called ergosterol, is transformed to D_2 or calciferol in the body and is essentially not available to birds. The animal provitamin 7-dehydrocholesterol is present in the skin of animals and birds and is converted to D_3 (cholecalciferol) by irradiation of the skin with ultraviolet light. Work by DeLuca (1971) revealed a more biologically active form as 25-hydroxycholecalciferol. Other synthesized analogues and related products have considerably more antirachitic activity than D_3.

STANDARD. The International Chick Unit (ICU) of vitamin D was defined by the World Health Organization as the biological activity of 0.025 μg pure D_3. The U.S. Pharmacopeia has also adopted this as the USP unit. Prior to 1951 vitamin D was measured in Association of Official Analytical Chemists (ADAC) units. The ICU is one-third larger than the old AOAC unit (Standard Brands 1954; Siegmund 1979).

FUNCTION. Vitamin D is indispensable to the normal formation of bone. Its absence results in rickets or osteomalacia. The vitamin is essential for the proper metabolism of calcium and phosphorus (Scott and Norris 1965) and for the absorption of calcium (Keane et al. 1956). The vitamin promotes growth and health and it increases vitality and livability. It is essential for eggshell formation and for hatchability of eggs. Bellin et al. (1954) showed that vitamin D increases the urinary excretion of citrate; thus it also affects citric acid metabolism.

UTILIZATION. D_3 is transported by the blood to the liver, where it is stored (Scott and Norris 1965). DeLuca (1971) reported that the active form of vitamin D is formed by the hydroxylation of cholecalciferol to 25-hydroxycholecalciferol in the liver and then to 1,25-dihydroxycholecalciferol in the kidney.

SIGNS. Vitamin D deficiency is characterized by muscular or leg weakness; paralytic signs; weak beady rib structure; rubbery beaks, legs, and keel; decreased egg production; thin shells; and/or cracked eggs and poor hatchability. Birds often rest on their hocks and fail to move. In red or buff-colored breeds of chickens, black feather pigmentation has been noted.

NECROPSY. In vitamin D deficiency, joint swelling with soft poorly calcified bones that lack a brittle break, beading of the ribs at the costochondral junction, and rib bending at the juncture with the spine may be noted. Cartilage increases at the joints in an attempt to compensate for an absence of calcification.

MICROPATHOLOGY. Enlargement of growing zones without tissue calcification, a lack of orderly bone development, failure of cartilage cells to develop normally, and irregular destruction of cartilage cells by advancing capillaries are observed in vitamin D deficiency.

TOXIC MANIFESTATIONS. Excessive vitamin D levels cause renal damage and calcification in kidney tubules. Calcification may also occur in other tissues (Austic and Scott 1984).

STABILITY. Vitamin D is relatively resistant to heat but is destroyed by mixing with manganese.

DIFFERENTIAL DIAGNOSIS. An imbalance or a deficiency of calcium or phosphorus may create the same clinical pattern.

PREVENTION. To ensure sufficient vitamin D, lofts must provide sunlight. A maintenance level of approximately 400–500 ICU D_3/lb or 900–1100 ICU/kg feed is necessary. These are approximately the levels given by the NAS-NRC (1977) for quail. Approximately two parts of calcium to one of phosphorus should be provided in the diet, or 20 lb dicalcium phosphate/t feed.

TREATMENT. Hooper at al. (1942) found that one single large dose of 15,000 IU/chick was more effective than continued feeding of higher levels. Where a deficiency is diagnosed this method of treatment should apply to pigeons as well.

Vitamin E

DEFINITION. A deficiency of fat-soluble, unstable organic compounds known as tocopherols can cause an uncommon disease of pigeons that may be characterized by encephalomalacia or exudative diathesis.

SOURCE. D-alpha-tocopherol acetate produces about four times the vitamin E activity of L-alpha-tocopherol acetate (Scott et al. 1969). Tocopherols are present in the oil of the germ of grains, including peas, corn, wheat, and soybeans, and in alfalfa leaf meal (Scott and Norris 1965).

STANDARD. One IU vitamin E is equivalent to the activity of 1 mg unesterified DL-alpha-tocopheryl acetate (Scott and Norris 1965).

FUNCTION. Purkinje cell degeneration and brain edema or encephalomalacia can be prevented in chicks by vitamin E. In the alcoholic form vitamin E is an effective natural antioxidant and as such protects essential fatty acids, vitamin A, D_3, the carotenes, and xanthophyll (Scott and Norris 1965). It is necessary for embryo development and hatchability (Adamstone 1931; Jensen et al. 1956). Selenium and vitamin E are involved in preventing capillary permeability and exudative diathesis in chicks (Dam and Glavind 1939; Scott et al. 1969). Scott (1953) also reported on hock enlargement and leg bowing in turkeys on a vitamin E–deficient diet, and when vitamin E deficiency is accompanied by a sulfur amino acid deficiency, muscular degeneration occurs in chicks. Furthermore, in vitamin E–deficient ducks a similar skeletal muscle problem occurs (Austic and Scott 1984). In vitamin E– and selenium-deficient chickens and turkeys, gizzard and heart lesions develop (Scott et al. 1967). Vitamin E also participates with selenium and cystine in preventing muscular degeneration (Austic and Scott 1984). It is further theorized that vitamin E and the enzyme glutathione peroxidase, which contains selenium, act to prevent peroxidative injury to the capillary endothelial lining (Noguchi et al. 1973). It also appears that the tocopherols are active in the NADH cytochrome C reductase system in the oxidative production of energy (Phillis 1976).

UTILIZATION. Bile aids the absorption of oil-soluble vitamin E and selenium also aids in the absorption, retention, and utilization of vitamin E (Scott et al. 1969; Ensminger and Olentine 1978).

SIGNS. In vitamin E deficiency, young birds appear droopy and inactive and may have clonic spasms with downward or backward head-curling. Backward propulsion may occur while the bird rests on its hocks. Adults may have no visible symptoms, but embryonic mortality generally occurs during the first week of incubation because a lethal mesodermal ring chokes off vitalline vessels.

NECROPSY. Gross brain changes are not obvious in vitamin E deficiency. Muscle degeneration such as white fish flesh, elongated muscle bundles, or streaks in breast muscle or other muscles may be noted. Exudative diathesis is characterized by edema of subcutaneous tissue, which results from abnormal capillary permeability. This may be observed under the skin of the breast and legs.

MICROPATHOLOGY. In vitamin E deficiency, the cerebellum of the brain is most commonly affected. Edema with destruction of Purkinje cells predominates. Other portions of the brain are also affected with ischemic necrosis of tissue. Muscle degeneration is characterized by Zenker's hyaline degeneration.

TOXIC MANIFESTATIONS. Toxic signs for vitamin E have not been observed and reported.

STABILITY. Vitamin E is easily destroyed by oxidation. Unsaturated oils, such as corn oil, increase the need for vitamin E. If such oils become rancid prior to ingestion, the vitamin E is destroyed. If oxidation occurs when ingested, body stores of vitamin E will also be depleted (Scott et al. 1969). Minerals such as ferric chloride destroy vitamin E, and mineral oil interferes with its absorption. Exposure to ultraviolet light and heat also is detrimental.

DIFFERENTIAL DIAGNOSIS. Vitamin A deficiency and selenium deficiency must be excluded.

PREVENTION. Diets should contain 10–12 IU D-alpha-tocopherol/kg feed (Griminger 1981). Singsen et al. (1953) discovered that the antioxidant diphenyl-*p*-phenylenediamine had a sparing effect on vitamin E. Presently 6-ethoxy-1,2-dihydro-2,2,4-trimethylquinoline (ethoxyquin) is widely used to protect vitamin E, A, carotenes, and xanthophylls from oxidation.

Selenium together with vitamin E prevents exudative diathesis. An effective diet for chicks contained 0.1 ppm selenium as $NaSeO_3$ (Gries and Scott 1972a). Previously Scott (1962) showed that 0.04–0.1 ppm selenium would prevent or cure exudative diathesis in chicks. The sulfur-bearing amino acids cystine and methionine should be present in the diet to prevent muscle degeneration. Based on the standards of the NAS-NRC (1977) feeding of 0.5% methionine plus cystine is recommended.

TREATMENT. The oral administration of a single dose of 300 IU vitamin E is suggested for individual birds (Siegmund 1979). Brain damage produced by the deficiency will not be altered by treatment. Capillary permeability or muscular degenerations may respond to dietary supplements of methionine, vitamin E, and selenium if the changes are not excessive.

Vitamin K

DEFINITION. Hemorrhagic disease is a nutritional disease caused by the absence of or interference with the fat-soluble vitamin K. It is characterized by prolonged blood clotting time and hemorrhages. The condition is observed in pigeons that have consumed toxic drugs.

SOURCE. Vitamin K_1, phylloquinone, is a yellow oil found in green vegetable leaves. K_2, menaquinone, is produced by bacteria in animals (Scott et al. 1969). Bacteria in the intestine synthesize the vitamin, but it is unlikely that this source furnishes all the body needs for chicks (Scott and Norris 1965). Synthetic menadione sodium bisulfite or diphosphate are reliable sources.

STANDARD. Synthetic menadione, K_3, represents the major feed supplement and essentially serves as a standard. It is a bright yellow crystalline powder several times more active than natural K_1.

FUNCTION. Vitamin K is necessary for the synthesis of prothrombin, an essential part of the blood clotting mechanism that prevents hemorrhages and embryo mortality late in incubation (Austic and Scott 1984). In 1978 Garvey and Olsen reported that during a deficiency of vitamin K the liver produces prothrombin that lacks gamma-carboxylglutamic acid. This faulty product increases blood clotting time and hemorrhages.

UTILIZATION. Fat-soluble forms require bile salts for absorption. Mineral oil apparently interferes with absorption of the vitamin if it is applied to the grain or feed during treatment. The vitamin is not stored.

SIGNS. In vitamin K deficiency, birds appear weak and listless from anemia.

NECROPSY. Intramuscular hemorrhages appear in the muscles of the breast, legs, wing, and heart in vitamin K deficiency. Hemorrhages may appear beneath the serosa of the intestine and mucosa of the proventriculus, in liver and kidney tissue, and under the gizzard lining. The blood fails to clot. Hemorrhages may be noted in or under the skin and along the shanks of the legs within a week of the deficiency.

MICROPATHOLOGY. In vitamin K deficiency, hemorrhages are evident in most tissues.

TOXIC MANIFESTATIONS. When excessive vitamin K levels are given, scaly skin and cere may be observed, together with skin and leg dehydration.

STABILITY. Vitamin K is destroyed by ultraviolet light and oxidation.

DIFFERENTIAL DIAGNOSIS. The use of drugs such as sulfaquinoxaline must be considered as a cause.

PREVENTION. The routine use of additional vitamin K_3 in the diet is not indicated; however, 0.5–1.0 mg K_1/kg feed should be present in the ration (Griminger 1981).

TREATMENT. Treatment is seldom necessary for vitamin K deficiency. When excessive medication of other drugs has been given, vitamin K should be given if clinical signs are present. Supplemental menadione sodium bisulfite may be provided in the feed at the rate of 0.5–2.0 g/t feed. Very careful and thorough mixing must be practiced to guarantee even distribution. Corn oil on the grain enables adherence of the powder to the grain. A gram scale must be used to weigh the vitamin. Water medication is not recommended because of the risk of poisoning.

Vitamin B_1

DEFINITION. Vitamin B_1 (thiamine) is a water-soluble, crystalline nutritional compound. The deficiency is characterized by polyneuritis. The disease condition is seldom observed in pigeons except on experimental diets.

SOURCES. Thiamine occurs in plant and animal tissue, including rice hulls, cereal grains, yeast, liver meal, milk, and green leaves.

STANDARD. One IU vitamin B_1 is equivalent to 3 μg crystalline thiamine hydrochloride (Scott et al. 1982). Levels are calculated in milligrams.

FUNCTION. Vitamin B_1 is necessary for carbohydrate metabolism and for normal appetite, growth, and nerve function.

UTILIZATION. Thiamine is converted to thiamine pyrophosphate, which is a coenzyme factor in oxidative carboxylation and aldehyde exchanges (Austic and Scott 1984).

SIGNS. In vitamin B_1 deficiency, pigeons exhibit loss of appetitie, loss of weight, ruffled feathers, incoordination, and finally paralysis with the head reflected backward. Skin edema may also be noted.

NECROPSY. In vitamin B_1 deficiency, the adrenals become enlarged, but shrinkage of genital organs may occur (Austic and Scott 1984).

MICROPATHOLOGY. Degeneration and dilation of the duodenal crypts, together with cytoplasmic vacuolation of the pancreas, may be present in vitamin B_1 deficiency (Austic and Scott 1984).

TOXIC MANIFESTATIONS. High vitamin B_1 levels have been given without evidence of toxicity (Scott et al. 1969). Excess vitamin B_1 in humans often results in red facial flushes.

STABILITY. Vitamin B_1 is destroyed by alkali.

DIFFERENTIAL DIAGNOSIS. The condition must be differentiated from vitamin E deficiency.

PREVENTION. Pigeons should receive 1.8–2.0 mg/kg feed (Griminger 1981).

TREATMENT. Seriously vitamin B_1–deficient birds can not ingest food on their own; thus individual birds should be treated orally with 1.0–2.0 mg thiamine hydrochloride.

Vitamin B_2

DEFINITION. A deficiency of the yellow-green, fluorescent, water-soluble B_2 or G vitamin (riboflavin) is characterized by curled toe paralysis. The problem seldom occurs in pigeons.

SOURCE. Good sources of vitamin B_2 include milk products, green plants, yeast, liver, and eggs. In addition, intestinal microflora synthesize the vitamin (Scott et al. 1969). Synthetic riboflavin is widely used as a feed supplement.

STANDARD. Riboflavin, 6,7-dimethyl-9-(1'-D-ribityl) isoalloxazine, is a crystalline powder measured in milligrams or micrograms.

FUNCTION. Vitamin B_2 is essential for growth and health, tissue repair, hatchability, and the prevention of curled toe paralysis. It is an active component of several body enzymes essential in cellular oxidation-reduction activities (Austic and Scott 1984). The deficiency affects primarily the epithelium and the myelin sheaths of some large nerve trunks, notably the sciatic (Scott et al. 1982).

UTILIZATION. Vitamin B_2 is stored in the liver, kidney, and heart. In tissues it occurs as riboflavin-phosphoric acid and this is combined with adeninedinucleotide (Stecher 1960). A number of compounds, including atabrine with its isoalloxazine nucleus, have an antiriboflavin effect (Scott et al. 1982).

SIGNS. Retarded growth is noted in vitamin B_2–deficient young birds. Squabs exhibit inward curled toes and walk with difficulty. Complete leg paralysis may develop, and shanks and feet may be swollen from edema. In adults decreased egg production and hatchability may be observed. Embryos may have clubbed down because the feathers fail to rupture the sheath.

NECROPSY. The livers may increase in fat content in vitamin B_2 deficiency.

MICROPATHOLOGY. In vitamin B_2–deficient chicks, degeneration of the myelin sheath covering the nerves to the extremities often occurs. In severe cases the sciatic and brachial nerves are soft and swollen (Austic and Scott 1984).

TOXIC MANIFESTATIONS. Signs of excessive vitamin B_2 are not readily evident.

STABILITY. Riboflavin is particularly sensitive to light and alkaline solutions. It is stable in mineral acids in the dark (Scott et al. 1969).

DIFFERENTIAL DIAGNOSIS. Early stages of the deficiency may not be recognized, but as the signs develop it is quite distinctive.

PREVENTION. The diet should contain 3.5–4.0 mg/kg or 1.6–1.8 mg/lb feed (Griminger 1981; NAS-NRC 1977).

TREATMENT. If vitamin B_2 deficiency has been present for some time, curled toes will not respond to vitamin therapy. Where a partial deficiency occurs birds may recover as they get older, indicating that the requirement is less for older birds (Austic and Scott 1984). Where a deficiency is present fresh whole milk may replace the water for birds to drink for 2–3 days, and during this period the diet should be adjusted to the preventative level to overcome the condition. Also, according to Scott et al. (1982), individual chicks have been treated with 100 μg.

Vitamin B_5

DEFINITION. The absence of water-soluble vitamin B_5 (pantothenic acid) is characterized in the chick by crust formation in the corners of the mouth and rough feathering. It is an uncommon nutritional disease of pigeons.

SOURCE. Pantothenic acid occurs in animal and plant tissues and is concentrated in liver meal. Rice bran, yeast, and molasses are good sources, but seeds are relatively poor sources. Synthetic pantothenic acid is usually included in vitamin supplements.

STANDARD. One g pantothenic acid equals 70,000–75,000 ICU (Stecher 1960). Free pantothenic acid is a highly hydroscopic, viscous oil, but the pantothenate form is a white needle-shaped crystal.

FUNCTION. Vitamin B_5 is essential for growth and health, for normal hatchability, and for the prevention of early squab mortality

(Gillis et al. 1948). Pantothenic acid is a component of coenzyme A, which participates in the metabolism of carbohydrates, proteins, and fats. It is involved in the synthesis of acetylcholine, fatty acid and keto acid oxidation, citrate formation, and the removal of nitrogen from amino acids (Austic and Scott 1984).

UTILIZATION. Only the dextro rotary forms of the salt are utilized (Stecher 1960). The chick B_5 requirement is greater when breeders are deficient in B_{12} (Yacowitz et al. 1951).

SIGNS. In vitamin B_5 deficiency, granular crusts may appear in the corner of the mouth and eyelids, and small cracks may appear on the bottoms of the feet. Uneven, retarded growth and rough feathering may occur. Loss of feathers from the head have been observed. Hatchability of fertile eggs is lowered and embryo mortality increases during the last 2–3 days of incubation. Stunted squabs hatch weak and may die during the first week of life.

NECROPSY. Fatty degeneration of the liver with enlargement may be found, together with kidney swelling, in vitamin B_5 deficiency.

MICROPATHOLOGY. Fatty infiltration of liver cells occurs in vitamin B_5 deficiency.

TOXIC MANIFESTATIONS. Adrenal and kidney changes may be expected with excessive medication of vitamin B_5.

STABILITY. Free pantothenic acid is readily destroyed by acids, bases, and heat (Stecher 1960). Heated kiln-dried corn generally has less B_5. Calcium pantothenate is fairly stable and only small amounts are lost in pelleting (Scott et al. 1969).

DIFFERENTIAL DIAGNOSIS. Winter chilling of squabs, omphalitis, and salmonellosis must be excluded as a cause of squab losses. Biotin deficiency signs are also similar. Interrelationships with B_{12} and folic acid must be considered.

PREVENTION. Based on NAS-NRC (1977) suggested levels for poultry, diets for young birds and breeders should contain 10 mg pantothenic acid/kg or 4.5 mg/lb feed.

TREATMENT. The water-soluble calcium chloride complex of pantothenic acid permits subcutaneous injection of individual birds with 1–2 mg/bird. Flock medication in the drinking water with 2.0 g calcium chloride complex and 0.5 g riboflavin in 50 gal for 5–7 days has been effective. The addition of 5.0–5.5 mg calcium pantothenate/kg feed has also been suggested (Siegmund 1979).

Vitamin B_6

DEFINITION. Vitamin B_6 (pyridoxine), the rat antidermatitis factor, is a group name for several pyridine derivatives, including coenzymes pyridoxal phosphate and pyridoxamine phosphate (Phillis 1976). A deficiency of pyridoxine is very unlikely when pigeons are fed a pigeon grain diet.

SOURCE. Pyridoxine hydrochloride is the synthetic source of vitamin B_6. Corn, soybeans, and other whole grains are plant sources, in which the vitamin takes the form of pyridoxol. Meat and liver meals are the best animal sources, but in these it takes the form of pyridoxal and pyridoxamine phosphate (Scott et al. 1969).

STANDARD. Vitamin B_6 activity is calculated in milligrams per kilogram of feed.

FUNCTION. Pyridoxine derivatives serve as coenzymes in the transamination, decarboxylation, and desulfuration of amino acids. Pyridoxine is involved in active transport of amino acids and certain metallic ions across cell membranes. Pyridoxal phosphate also acts as a coenzyme in changing linoleic to arachidonic acid (Phillis 1976). In chickens, egg production, hatchability, and feed intake are reduced by the deficiency, and it causes a rise in serum iron and a fall in copper, resulting in anemia (Scott et al. 1969).

UTILIZATION. Pyridoxine requirements increase when protein intake is high or when the amino acid balance is poor (Phillis 1976). High-tryptophan diets increase the need for pyridoxine (Fuller 1964).

SIGNS. Poor growth, loss of weight, anemia, perosis, and nervous signs may characterize vitamin B_6 deficiency. Exhaustive spasms often result in death as a result of insufficient energy for the brain. A severe deficiency in breeders will result in rapid cessation of ovarian and testicular activity (Scott et al. 1982).

NECROPSY. In vitamin B_6–deficient chicks and turkey poults, anemia is seldom observed, but severe anemia is reported in ducklings (Scott et al. 1969). An examination of squabs should be made for anemia.

MICROPATHOLOGY. Chick perosis caused by intermediate levels of pyridoxine, 2.5–2.8 mg/kg diet, and 31% protein causes disorganized cartilage formation at the tarsometatarsal joint (Gries and Scott 1972b).

STABILITY. Pyridoxine is stable to heat, acids, and alkalies, but some losses occur in feed processing (Scott et al. 1969).

DIFFERENTIAL DIAGNOSIS. Any deficiency diagnosis involving perosis must exclude other causes, such as a deficiency of manganese, zinc, biotin, choline, folic acid, and niacin. When nervous signs are present, vitamin E deficiency must be considered, but in pyridoxine deficiency the nervous signs are usually intensified.

PREVENTION. Pyridoxine is required at a

level of 2.6–4.5 mg/kg feed (Griminger 1981). Breeders require the higher level. In experiments with chicks, Gries and Scott (1972b) reported that high protein levels increased the requirement for pyridoxine.

TREATMENT. In the absence of natural clinical disease, treatment is not indicated.

Vitamin B_{12}

DEFINITION. Vitamin B_{12}, the animal protein factor, is a water-soluble, dark red, crystalline growth substance consisting of a macrocyclic, porphyrin-type, cobalt-containing nucleus with a nucleotide attached (Phillis 1976). The deficiency may occur in pigeons.

SOURCE. Bacteria in the intestine synthesize vitamin B_{12} if cobalt is available. Vitamin B_{12} occurs naturally in foods of animal origin, including meat, milk, fish, and eggs, but not to any extent in plant materials (Scott et al. 1982). Fermentation products are available for feed supplementation.

STANDARD. Cyanocobalamin and hydroxycobalamin are the usual forms of vitamin B_{12} used as feed additives. They are calculated in milligrams.

FUNCTION. Vitamin B_{12} is essential for good food utilization, growth, and hatchability. It is involved in the formation of nucleic acid, methyl groups, and fat and carbohydrate metabolism (Scott et al. 1978). It is a cofactor for homocysteine transmethylase and methyl malonyl coenzyme A isomerase, and it is interrelated with folic acid, pantothenic acid, choline, and methionine (Scott et al. 1982).

UTILIZATION. For absorption vitamin B_{12} requires an intrinsic factor from the proventriculus and transcobalamin II formed in the liver (Scott et al. 1982) but only about 3% of the true B_{12} is absorbed (Phillis 1976). It is retained in the liver for long periods but is depleted rapidly when high protein levels are fed (Scott et al. 1982). The cobalt fraction is quickly removed from the body by the kidneys (Harrow 1946). Methylmalonic acid is found in the urine and is the basis for the B_{12} deficiency test (Scott et al. 1982).

SIGNS. Slow growth and poor feed utilization, poor hatchability and mortality are signs of a vitamin B_{12} deficiency. Perosis may occur if the diet is also deficient in choline, methionine, or methyl groups (Scott et al. 1969). Pernicious anemia develops in humans when B_{12} is not absorbed (Scott et al. 1982).

NECROPSY. Pale flesh and watery blood may be present in vitamin B_{12} deficiency.

TOXIC MANIFESTATIONS. Toxic signs of vitamin B_{12} have not been reported.

STABILITY. Aldehydes, ferrous salts, alkali, ascorbic acid, and oxidizing and reducing chemicals destroy vitamin B_{12} (Scott et al. 1982). It is stable at a temperature of 121°C for a short time.

DIFFERENTIAL DIAGNOSIS. Other nutritional problems and infectious diseases must be considered when poor growth and poor hatchability occur.

PREVENTION. As little as 0.003–0.009 mg vitamin B_{12} (cyanocobalamin)/kg feed is sufficient to prevent a deficiency (Scott et al. 1969).

TREATMENT. Injectable vitamin B_{12} is available for subcutaneous use; 2–5 mg/bird may be given if a definite diagnosis is established. Repeated injection is of questionable value.

Biotin

DEFINITION. Vitamin H (biotin) is a water-soluble, crystalline, sulfur-bearing vitamin that is active in carboxylation and decarboxylation reactions, which fix carbon dioxide (Phyllis 1976). The natural clinical deficiency can occur in pigeons.

SOURCE. Intestinal bacteria normally synthesize biotin in adequate amounts. Liver meal, yeast, egg yolk, and blackstrap molasses are good sources, but the vitamin is present in all living plant and animal cells (Stecher 1960). On the other hand, cereals and fish and meat meal are poor sources (Scott et al. 1969).

STANDARD. Vitamin H is calculated in milligrams per kilogram of diet.

FUNCTION. Biotin is involved in carbohydrate, lipid, and protein metabolism. It is needed in chicks for the prevention of dermatitis, particularly involving the feet and the skin adjacent to the eyes and beak. It is essential for hatchability. Enlarged hocks or slipped tendon has also been reported in vitamin-deficient turkeys (Scott et al. 1978).

UTILIZATION. Biotin is inactivated by avidin of raw egg white. When this occurs, graying and loss of hair has been reported (Phillis 1976). According to Scott et al. (1982), much of the biotin present in natural ingredients is not biologically available.

SIGNS. Crusts form at the corner of the mouth, along the skin of the beak, and on the eyelids in biotin deficiency. Cracks may be noted on the bottoms of the feet. According to Cravens et al. (1944), deficient embryos develop parrot beaks, are small, and fail to hatch, with death occurring during the first and last week of the hatch. They develop webbing between the third and fourth toes. Couch et al. (1948) found that the embryo tibiotarsus is shortened and bent posteriorly and the tarsometatarsus is abnormal and short.

NECROPSY. Internal signs of biotin deficiency are not significant. Minor subcutaneous inflammation is present with dermatitis. Joint changes may be evident with perosis.

TOXIC MANIFESTATIONS. Signs of excessive biotin dosage have not been reported.

STABILITY. Biotin is reasonably stable (Stecher 1960), but it is destroyed by oxidizing agents, rancidity, and strong acids and bases (Scott et al. 1969).

DIFFERENTIAL DIAGNOSIS. The skin condition is similar to pantothenic acid deficiency, and the hock problem is like perosis. A diagnosis of perosis must consider the other causes.

PREVENTION. Jukes and Bird (1942) injected 2 μg biotin daily into young chickens and partly overcame the deficiency signs. A suggested feed level is 0.15–0.2 mg/kg (Griminger 1981). This will prevent a deficiency, but one may be created by the excessive use of antibiotics, which destroy intestinal bacteria.

TREATMENT. The preventative level of biotin is adequate for treatment.

Niacin

DEFINITION. Niacin (nicotinic acid) is a water-soluble, white, crystalline vitamin component of dehydrogenase coenzymes: nicotinamide adenine dinucleotide phosphate (NADP) and nicotinamide adenine dinucleotide (NAD). These coenzymes and others participate in energy intake (Austic and Scott 1984). This deficiency is not clinically recognized in pigeons.

SOURCE. Cereal grains, legumes, greens, yeast, milk, and liver meal are sources of niacin.

STANDARD. The activity of nicotinamide, the metabolic form in animals, and nicotinic acid in plants is expressed in milligrams or micrograms of the chemicals (Scott et al. 1969).

FUNCTION. Niacin prevents hock disorders, poor feathering, irritability, and pellagra. Interaction of nicotinic acid and vitamin E prevents enlarged hocks in turkeys (Scott 1953). The coenzymes participate in the metabolism of carbohydrates, fats, and protein and in rhodopsin synthesis (Scott et al. 1982).

UTILIZATION. Nicotinic acid requirements of chicks and hens depend on the amount of tryptophan in the ration (Briggs et al. 1946). Pyridoxine is required to transform tryptophan to nicotinic acid (Phillis 1976; Austic and Scott 1984).

SIGNS. In niacin-deficient chicks, ducks, and turkeys, bowed legs and hock enlargement have been reported (Scott and Heuser 1954). This is similar to perosis except that the tendon seldom escapes from its condyles. Poor feathering and flightiness have been noted (Austic and Scott 1984). Retarded growth, loss of weight, and anemia can occur.

NECROPSY. Gross internal changes in niacin deficiency are not found except as related to the hock.

TOXIC MANIFESTATIONS. High levels of niacin result in vasodilation.

STABILITY. Niacin is unusually stable to heat, air, light, and alkali.

DIFFERENTIAL DIAGNOSIS. The other causes of perosis must be considered.

PREVENTION. Pigeon rations normally contain an adequate level of niacin, but Griminger (1981) has indicated that birds need 27–70 mg/kg feed.

TREATMENT. Supplemental niacin is considered unnecessary.

Folic Acid

DEFINITION. Folic acid (folacin; pteroylglutamic acid) is the name for several slightly water-soluble, yellow-orange compounds that contain a pteridine nucleus, para-amino benzoic acid, and glutamic acid (Stecher 1960). Folic acid deficiency is an uncommon clinical disease of pigeons.

SOURCE. The folic acid compounds are widely distributed in nature and occur in natural materials conjugated with two or more glutamic acid residues (Scott et al. 1969). Greens, yeast, and liver meal are good sources. Some intestinal microorganisms require only para-amino benzoic acid to form the rest of the folic acid molecule; thus where folic acid is deficient para-amino benzoic acid would be beneficial (Phillis 1976).

STANDARD. Folic acid is calculated in micrograms.

FUNCTION. Folic acid participates in the interconversion of glycine and serine, the synthesis of nucleic purines and pyrimidines, the breakdown of histidine, and the synthesis of methyl groups for choline, thiamine, and methionine (Phillis 1976). By supplying nucleic acid it also prevents macrocytic anemia (Scott et al. 1982). In chickens and turkeys, it prevents the reduced hatchability occurring after pipping, caused by an abnormal mandible and tibiotarsus in the embryo (Sunde et al. 1950a,b).

UTILIZATION. Folic acid appears in the intestinal wall conjugated with a nitrogen compound. This form is found in the blood (Scott et al. 1969). When this blood folic acid conjugate is deficient in perosis, the choline requirement is increased (Young et al. 1955).

SIGNS. Poor growth and feathering, anemia, and perosis are signs of folic acid deficiency. In colored chickens, such as Rhode Island Reds and Black Leghorns, folic acid, lysine, and iron are needed for feather color (Austic and Scott 1984).

NECROPSY. Internal evidence of folic acid deficiency is not significant. When anemia is present, the flesh and blood will be pale.

MICROPATHOLOGY. Blood smears may identify the type of anemia in folic acid deficiency.

TOXIC MANIFESTATIONS. Toxic signs of folic acid have not been reported, but like other vitamins, levels should not exceed treatment amounts.

STABILITY. Folic acid is quite stable (Scott et al. 1982).

DIFFERENTIAL DIAGNOSIS. A diagnosis of perosis must consider the other causes.

PREVENTION. Griminger (1981) has suggested a level of 0.55-1.0 mg folic acid/kg feed.

TREATMENT. Recovery of affected chicks has been demonstrated using folic acid at 500 μg/100 g feed (Austic and Scott 1984). Intramuscular injection of 50–100 μg pure folic acid improved anemia in chicks in 4 days (Robertson et al. 1947).

Choline

DEFINITION. Choline (choline chloride) is a water-soluble, white, crystalline compound. The pure vitamin is a strongly alkaline, viscid liquid that is a component of lecithin and acetylcholine. It is an indirect source of labile methyl groups for such compounds as creatine and methionine (Scott et al. 1982). A deficiency of choline is not clinically recognized in pigeons.

SOURCE. Choline chloride is the form usually added to feed. Choline is also found in plant and animal tissues. Chicks and other mammals can synthesize choline but often in insufficient amounts to meet their needs. (Scott et al. 1982). Pigeons may obtain choline from soybean meal, fish meal, liver meal, and yeast.

STANDARD. Choline is calculated in milligrams per pound of diet.

FUNCTION. Choline is essential for growth and perosis prevention (Jukes 1947). It helps prevent fatty liver, hemorrhagic degeneration of the liver, anemia, and low blood protein (Phillis 1976). Without choline, the oxidation of long-chain fatty acids is reduced and phospholipid turnover is curtailed (Scott et al. 1978). It is a source of methyl groups via betaine for such compounds as creatine and methionine and it is also required for acetylcholine (Scott et al. 1982). The latter agent, released at the end of parasympathetic nerves, slows the heart, causes contraction of the oviduct, and enables crop emptying (Sturkie 1954). Eggs also require a high level of choline in the yolk phospholipids.

UTILIZATION. Most of the body choline is found as phospholipid, and it is in this form that it aids the maturation of cartilage matrix of bone in the prevention of perosis (Scott et al. 1969). This requirement for choline can be reduced by vitamin B_{12} (Siegmund 1979) and folic acid (Scott et al. 1969).

SIGNS. Poor growth and perosis have been reported in choline-deficient birds (Scott et al. 1978). Swelling and tiny dot hemorrhages may occur at the hock. Lowered egg production may likewise be expected.

NECROPSY. Enlarged yellowish fatty livers may be observed in choline deficiency.

MICROPATHOLOGY. Fatty infiltration of liver cells occurs with a deficiency of phospholipid, which is necessary for the normal transport of fat.

TOXIC MANIFESTATIONS. Toxic signs of choline have not been reported.

STABILITY. Choline stability is not a problem.

DIFFERENTIAL DIAGNOSIS. When perosis is observed other causes must be considered.

PREVENTION. The choline level should approximate 1300–2000 mg/kg feed (Griminger 1981). The NAS-NRC (1977) suggested quantity is 1000–2000 mg/kg feed for quail.

TREATMENT. Without clinical disease, treatment for choline deficiency is not indicated.

Vitamin C. Vitamin C (ascorbic acid) is synthesized by poultry; thus it is not required in the ration (Austic and Scott 1984). It is necessary for the formation and maintenance of connective tissue, for metabolism, for the mobilization of iron, and as a hydrogen transport agent (Phillis 1976). Chaudhuri and Chatterjee (1969) reported that pigeons form ascorbic acid in the kidneys.

Inositol. The role of inositol other than as a component of phosphatides is unknown. It is not required in a poultry diet (Austic and Scott 1984).

Mineral Requirements and Imbalances

At least fifteen essential minerals or inorganic elements are required by poultry for growth, maintenance, and production. These include calcium, phosphorus, magnesium, potassium, sodium, chlorine, manganese, iron, copper, zinc, iodine, molybdenum, selenium, fluorine, and sulfur. Calcium, phosphorus, magnesium, and zinc are largely found in the bones. Most of the other mineral elements are found in body fluids, soft tissues, and muscles.

Calcium and Phosphorus

DEFINITION. Since bone formation involves the metabolism of calcium and phosphorus, the two structural minerals can be considered together. A deficiency or imbalance of either calcium or phosphorus and/or vitamin D results in rickets in young birds. In older birds deficient calcium and/or phosphorus leads to resorption, resulting in osteomalacia or bone softening. Both are nutritional problems of pigeons.

SOURCES. Pigeon grains are low in calcium; thus supplemental grit must be fed, such as medium-sized oyster shell and/or crushed limestone. Other sources of calcium and phosphate are fish meal, steamed bonemeal, steamed meat-and-bone meal, inorganic mono-, di-, and tri-basic calcium phosphate. Inorganic rock phosphates found in soil are essentially only available to pigeons after heat treatment (Gillis et al. 1954).

STANDARD. Calcium and phosphorus are each calculated as a percentage of the diet.

FUNCTION. Over 70% of the ash content of the body is calcium and phosphorus. Calcium is essential in the blood clotting mechanism. It is involved with the acid-base balance and acts with potassium and sodium in maintaining normal heart rhythm. It is required for strong eggshells and normal hatching. Phosphorus is vital for bone and shell formation, health, acid-base balance, nerve tissue phospholipids, sugar and fat metabolism, and several intermediary enzyme systems (Mineral Nutrition 1967).

UTILIZATION. A limited supply of either calcium or phosphorus reduces the nutritive value of the other. The ratio of calcium to phosphorus in a pellet ration should approximate 1.25–1.4:1, utilizing a supplemental mineral mix to provide the needed additional calcium (Smith 1985). In the complete diet for young birds the ratio should be about 1.5–2.0:1 and for breeders 2.5:1 (Scott et al. 1969).

Vitamin D_3 is required for the absorption of calcium but not phosphorus (Sallis and Holdsworth 1962; Scott et al. 1969), but less vitamin D is needed when the calcium-phosphorus ratio is within limits (Ensminger and Olentine 1978). Oxalic acid chelates calcium and forms an insoluble complex that prevents absorption. Tetracycline antibiotics form calcium salts that interfere with the utilization of either mineral in the blood. Lactose, on the other hand, improves calcium and phosphorus absorption (Scott et al. 1969). Phytin phosphorus, an organic phosphorus compound, forms perhaps two-thirds of the phosphorus in seeds and their by-products. This form is not well assimilated because of a deficiency of the enzyme phytase, which splits the compound. Excess fat in the ration may combine with calcium to some extent, forming insoluble soaps that prevent absorption of calcium. This reaction normally occurs at a pH of 8.0 or above (Mineral Nutrition 1967). The ingestion of magnesium also interferes with phosphorus availability in the intestine and large quantities of aluminum salts retard the assimilation of phosphorus (Harrow 1946).

Over 90% of body calcium is in the bones. The rest is in soft tissues and blood. On the other hand, 80% of body phosphorus is found in bones and 20% in soft tissues and blood. Blood calcium is governed by the parathyroid glands. Organic phosphates in the blood are regulated by alkaline phosphatase produced in the bone marrow or cartilage. It lays down soluble phosphorus in developing bone (Mineral Nutrition 1967).

SIGNS. A lack or an imbalance of either calcium or phosphorus may cause signs including a drop in food intake, retarded growth, reduced activity, abnormal posture, weak rib and bone structure, thin shells, an increase in cracked eggs, reduced egg production, and poor hatchability.

NECROPSY AND MICROPATHOLOGY. The evidence of calcium or phosphorus deficiency are similar to those for vitamin D deficiency.

TOXIC MANIFESTATIONS. Too much calcium in pellets interferes with palatability and the utilization of other minerals and fat (Siegmund 1979). Pullets 8–20 wk old fed 2.5% calcium for 1–2 wk developed nephrosis and gout with 10–20% mortality. This was reduced by increasing the available phosphorus to 1%. Pullets over 20 wk of age were not harmed by the high calcium (Scott et al. 1969).

STABILITY. Calcium and phosphorus are stable minerals, but vitamin D can be destroyed by various substances.

DIFFERENTIAL DIAGNOSIS. In arriving at a diagnosis, it is often difficult to ascertain which is at fault, an imbalance or a deficiency of calcium, phosphorus, or vitamin D.

PREVENTION. Diets must contain 0.6–0.7% phosphorus; 0.2% may be phytin phosphorus but 0.4% must be inorganic phosphorus (White Stevens 1956). A suggested feed level for calcium is 0.6–1.2% and for phosphorus 0.6–0.8% (Griminger 1981). The NAS-NRC (1977) levels for quail are 0.6–2.3% for calcium and 0.6–1.0% for phosphorus. Dicalcium phosphate 1.5% together with 1% ground limestone may be incorporated into pellets as provided in the Cornell ration (see Pigeon Diets later in this chapter).

TREATMENT. Treatment usually involves the temporary daily addition for 1 wk of ½ tsp USP cod liver oil containing oil-soluble vitamins A and D. The oil floats on the surface of the water for the birds to ingest. In addition

5% steamed bonemeal or dicalcium phosphate may be added to the mineral mixture to supply calcium and phosphorus. The ration must be reevaluated and the source of the trouble determined before further treatment is given.

Magnesium

DEFINITION. Magnesium is a trace mineral somewhat related to calcium in its metabolism. It is required for normal bone, muscle, and nerve health. The deficiency disease seldom occurs in pigeons, but excessive levels in mineral mixtures are possible.

SOURCE. Magnesium is considered to be sufficient in feedstuffs, but dolomitic limestone (magnesium carbonate) may be used if the ration is considered deficient.

STANDARD. Magnesium is calculated in milligrams per kilogram of ration.

FUNCTION. Erdtmann in 1927 observed that alkaline phosphatase is activated by magnesium, which is necessary in bone formation. It is present in bone and eggshells as magnesium carbonate (Austic and Scott 1984). It also acts in conjunction with adenylic acid as a coenzyme in carbohydrate metabolism (Harrow 1946).

UTILIZATION. The magnesium level is critical in the diet because excessive amounts interfere with calcium utilization (Mineral Nutrition 1967). About 70% of the total body magnesium is in the bones (Phillis 1976), but like potassium, it is found to a large extent within the soft tissues of the liver, striated muscle, kidney, and brain. The blood serum contains about 50 mg/l or 10% of other tissues (Scott et al. 1969). Magnesium is excreted by the kidneys, which also reabsorb and conserve the body reserves (Ensminger and Olentine 1978).

SIGNS. Lack of vitality and poor growth, together with muscle spasms when stimulated, characterize magnesium deficiency. Comatose birds may die following a prolonged deficiency, and chronic deficiencies result in skin lesions (Austic and Scott 1984).

NECROPSY. No gross evidence of magnesium deficiency may be noted.

MICROPATHOLOGY. In magnesium deficiency, inflammatory necrotic foci have been observed in small blood vessels (Scott et al. 1969).

TOXIC MANIFESTATIONS. Excess dolomitic limestone can be harmful, causing diarrhea, thin eggshells, and hypersensitive birds (Adler 1927).

STABILITY. Magnesium stability is not a problem.

DIFFERENTIAL DIAGNOSIS. Calcium metabolism must be considered.

PREVENTION. The chick diet should contain 0.04% magnesium during the first 4 wk (Almquist 1942). Griminger (1981) placed the level at 150–600 mg/kg diet as the level required for maximum growth. Scott et al. (1982) has suggested 500 mg/kg as the requirement.

TREATMENT. Treatment for magnesium deficiency is not indicated. Any excess must be removed from the diet.

Sodium Chloride

DEFINITION. Sodium chloride (salt) is an essential homeostatic mineral that helps to maintain the acid-base and ionic balances in the body fluids. Deficiencies are rare, but excesses in pigeons do occur.

SOURCE. Iodized salt is suggested as a source of sodium chloride.

STANDARD. Salt is calculated as a percentage of the total ration. The individual requirements for sodium and chlorine are also specified in percentages.

FUNCTION. Sodium helps to maintain the acid-base balance, along with calcium and potassium, and it is present mainly in extracellular tissue fluid. Without it, blood volume and tissue fluids decrease as osmotic pressure is lowered. The utilization of proteins and carbohydrates is also impaired by a sodium deficiency (Scott et al. 1978).

Chlorine is important in maintaining acid-base equilibria. It is also an important aspect of osmotic pressure. It serves as a constituent of hydrochloric acid of the proventriculus.

UTILIZATION. The level of sodium in the body is controlled by diet intake, absorption, and kidney excretion. Sodium ions can move in both directions across the intestinal mucosa and are actively transported out of the lumen of the small intestine and colon (Phillis 1976). Excess sodium in the diet causes excessive kidney excretion of potassium and vice versa. Chlorine, as chloride, is readily absorbed and can readily pass across cell membranes (Harrow 1946).

SIGNS. Birds crave salt and will strive to get it. Cannibalism may develop as a result. A prolonged deficiency results in dehydration, muscular shrinkage, poor growth and bone development, corneal thickening, nervous signs, and ultimate death. Decreased egg production and small testes occur in breeders.

NECROPSY. The adrenals enlarge with a deficiency of sodium (Scott et al. 1978). The skin, legs, and body tissues appear dryer and less pliable.

MICROPATHOLOGY. In salt deficiency, the adrenals show evidence of stress.

TOXIC MANIFESTATIONS. Excessive salt causes edema. Water follows salt into the tissues and places a burden on heart action. Signs include intense thirst, watery droppings, muscular weakness, convulsions, and death. Internally congestion of the viscera with tissue hemorrhages may be noted. The lethal level for chickens is 4 g/kg body weight (Scott et al. 1978).

DIFFERENTIAL DIAGNOSIS. Other causes of poisoning must be considered if toxic signs are present. If dehydration occurs, the water supply should be checked.

PREVENTION. The addition of 0.25–0.5% salt to pellets is adequate to prevent a deficiency. Griminger (1981) has placed the requirement for sodium at 0.08–0.15% of the ration. Scott et al. (1982) reports the level for chickens up to 20 wk old at 0.15% for both sodium and chlorine; for birds over 20 wk the levels are 0.12 and 0.1% respectively.

TREATMENT. When excesses occur, the level should be reduced and easy access to fresh water provided. To overcome a deficiency, 0.5% salt is added to the diet.

Potassium

DEFINITION. Potassium is an essential homeostatic mineral found in body fluids and soft tissues. It is concerned with the acid-base balance and ionic balances in body fluid. Pigeons may develop a clinical deficiency.

SOURCE. Potassium is found in both plant and animal feedstuffs. Supplemental potassium may be provided as potassium chloride, but it is seldom needed.

STANDARD. Potassium is calculated as a percentage of the diet.

FUNCTION. Potassium is necessary for membrane permeability, reducing contractility of the cardiac muscle, helping to provide an ionic balance in tissue fluids, and in acid-base balance. It depresses muscle irritability and acts in opposition to calcium ions.

UTILIZATION. Potassium is found chiefly within body cells.

SIGNS. Muscle weakness, kidney impairment, and heart damage occur with a potassium deficiency. Respiratory muscles become weak and eventually fail. Blue comb–like signs may occur, including dehydration, shriveled legs, and a sour, watery crop. Leach (1974) also noted a decrease in egg production.

NECROPSY. In potassium deficiency, lack of intestinal muscle tone with distention and blanching of the intestine may be observed.

MICROPATHOLOGY. Cellular changes in potassium deficiency primarily involve the kidneys, intestine, spleen, muscles, and adrenals.

TOXIC MANIFESTATIONS. An excess of potassium appears less harmful than an excess of sodium (Scott et al. 1969).

STABILITY. Potassium chloride is stable.

DIFFERENTIAL DIAGNOSIS. Other causes of muscle weakness and dehydration must be considered.

PREVENTION. A suggested level for potassium in the feed is 0.2–0.4% (Griminger 1981). Scott et al. (1982) indicates the level to be 0.4%.

TREATMENT. This area has not been studied adequately but certainly the potassium dietary requirement must be met.

Copper

DEFINITION. Copper is a dietary trace mineral essential for the production of hemoglobin. A copper deficiency is highly unlikely in pigeons.

SOURCE. Copper sulfate is a source of copper, but generally rations do not require supplementation.

STANDARD. The copper requirement is calculated in parts per million or in milligrams per kilogram of ration.

FUNCTION. Copper is required as a part of enzyme systems with oxidase functions. These include ascorbic acid oxidase, tyrosinase, and amine oxidase (Scott et al. 1969). It enables synthesis of hemoglobin and stimulates red blood cell formation (Phillis 1976). The enzyme polyphenoloxidase, which contains copper, can initiate the formation of melanin from L-tyrosine, and this appears to affect the melanin color of feathers (Scott et al. 1982). Further studies indicate that dissecting aneurysms of the aorta occur in chicks deficient in copper and deficient breeders produce chicks with no amine oxidase in the aorta or liver. Kim and Hill (1966) showed that amine oxidase used lysine in forming elastins in the aorta.

UTILIZATION. Copper is transported in blood in the protein-bound form ceruloplasmin (Phillis 1976). It is stored in the liver, red blood cells, and spleen and is excreted in the bile (Ensminger and Olentine 1978). When molybdenum is too high, copper utilization is decreased (Siegmund 1979), and when copper is too high, molybdenum deficiency can develop if it is already marginal (NAS-NRC 1977).

SIGNS. Signs of copper deficiency are related to the degree of the deficiency. Anemia with pale lemon-yellow skin color, weakness, and shortness of breath may develop if copper is deficient. Also poor growth, feather depigmentation in New Hampshire chickens, and bone weakness have been noted. Chicks may be unable to walk and become paralyzed (Scott et al. 1969).

NECROPSY. In copper deficiency, the flesh may be pale and the blood watery. Gelatinous pink fluid accumulations may appear about the heart, in the abdominal cavity, and under the skin. The bone marrow may be found to be quite active, red, and spongy.

MICROPATHOLOGY. In copper deficiency, blood changes may be found in bone marrow and heart blood smears.

TOXIC MANIFESTATIONS. High dietary levels of copper cause notable dehydration and liver and kidney changes.

STABILITY. Copper is quite stable, but it is antagonized by silver (Scott et al. 1982).

DIFFERENTIAL DIAGNOSIS. The cause of anemia, if present, must be determined. Iron may be deficient or a toxic factor may be involved.

PREVENTION. The NAS-NRC (1977) set the copper requirement at 2–3 ppm in the ration to prevent anemia. Scott et al. (1982) set the total suggested level in the ration at 10 ppm or 5 mg/kg. Griminger (1981) set the requirement at 4–6 mg/kg feed.

See *Trichomonas* (Chap. 9) for further information concerning copper.

TREATMENT. Without a clinical disease treatment for copper deficiency is not indicated.

Iron

DEFINITION. Iron is a trace mineral essential in the production of hemoglobin. The deficiency is uncommon in pigeons.

SOURCE. Iron (ferric) oxide or iron (ferrous) sulfate serve as a source. Alfalfa, fish meal, meat meal, and soybean meal are high in iron.

STANDARD. Iron is calculated in milligrams per kilogram of ration.

FUNCTION. About 57% of the total body iron is included in the hemoglobin of red blood cells, and 7% is present in myoglobin of muscle cells. Porphyrins bind over 90% of the iron. Some of these compounds are a part of heme, the nonprotein pigment portion of hemoglobin (Scott et al. 1969). Iron is a part of several oxidizing enzymes, including cytochrome, catalase, peroxidase, succinic dehydrogenase, and xanthine oxidase (Phillis 1976).

UTILIZATION. Because there is little iron lost from the body iron, absorption by the mucosal cells of the intestine is regulated to avoid harmful accumulations (Scott et al. 1969). High levels of phosphate, phytate, and dietary fat also reduce iron absorption (Phillis 1976) and ferrous iron is more readily absorbed than ferric iron (Sodeman 1950). Transferrin in the blood plasma receives absorbed ferritin iron from the intestine and iron from the death of red blood cells and the breakdown of hemoglobin and transports it wherever it is needed (Scott et al. 1969).

SIGNS. Hypochromic microcytic anemia develops primarily from a lack of iron hemoglobin. This means that pale, small red blood cells are formed. A reduction in the number of red blood cells also occurs. In New Hampshire chickens iron deficiency resulted in faulty feather pigmentation in red and black feathers (Scott et al. 1978).

NECROPSY. In iron deficiency, characteristic anemia signs as given for copper deficiency may occur.

MICROPATHOLOGY. The appearance of appreciable amounts of hemosiderin within cells of the reticuloendothelial system or in epithelial cells of liver or kidney is an indication of blood destruction. Anemic red blood cells are more fragile and are easily destroyed.

TOXIC MANIFESTATIONS. Excess iron may cause liver damage. In addition, phosphorus may be tied up, resulting in rickets. The iron phosphate can also interfere with the absorption of vitamins. With an excess of iron, the red blood cell may carry more pigment and hemosiderin may accumulate in the blood (Scott et al. 1969).

STABILITY. Iron salts are quite stable.

DIFFERENTIAL DIAGNOSIS. The cause of anemia must be determined. A deficiency of copper or some other factor may be involved.

PREVENTION. Iron is required at the rate of 60–80 mg/kg feed (Griminger 1981). Scott et al. (1982) set the level at 45–80 mg/kg feed, with younger birds requiring the higher level.

TREATMENT. Iron oxide (red hematite) is chiefly used in feed. For pigeons it can be included in the mineral mix at the rate of 10 g/100 lb until the total iron content of the ration is recalculated. The preventative level should be established.

Manganese

DEFINITION. Manganese is an essential trace mineral required for growth, reproduction, and the prevention of perosis or slipped tendon disease, which is characterized by an anatomical bowing of leg bones with the lateral tendon displaced at the joint. The deficiency may occur in pigeons.

SOURCE. Manganese sulfate is usually incorporated into feed, but manganous chloride, dioxide, or carbonate may be used.

STANDARD. Manganese is calculated in milligrams per pound of feed or in parts per million.

FUNCTION. Manganese is chiefly found in bone. It is essential to prevent perosis and to provide for normal bone matrix formation (Wilgus et al. 1937). Lyons and Insko (1937) found that a deficiency lowered hatchability. Manganese also is necessary for the activation of several enzymes and in amino acid complexes (Scott et al. 1969).

UTILIZATION. Manganese is poorly absorbed, and its absorption is apparently dependent on bile salts (Scott et al. 1969). It is removed from the body almost exclusively in bile in the feces (Harrow 1946). The mineral is interrelated with biotin, choline, and folic acid (Mineral Nutrition 1967).

SIGNS. In manganese deficiency, there is a distorted enlargement of the tibiometatarsal joint involving the lower end of the tibia and the upper end of the metatarsus. The gastrocnemius tendon slips laterally outward away from the joint condyles. There is inability to walk, and spasms may be stimulated by excitement.

NECROPSY. In manganese deficiency, embryos are small and have very short thickened legs, short wings, and parrot beaks. The abdomen protrudes with marked edema (Scott et al. 1978).

MICROPATHOLOGY. In manganese-deficient chicks, the intracellular cartilage matrix is unorganized or lacking at the epiphyseal plates (Scott et al. 1969). It is the cartilage matrix that becomes impregnated with calcium salts in the process of bone formation.

STABILITY. Manganese is very stable.

DIFFERENTIAL DIAGNOSIS. Hock disease must involve a consideration of other causes, such as a deficiency of zinc, biotin, choline, folic acid, niacin, or pyridoxine.

PREVENTION. A manganese deficiency may be offset by the feeding of 25–40 ppm manganese (Mineral Nutrition 1967). Griminger (1981) suggested 40–90 mg/kg feed. Scott et al. (1982) set the chick requirement at 50 mg and the older bird level at 33 mg/kg feed.

TREATMENT. Manganese sulfate 70% at the rate of 6–8 oz/t feed or 15–25 mg pure manganese/lb feed is recommended when a problem develops.

Zinc

DEFINITION. Zinc is a required trace mineral essential for growth and performance and is a vital part of a number of enzymes and insulin. Zinc deficiency is rare in pigeons.

SOURCE. Zinc oxide or carbonate may be used in feed. Zinc is also released from galvanized feed and water containers if reactive chemicals are given in the feed or water. Meat meal and fish meal are natural sources.

STANDARD. Zinc is calculated in milligrams per kilogram of ration.

FUNCTION. Zinc is necessary for the breakdown of carbonic acid to carbon dioxide and water. It is a component of insulin and of several enzymes involved in intermediary metabolism (Phillis 1976). The mineral is associated with the follicle-stimulating hormone and the luteinizing hormone and is a part of the enzyme that stimulates hydrochloric acid in the proventriculus, according to Ensminger and Olentine (1978). In chicks it prevents skin scaling, hock and leg deformities, corneal changes, and abnormal feathering (Scott et al. 1969).

UTILIZATION. Less than 10% of dietary intake of zinc is absorbed, but metallic zinc, zinc sulfate, carbonate, and oxide are absorbed equally. The mineral is excreted in feces in the bile and by the pancreatic duct (Ensminger and Olentine 1978). Zinc accumulates in the choroid of the eye, skin, muscle, heart, kidney, liver, and pancreas. It is absorbed and bound to protein in plasma for transport (Phillis 1976).

SIGNS. Skin dermatitis (particularly involving the feet), frayed feathers, and retarded feather development are reported (Mineral Nutrition 1967). Other signs include retarded growth, shortened and thickened leg bones, enlarged hocks, poor appetite, and lowered egg production (Scott et al. 1978). Chicks hatched from deficient breeders may be weak and unable to stand (Scott et al. 1969).

NECROPSY. In zinc-deficient embryos there is an absence of wings, body wall, legs, and spinal column. Only a head and complete viscera develop (Kienholz et al. 1961).

MICROPATHOLOGY. In zinc deficiency, changes may be observed in the skin, joints, and bones.

TOXIC MANIFESTATIONS. Zinc is relatively nontoxic for pigeons, but toxic levels in galvanized drinking water containers can interfere with water consumption. When this occurs, the birds should be provided fresh water.

STABILITY. Zinc is quite stable.

DIFFERENTIAL DIAGNOSIS. Other causes of enlarged hocks must be considered.

PREVENTION. A total suggested dietary level is 25–75 mg/kg (Griminger 1981). Scott et al. (1982) set the value at 30–40 mg/kg for young birds and 40–60 mg for older birds.

TREATMENT. Treatment is not indicated without clinical disease.

Iodine

DEFINITION. Iodine is a trace mineral essential for proper thyroid function. A deficiency very seldom occurs in pigeons.

SOURCE. Sodium iodide may be included in the salt mix.

STANDARD. Iodine is calculated in milligrams per kilogram of feed.

FUNCTION. Iodine is utilized by the thyroid in the production of thyroxin, which regulates heat production and basal metabolism. It influences lipid, and in particular, cholesterol metabolism. Thyroxin also affects neuromuscular function and reproductive activity (Phillis 1976). Potassium iodide specifically prevents the development of goiter and weak squabs (Hollander and Riddle 1946).

UTILIZATION. Absorption of iodide is rapid and almost complete. Excretion is by the kidney as iodide (Phillis 1976). Most iodine is bound as thyroglobulin in the thyroid and two molecules of 3,5-diiodotyrosine combine to form thyroxine (Harrow 1946).

SIGNS. Whenever iodine is absent in the diet goiter (an enlargement of the thyroid glands) may be noted as a bulge under the crop and from within the chest cavity. Depletion of body reserves of hens with a dietary deficiency of iodine lasting a year or more produces only slight follicular enlargement (Scott et al. 1969).

NECROPSY AND MICROPATHOLOGY. In iodine deficiency, only the thyroids are enlarged. The follicles or acini of the thyroid increase in size and number when iodine is depleted.

TOXIC MANIFESTATIONS. Excessive iodine intake results in poor hatchability.

PREVENTION AND TREATMENT. In addition to natural iodine in feedstuffs, 0.25% added iodized salt will prevent the problem (Scott et al. 1978). Morrison (1951) suggests ½ oz finely ground potassium iodide/300 lb salt. The NAS-NRC (1977) recommends 0.30 mg iodine/kg feed for quail. Ethylenediamine dihydroiodide at 0.156% in salt may also be fed (Maas et al. 1984).

Molybdenum

DEFINITION. Molybdenum is an essential trace mineral, but it is required only in very, very small quantities. Thus a deficiency is unlikely.

SOURCE. Molybdenum is present in most grains grown in the United States.

STANDARD. Molybdenum is calculated in milligrams per kilogram of feed.

FUNCTION. Molybdenum is involved as a cofactor for the enzyme xanthine oxidase, which converts purines to uric acid for excretion by the kidneys (Siegmund 1979). This enzyme also acts on other substrates (Scott et al. 1969).

UTILIZATION. Molybdenum is quickly absorbed as water-soluble molybdate and is also excreted as such. This action is altered by a normal level of sulfate in the diet, which reduces absorption (Phillis 1976) and increases the rate of excretion (Siegmund 1979). In addition, excessive quantities may decrease the utilization of copper (Siegmund 1979). Molybdenum increases fluorine absorption and retention, and when sodium tungstate is present in the diet, it is competitively excluded from its role in the formation of xanthine dehydrogenase (Scott et al. 1982).

SIGNS. Poor growth may result in birds on molybdenum-deficient diets.

TOXIC MANIFESTATIONS. Molybdenum is toxic in excess of 20 ppm (Ensminger and Olentine 1978).

PREVENTION AND TREATMENT. Molybdenum is needed in the diet at 2 ppm (Ensminger and Olentine 1978). Scott et al. (1982) set the level at 0.2 mg/kg ration (2 ppm). The mineral must come from the natural quantities in grains. It is not approved for addition to feeds.

Selenium

DEFINITION. Selenium is an essential trace mineral involved as an antioxidant and in the formation of antioxidants. The requirement is spared by high levels of vitamin E in the diet. Deficiencies and excesses can occur in pigeon diets.

SOURCE. Selenium may be supplied as sodium selenite. Fish meals and dried brewer's yeast are good feed sources. The level of selenium in plants is largely a reflection of soil content. In general plants are poor sources, but when selenium is present, it is highly available and is associated with plant proteins (Ensminger and Olentine 1978).

STANDARD. Selenium is calculated in parts per million.

FUNCTION. One-tenth ppm selenium will protect chicks from exudative diathesis and heart and gizzard muscle degeneration by enhancing the metabolic activity of vitamin E (Scott et al. 1978). It acts as an antioxidant on the lipid layer of lysosomes (the intracellular proteolytic enzymatic laminar bodies). It is involved in the formation of lipid antioxidants (Scott et al. 1969). It is a component of glutathione peroxidase, an enzyme that acts on peroxides in tissues, and it is involved with lipid absorption (Ensminger and Olentine 1978). Deficient chicks have altered growth and feathering and faulty fat utilization (Thompson and Scott 1969).

UTILIZATION. Selenium is absorbed in the duodenum and excreted by the kidneys and feces (Ensminger and Olentine 1978), but mechanisms are present to conserve selenium when the dietary level is low (Scott et al. 1969). Selenomethionine is transported across the intestinal mucosa, and selenium becomes associated with serum albumins and later with globulins. It is also incorporated into the proteins of red blood cells, liver cells, and other tissues. It combines with sulfur-bearing amino acids: cysteine, cystine, and methionine (Scott et al. 1969).

SIGNS. Signs of selenium deficiency may include those for vitamin E deficiency. Other signs in seriously deficient chicks are poor growth, rough feathering, faulty lipid utilization, and pancreatic degeneration (Thompson and Scott 1969).

NECROPSY. Pancreatic degeneration is reported in selenium deficiency by Gries and Scott (1972a).

MICROPATHOLOGY. In selenium deficiency, the cells of pancreatic acini degenerate and interstitial fibrosis occurs.

TOXIC MANIFESTATIONS. Soils containing selenium over 0.5 ppm are potentially dangerous because of the grains grown on the soil (Ensminger and Olentine 1978). In large animals toxic levels produce hemorrhagic degeneration, cirrhosis, and anemia (Phillis 1976). Selenium toxicity appears to be related to its affinity for sulfur in the formation of sulfur selenium complexes and in its ability to inhibit or inactivate enzymes. Excess levels lower growth rates, egg production, and hatchability in chickens (Scott et al. 1969). As little as 8–10 mg/kg feed is toxic for poultry according to Siegmund (1979). Earlier, Poley et al. (1937) reported 5 ppm toxic in chicken feed.

STABILITY. Selenites are quite stable under normal conditions, but they can be reduced to selenium by reducing agents such a vitamin C (Scott et al. 1969).

PREVENTION. In 1957 it was shown that 0.05–0.2 ppm of selenium was essential in the diet, but 0.15–0.2 ppm is now considered the practical level (Scott et al. 1982).

TREATMENT. If 0.1 ppm selenium is available in the absence of vitamin E or if 100 IU vitamin E/kg diet is present when selenium is deficient, exudative diathesis is prevented (Thompson and Scott 1969). The preventative level may be used for treatment, but Scott et al. (1978) reported that the addition of 0.1 ppm sodium selenite to a chick diet for 2 wk caused complete recovery from pancreatic degeneration.

Fluorine

DEFINITION. Fluorine is a trace mineral present in small quantities in bone and most tissues. A deficiency is unlikely; an excess is more often the problem.

SOURCE. Fluorides may be found in phosphates, phosphatic limestone, industrial pollution, brick manufacture using fluorine-containing clay, and underground water supplies. Most grains contain an adequate supply.

STANDARD. Fluoride is calculated in parts per million.

FUNCTION. In animals, fluoride is absorbed in small quantities by hydroxyapatite of tooth enamel; thus it aids in cavity prevention. In pigeons its function is not clearly defined.

UTILIZATION. Most fluoride is deposited in the bones. The kidneys readily excrete fluorine (Phillis 1976).

SIGNS. No fluoride-deficiency signs have been reported for birds.

TOXIC MANIFESTATIONS. Toxic signs of fluoride include lameness, dehydration, loss of appetite, diarrhea, and soft thickened bones that develop bony enlargements (Siegmund 1979). Only 5–10 ppm acts as an enzyme inhibitor by substituting for manganese in oxidative enzymes (Mineral Nutrition 1967). According to Siegmund (1979), 300–400 ppm are toxic for chickens.

PREVENTION AND TREATMENT. Fluoride supplementation is not indicated.

Sulfur

DEFINITION. Sulfur is an essential trace mineral that is naturally incorporated into organic molecules of three amino acids. Pigeons are unlikely to show evidence of such a deficiency, and they cannot use elemental sulfur if added to a ration.

SOURCE. Methionine, cystine, and cysteine are sulfur-bearing amino acids and are natural ingredients of protein feedstuffs. These amino acids are adequate in feed grains.

FUNCTION. Small quantities occur in blood and tissues. Cystine is a principal constituent of skin keratin and feathers. Aside from this, sulfur is also found in the body in coenzyme A, glutathione, insulin, thiamine, ergothionine, taurocholic acid, sulfocyanide, chondroitin, and melanins (Harrow 1946). Sulfur is essential in several hormones and enzymes and in bile salt molecules (Mineral Nutrition 1967).

UTILIZATION. The metabolism of sulfur is associated with the metabolism of proteins and inorganic sulfur is largely oxidized to inorganic sulfate in the body. Neutral and ethereal sulfate compounds are formed and excreted by the kidneys (Phillis 1976).

TOXIC MANIFESTATIONS. Continued ingestion of elemental sulfur results in scaly skin.

PREVENTION AND TREATMENT. A sulfur deficiency is very unlikely.

Feeding Factors

Mineral Mixtures. Since the grains fed to pigeons are low in mineral content, a mineral mixture should be provided free choice in a separate hopper. The U.S. Department of Agriculture (1960) suggests as a mineral mixture 45 parts medium-sized crushed oyster shell, 40 parts limestone or granite grit, 5 parts steamed ground bonemeal, 5 parts ground limestone, 4 parts salt, and 1 part venetian red.

Medium- or chick-sized limestone grit or granite grit helps to grind the grains in the gizzard. Limestone grit serves to provide calcium in addition to its grinding function. Medium-sized crushed oyster shell is a good source of calcium. It contains about as much calcium as high-calcium limestone. Ground limestone is mostly calcium carbonate. High-calcium limestone or calcitic limestone is usually about 38.5% calcium. Dolomitic limestone contains magnesium carbonate, which is not recommended. Ground bone must be steamed; it contains about 31.74% calcium, 15.0% phosphorus, 7.1% protein, and 3.3% fat (Morrison 1951). Salt should contain sodium iodide. Excess salt will result in increased water consumption and diarrhea. Venetian red is supplied as an iron-containing mineral.

ANTIOXIDANTS. Antioxidants are essential in a pellet ration to prevent the oxidative breakdown of ingredients. Several are in use and include diphenyl-*p*-phenylenediamine, santoquin, and ethoxyquin. Ethoxyquin can be used at the rate of 113.5 g/t feed (Smith 1986).

GRAINS

Types of grains. Since pigeons will not eat mash, the rations must be composed of all grain, grain and pellets, or all pellets. Grains include barley, buckwheat, corn, hempseed, millet, peanuts, peas, rapeseed, rice, rye, sesame, sunflower, sorghum, soybeans, vetch, and wheat. A brief description of each grain follows. For further information Levi (1941) has given an excellent evaluation of pigeon grains.

Barley (*Hordeum vulgare*)–Winter and spring barley is widely grown; these varieties may be beardless or bearded varieties with awns. Both types are avoided by pigeons if they can get grain without hulls. Hull-less or bald barley is a good pigeon grain when available, but little is grown. Barley is generally too expensive because it is largely used in malt production. Pearled barley, the rounded kernel free of hull and bran, is excellent but it too is costly.

Buckwheat (*Fagopyrum esculentum*)–Buckwheat is not a cereal grain; it belongs to a different group of plants. Not more than 5% of the whole grain should be used since the hull has a fiber content of about 10%. Pigeons like the whole grain, and it is a good winter supplement if it becomes available at reasonable cost. In addition, buckwheat groats are becoming more plentiful and will in time be used more widely.

Dent corn (*Zea mays indentata*)–Dent corn, the most commonly grown corn in the United States, generally constitutes 50% or more of a ration. It contains both hard and soft starch. The central soft starchy endosperm extends to the crown of the kernel and becomes indented as it dries. The sides of the kernel are hard. The color of the starchy endosperm is independent of the seed coat and varies from off-yellow to white. The outer aleurone layer may be colorless, purple, or bluish black. In red corn the color appears in the seed coat. Only the yellow starch contains vitamin A; thus white corn in a ration must be supplemented with vitamin A. A mutant gene in hybrid strains increases the lysine level. Whole corn is better than cracked unless it is freshly cracked. Molds easily penetrate the tip of whole field shelled corn that is not dried to under 13% moisture within 24 hr of harvesting. Damaged or cracked corn likewise is easily penetrated by mold spores and is likely to develop mycotoxins.

Flint corn (*Zea mays indurata*)–Flint corn is suited for growing in cool climates at high altitudes. Very little is grown in the United States. The rounded kernels are of various colors. The central endosperm is soft and starchy, but the outer layer is very hard.

Popcorn (*Zea mays everta*)–Popcorn is higher in protein and fat than flint or dent corn. The kernel is smaller in size and harder and it lacks vitamin A. It is generally not used in pigeon feed because of its greater cost.

Hemp seed (*Cannabis sativa*)–Because hemp is marijuana, only sterilized seed may be legally sold in the United States. The seed is high in protein (31%) but low in fat. Pigeons like the seed, and if given free choice, they would eat it to the exclusion of other grains. If fed it should be given separately and as a delicacy. Not more than 5% should be used, and its feeding value should be calculated as a part of the ration.

Millet proso (*Panicum miliaceum*)–Hog millet or broom corn millet is distinct from the sorghums, but the seed is somewhat similar to that of sorghum broom corn. The seed is dark in color, oblong, and has a short awn.

Hulled oats (*Avena sativa*)–Oat groats or kernels contain about 16% protein, 6% fat and are low in fiber. This is an excellent feedstuff and highly desirable, but the cost may be high. The threshed whole oat grain has a long, thick, fibrous hull with pointed ends. Pigeons will discard this whole grain unless forced to eat it. In some areas naked oats are grown, but because of poor yields, very little is available (Justin 1985). It is, however, a suitable replacement for corn and soybean meal.

Peanuts (*Arachis hypogaea*)–Peanut kernels sometimes replace a percentage of peas in the ration. Even though they are called earth nuts, they are in fact related to peas and beans. Shelled peanuts contain 13% fat, which is desirable for squabs but less so for breeders. Peanuts pose a problem because the fat becomes rancid in shelled peanuts that are stored for a length of time and mold mycotoxins often form.

Southern field peas, or cowpeas (*Vigna sinensis*)– Cowpeas are usually cheaper than Canada peas and are a different legume than actual field peas. They are grown chiefly in southern and central states. Cooked cowpeas give better results than raw cowpeas (Levi 1941). Ewing (1963) reported slight or no improvement in feeding value when cooked at 15 lb pressure for 20 min.

Maple peas or New Zealand peas (*Pisum sativum arvense*)–These brown peas are similar to cowpeas in size. They are not grown in the United States.

Canada peas, or field peas (*Pisum arvense*)–Field peas are relished by pigeons and are high in digestible nutrients. They are grown in colder areas of the United States and Canada. They have about 22% protein and can be used to supply half the animal protein supplement. Because of cost, grain rations seldom include more than 20%.

Garden peas (*Pisum sativum*)–Garden peas are distinguished by a wrinkled surface. They are higher in vitamin A than field peas or cowpeas and are generally more costly. Pigeons adjust to the larger size and eat them quite well.

Rape seed (*Brassica napus*)–Rape seed oil meal contains about 35% protein, 7% fat, and 12% fiber. Tower, grown in Canada, is the best variety because it is low in both erucic acid and glucosinolate, which are toxic for birds. Compared with soybean meal, Tower meal has more calcium, phosphorus, niacin, choline, methionine, and crude fiber and less protein, lysine, and energy (Canadian Tower Rapeseed Meal 1977). Eggs from birds fed rape seed may have a fishy flavor (Scott et al. 1969).

Rice (*Orysa sativa*)–The feeding value of each of the rice varieties is similar. Rice is fed as polished rice with the hull removed and is usually only included in pigeon grains when it is off-grade. Pigeons find the whole grain with the hull quite unpalatable. Not more than 5% is included in a ration.

Rye (*Secale cereale*)–Rye is largely grown for human consumption. It can be substituted for wheat in a pigeon ration, but like barley, it contains betagluconase, which limits its usefulness. Much beyond 5% in a ration is constipating.

Sesame seed (*Sesamum indicum*)–Sesame seed, grown in India, averages 42.8% good-quality protein. It lacks lysine, but when blended with soybean meal in pellets, the deficiency is easily overcome (Morrison 1951).

Sorghums (*Andropogon sorghum*)–Sorghums were introduced into the United States from Japan in 1854 (Hutcheson et al. 1936). For years several distinct groups of grain sorghums were grown, but now only hybrid grain sorghums are available. The grain is round and smaller in size than most grains. Strains have been developed with high tannin levels to reduce field pillage by wild birds. Most of this tannin is dissipated in dried mature grain, but pigeons may discard dark brown seeds containing higher tannin levels. Because the seeds are hard-coated, the grains may contain more than 13% moisture when apparently dry; thus they may heat in storage and decrease in quality. The grain is low in vitamin A, but it has feeding value almost equal to corn.

Sunflower seeds (*Helianthus annus*)–The sunflower groats (kernels without the hulls) have increased in popularity. They are an excellent feedstuff containing about 37% protein and 5% fat. The protein is quite adequate in amino acid composition (Schaible 1970). Sunflower meal with lysine and methionine can completely replace soybean meal (Michel and Sunde 1985). The whole seed, however, has too much fiber, and pigeons will discard it.

Soybeans (*Soja max*)–Raw soybeans contain up to 50% protein but are not desirable for pigeons because thorough cooking is required to destroy trypsin inhibitor and to make cystine and methionine fully available. It has also been observed that the proteolytic inhibitor not only affects the speed of digestion of soy protein but also affects other protein in the same diet (Almquist 1957). Too high cooking of the meal (above 180°F) or too prolonged cooking causes permanent nutritional destruction of lysine, arginine, and tryptophan by apparent reaction with carbohydrates (Riesen et al. 1947). Properly cooked soybeans or soybean oil meal is equal in value to the protein in milk or fish (Morrison 1951).

Vetch seeds (*Vicia astiva* and *V. hirsuta*)–Vetch seed is similar to peas in composition and contains about 29% protein. The seed is small, round, and grayish black. It is not widely grown; thus it is generally more expensive.

Wheat (*Triticum sativum vulgare*)–Wheat is an excellent grain and generally makes up 15–25% of the diet. Hard red spring and winter wheats and hard spring Durum wheat are the predominant types grown in the United States. Red wheat is more commonly used than soft white varieties. New wheat, which contains a higher moisture content, may cause digestive problems, but wholesome, well-cured grain can be fed with safety and confidence.

Grain quality. Grain should be clean, wholesome, and free from broken kernels. High-moisture grain (over 13%) and broken or insect-damaged kernels may have a higher level of mold toxins. Only 1–2 ppm of mycotoxin reduces the bird's ability to utilize vitamins and minerals (Brown 1983). Cracked grains, unless dried and properly stored, can have higher levels of mycotoxins. Pigeons will often discard unwholesome seeds, and they also show a dislike for bitter-tasting grains, such as brown sorghum grains containing tannin (Yacowitz 1976).

Carr and James (1931) quoted the following table from a graduate thesis from the University of Washington. It presents some interesting data on the weekly gains in squab weights when the squabs were fed different grains in addition to crop milk, which was provided for 1/2–2 mo.

	Weight in Grams by Week				
Grain	1	2	3	4	5
Hemp	80	194	210	240	350
Kafir	78	154	178	220	280
Corn	48	96	140	166	180
Wheat	74	186	220	280	310
Soybean	80	164	210	225	290
Oats	70	168	320	355	380

From the standpoint of grain quality, oats, wheat, and hemp seed gave the greatest weight gains.

Pigeon diets. Feral pigeons are adaptable and will consume what is seasonably available. It is obvious that pigeons meet their requirements for survival by foraging for a variety of foods. The crop contents of scavenger pigeons have been examined by several investigators, who found corn, barley, oats, clover, grass and weed seeds, vetch, peas, broad beans, rye, black nightshade berries, earthworms, slugs, snails, and other dry matter (Griminger 1983). Given free choice, pigeons are independently selective and will exhibit individual preferences for different plant foods. Moon and Zeigler (1979) provided only kafir, maple peas, and hemp seed in a controlled test of 19 loft pigeons. They observed that all 19 ate kafir, 7 ate peas, 10+ ate hemp seed, and 4 ate all three grains.

Aside from these findings, it is well established that balanced formulated grain or pellet rations are consistently superior for growth and squab production. Platt (1951) raised White King pairs for 18 mo on a diet of 35% yellow corn, 25% milo, 20% wheat, 20% green peas, and a mineral mixture composed of 85% oyster shell, 10% charcoal, and 5% sodium chloride. This was adequate for squab production.

The all-grain ration suggested by the U.S. Department of Agriculture (1960) in Bulletin 684 includes 35% whole yellow corn, 20% kafir or milo, 20% cowpeas or field peas, 15% hard red wheat, 5% oat groats, and 5% hemp seed. It is suggested that the corn be reduced to 25% during the summer months. The ration provides approximately 14.2% crude protein, 66.9% nitrogen free-extract, 2.8% fat, and 2.6% crude fiber. The USDA bulletin further places squab requirements at 13.5–15% crude protein, 60–70% carbohydrate, 2–5% fat, and under 5% fiber.

Pellets have been formulated to ensure that pigeons receive all the ingredients, including minerals and vitamins, that are essential at different ages and under changing conditions. Pellets can be concocted to be fed with grain or as an all-pellet ration.

The Cornell University pigeon ration formulated by Dr. M. L. Scott provides a pellet with 22.7% protein and 4.7% added fat to be fed flying birds 1:3 with scratch grains (1:1 cracked corn and soft white wheat). This combined ration provides 13% crude protein, 3.4% crude fat, and 2.7% crude fiber. Brown (1983) reported good success with this ration on flying birds. Their breeders were fed ad libitum 2 parts pellets, 1 part whole corn, and 1 part scratch (50% cracked corn, 50% soft white wheat).

Pellet Ingredients

	lb/t	%
Corn meal	1100	55.00
Soybean meal, 50% protein, heat-processed, dehulled	700	35.00
Alfalfa meal, 17% protein	100	5.00
Limetone, ground	20	1.00
Dicalcium phosphate	30	1.50
Iodized salt	5	0.25
Tallow, fat	40	2.00
Vitamin-mineral premix	5	0.25

Vitamin-Mineral Premix

	g/t	units/t
Stabilized vitamin A	30.89	10,000,000
Vitamin D_3	667.00	2,000,000
Vitamin B_{12}	0.02	
Stabilized vitamin E	90.50	20,000
Riboflavin	8.00	
Manganese sulfate	200.00	
Zinc carbonate	100.00	
Niacin	60.00	
Calcium pantothenate	20.00	
DL-methionine	500.00	
Sodium selenite	0.20	
Santoquin antioxidant	125.00	

Once birds become adjusted to eating pellets, they can be fed with confidence. Special machines compact the formulated ground concentrate mixtures into small cylindrical bits 3/16 in. in diameter and 1/4 in. long. The moisture content is reduced from 17% to 11%. Heat of 160–180°F and pressure of 6000–8000 psi are applied to form the pellet. The heat reduces the vitamin potency of the pellet, but this loss can be compensated prior to pelleting. The heat also lowers the bacterial count of the feed, which is desirable.

Formulating grain mixtures. To increase the protein of the USDA grain formulae above the present 14.2%, for example, an easy way would be to add high-protein field peas. Using the "square technique" of Yacowitz (1976), place the protein percentage of the USDA grain in the upper left corner of a square. Place the protein content of Canada peas in the lower left. In the center of the square place the desired protein level. Subtract diagonally and place 4 in the upper right corner. This indicates that 4 lb of the present USDA mixture blended with 3.8 parts of additional field peas will give a mixture containing 18% protein (see an example of this calculation below).

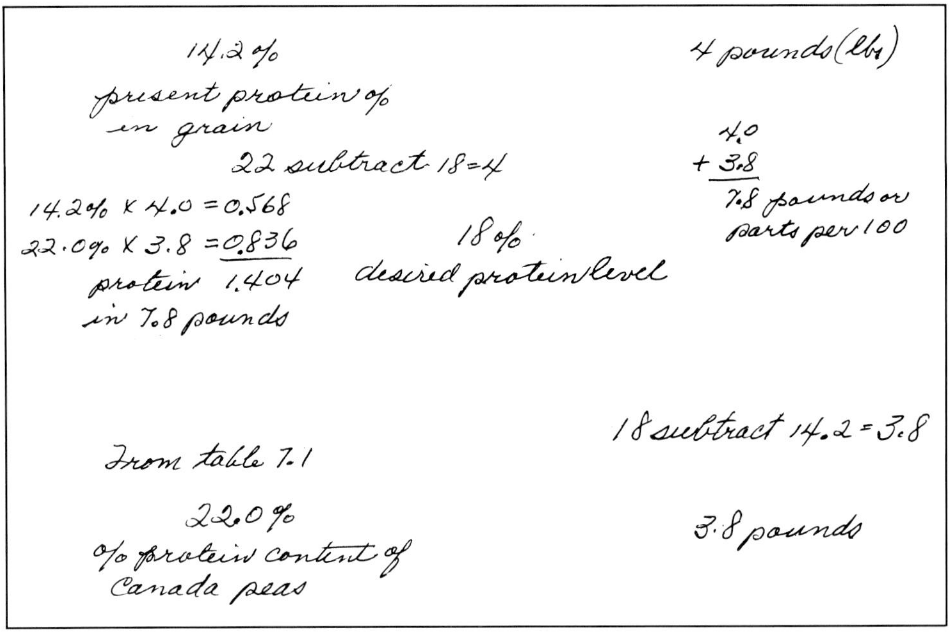

This means that 4 lb of the present grain yields 0.568 lb of protein when 3.8 lb of peas are added. The combination gives a total of 1.404 lb of protein in 7.8 lb of grain. This 1.4 lb is to 7.8 lb as 18% is to 100%. Dividing 100 by 7.8 and multiplying by 1.4 equals 17.94%, or 18%. The same method may be used to prepare a ration with a different fat or protein content.

By consulting Table 6.1 the total protein value for the USDA ration can easily be determined. Yellow corn contains 8.6% protein. The ration contains 35% or 35 lb corn. By multiplying the pounds times the percentage (35 × 0.086) 3.01 lb of protein are found to be present in the corn of the ration. By computing the value for each ingredient and totaling the amounts, the final figure should be 14.2% or their figure for crude protein. This same procedure can also be used to calculate fat, fiber, energy, amino acid, vitamin, and mineral levels. Minor differences occur in calculations because of differences in table percentages.

ENERGY. Pigeons convert chemical molecular energy of feedstuffs into kinetic energy of work and heat. The caloric value of a ration must supply sufficient metabolizable energy to equal the basal and extraneous metabolic requirements if the body weight of the pigeon is to be maintained. The diet must first satisfy the basal life maintenance (resting) requirement. This is calculated in pigeons after a 28-hr fast (Benedict and Riddle 1929). Brody (1945) quoted two investigations pertaining to the basal energy metabolism of pigeons. One report gave 27.2 Kcal/day or 3.3 Kcal/kg/hr for a 340-g male pigeon. Another gave 6.5 Kcal/kg/hr. Extraneous energy needs vary depending on age, disease, environmental temperature, and work activity. For example, long races require a greater expenditure of muscular energy; thus the diet must be formulated to satisfy this demand. Muscular activity is largely supported by oxidative processes of noncarbohydrate substances. Fatty acids are used preferentially (Hazelwood 1986).

In order to calculate energy requirements of a pigeon one must understand the general aspects of energy utilization and associated terminology. Ingested feedstuffs are temporarily stored and softened in the crop before passing to the proventriculus (stomach) and gizzard. This begins the process of digestion, which is the sum of physical and chemical changes in the digestive tract that breaks down feedstuffs to absorbable molecules. In the muscular gizzard, grit serves to grind and pulverize the food prior to further digestion and absorption in the intestine. A portion of the food remains undigested and thus energy is lost in the feces. The remaining food is absorbed as digestible energy and is used except for a small portion lost through the kidneys as uric acid or as other unoxidized compounds. This retained metabolizable energy serves to maintain body heat; the remainder, net productive energy, is used for the work of the body and any excess may be stored in body tissue. Work includes growth; tissue repair; maintenance of vital activities; production of crop milk, body fat, eggs, and feathers; and locomotion (flight).

Carbohydrates, fats, and proteins all contribute energy. Of the carbohydrates, starch is the major source of energy for pigeons. In chickens, such carbohydrates as sucrose, maltose, glucose, fructose, mannose, and galactose also provide energy (Scott et al. 1969).

In the process of digestion, starch is broken down to D-glucose. It undergoes metabolism in several ways interrelated with fatty acid and amino acid metabolism. The formation and storage of glucose as glycogen occurs in all tissue cells but mostly in liver and muscle. When required by the body, glycogen reserves are reconverted to glucose. As glucose is oxidized in the cell to carbon dioxide and water via pyruvic acid, energy is released to high-energy phosphate linkages. Adenosine diphosphate converts to adenosine triphosphate, and this energy is stored as phosphocreatine. Most energy for body activities is derived from these reserve energy compounds (Phillis 1976).

Dietary lipids chiefly in the form of triglycerides also contribute energy. They are emulsified in the upper intestine by conjugated bile salts. Fat triglycerides are hydrolized by pancreatic lipase to two alpha fatty acids and one beta fatty acid (beta monoglyceride), which remains attached to the glycerol. These fatty compounds form tiny micelles that permit absorption by epithelial cells of the microvilli. Bile salts remain in the lumen of the intestine and are resorbed in the ileum and reused. Triglycerides are reesterified within mucosal cells, and fat chylomicrons are transported to the liver. The portal vein carries 15% and the lymph carried 85% (Griminger 1986). Fat transport in chickens differs from that in mammals. Neutral fat and other lipids are stored to a limited extent in the liver. The liver also serves to convert excess carbohydrate into fat, but high levels of dietary fat reduce fat storage (Leveille 1969). This liver fat is carried by the blood and is rapidly removed from the plasma for storage by fat cells in adipose tissues (Scott et al. 1982). These energy reserves may be found in many parts of the body, but accumulations in excess of about 15% dry weight in the liver or kidney are considered harmful.

Since fatty acids are not excreted in the urine, the metabolizable energy is directly related to the intestinal absorption, and this rate varies for individual fatty acids and glycerides. Chickens, and presumably pigeons, can use appreciable levels of fat as sources of energy provided the diet is formulated to supply all other nutrients in proportion to the caloric increase, but growth rates are reduced on diets with more than 20% free fatty acid. In general the efficiency of energy utilization by squabs is higher when fat is part of the diet as opposed to a low-fat diet (Scott et al. 1982).

If the caloric intake is inadequate, dietary proteins are burned for calories, which is an expensive and inefficient source of energy. Special metabolic pathways are required to rid the body of nitrogenous compounds that accumulate when excess protein is used for energy. For optimum efficiency of feed utilization, an interrelationship must be present between dietary calories and protein.

The metabolism of carbohydrate or protein, on the other hand, yields 4.15 Kcal/g whereas fats release 9.4 Kcal/g. For this reason fats are used as energy supplements. They supply essentially twice the energy of carbohydrates and proteins. One kg starch when completely metabolized provides 3650 Kcal net metabolizable energy, or 1659 Kcal/lb. One kg corn yields 3370 cal metabolizable energy or 1530 Kcal/lb corn. Some of this energy comes from the fat and protein of the corn, but most of it comes from the starch. Fats and oils provide 3200–3900 cal metabolizable energy/lb, which is essentially twice the energy available from corn.

Energy requirements are calculated in therms, calories, and British thermal units (BTUs). A therm is the amount of heat required to raise the temperature of 1000 kg water 1°C. This is equal to 1000 large Calories or 1,000,000 small calories. One Calorie represents 4 BTUs (Morrison 1951).

The NAS-NRC (1977) places the requirement of metabolizable energy of quail at about 2800 Kcal/kg feed. The level is not specified for pigeons, but this amount should approximate the need, depending on the time of year and activity. Griminger (1981) suggested 2800–3200 Kcal/kg feed.

Goodman (1969) and Goodman and Griminger (1969) reported that high-fat-energy diets sustained racing pigeons better than commercial pellets or grain diets on short and long races. This work supports George and Jyoti (1955), who reported that pigeons during sustained muscular flight receive 77% of the energy from fat oxidation, and Visscher (1938), who showed that the heart obtained nearly all of its energy from fat. This indicates that pigeons in races are sustained better by storage fat than by carbohydrate energy.

Tables 6.1–6.3 give the average composition of commonly used feedstuffs for pigeons. These figures are provided to enable the formulation and computation of pigeon rations.

WATER. Water is essential for life. About 55% of the adult pigeon is water. Squabs' tissues are softer and contain a higher percentage. Loss of 10% of body moisture results in malfunction of the kidneys and other body activities. The loss of 20% can cause death (Scott et al. 1969). Zeigler et al. (1972) found that White Carneau pigeons averaging 542 g consumed ad libitum 27 g feed and 44 g water. When deprived of food for 3 days the birds lost 5% of their body weight, which was regained in 5 days. More appreciable losses of 15% and 25% required 8 and 19 days, respectively, for weight losses to return. This provides an indication of the value and quantity of feed and water required. Most of the water is consumed during or right after eating as an aid to swallowing and digestion.

Although electrically neutral, water is a polar compound and is thus an excellent solvent for salts. More substances will dissolve in water than in any other solvent. Within the body in intra- and extracellular fluids and in the blood, nutrients and waste products are dissolved, absorbed, and transported in water-based fluid. All body reactions and activities take place in a water solution. Water is a major constituent of fluids that lubricate tissues, joints, and muscles. Water aids in the control of body functions, such as osmotic pressure, pH, and electrolyte balance. It softens and lubricates the feed on its way to the crop and gizzard; without it, food could not be swallowed. Since water has the ability to absorb or give off heat with a relatively small change in temperature, it helps to control body temperature as it conducts metabolic heat from the body. Actually much of the body heat is dissipated by the release of water vapor during respiration.

It is important to provide an easily accessible, constant supply of pure, fresh, unfrozen, cool water. Birds prefer water at a temperature of 50–55°F, and water consumption will drop when its temperature is near freezing or above 80°F. The addition of anything to water may reduce intake. Hardness of water below 2000 ppm will not be detrimental unless the hardness is due to magnesium sulfate and the sulfate exceeds 50 ppm, in which case a laxative effect occurs. The normal magnesium concentration is 14 ppm (Carter 1985). Salinity or total dissolved solids must be under 1000 ppm (Shapiro 1985). Sodium alone should be under 32 ppm to prevent diuresis. Water normally contains no arsenic, lead, manganese, zinc, or fluoride, and respective levels of: 0.2 ppm, 0.1 ppm, 0.05 ppm, 25.0 ppm, and 2.0 ppm are considered potentially toxic (Carter 1985). Nitrates (NO_3), which suggest fertilizer contamination, must not exceed 50 ppm. Nitrites (NO_2) should not exceed 4 ppm. Sulfur in the form of sulfates should not exceed 250 ppm, or a laxative effect is produced (Shapiro 1985). Copper in excess of 10 ppm may also be considered harmful (Scott et al. 1969).

Birds prefer water with a pH about neutral (pH 7.0). Water with a pH on the acid side is more acceptable than that on the alkaline side. Pigeons should not be provided water with a pH of 8 or above. Disinfectants added to waterpans to kill bacteria should be rinsed away, but sodium hypochlorite added to water up to 5 ppm may be used without harm. Hypochlorite as a disinfectant may be used at 15 ppm in water, but it should be rinsed away.

Water is often used as a means of medication of birds. Care must be taken to ensure proper dosage and also to guarantee that the birds drink about the usual amount. Inadequate intake leads to dehydration and death.

References

Adamstone, F. B. 1931. J Morph Physiol 52:47.

Adler, B. 1927. Proc 3d World Poultry Congress, pp. 231–34.

Almquist, H. J. 1942. Proc Soc Exp Biol Med 491:544–45.

______. 1957. *Proteins and Amino Acids in Animal Nutrition.* 4th ed. New York: Reprinted courtesy US Industrial Chemical Corp.

Austic, R. E., and M. L. Scott. 1984. In *Diseases of Poultry,* 8th ed., ed. M. S. Hofstad, pp. 38–64. Ames: Iowa State University Press.

Bellin, S. A., et al. 1954. Arch Biochem Biophys 50:18.

Benedict, F. G., and O. Riddle. 1929. J Nutr 1:497.

Best, C. H., and N. B. Taylor. 1945. *Physiological Basis of Medical Practice.* 4th ed. Baltimore: Williams & Wilkins.

Briggs, G. M., A. C. Groschke, and R. J. Lillie. 1946. J Nutr 32:659–75.

Brody, S. 1945. *Bioenergetics and Growth.* New York: Reinhold.

Brown, I. 1983. Am Racing Pigeon News 99:28–30.

Canadian Tower Rapeseed Meal. 1977. Dawe's Frontiers in Nutrition (Dawe's Laboratories, Chicago) 279:1084.

Carr, R. H., and C. M. James. 1931. Am J Physiol 97:227–31.

Carter, T. 1985. Poult Dig 2.

Chaudhuri, C. R., and I. B. Chatterjee. 1969. Science 164:435-36.

Couch, J. R., et al. 1948. Anat Rec 100:29–48.

Cravens, W. W., W. H. McGibbon, and E. E. Sebesta. 1944. Anat Rec 90:55–64.

Dam, H., and J. Glavind. 1939. Skand Arch Physiol 82:299–316.

DeLuca, H. F. 1971. Nutr Rev 29:179–81.

Ensminger, M. E., and C. G. Olentine. 1978. *Feeds and Nutrition.* Clovis, Calif.: Ensminger.

Erdtmann, H. 1927. Z Physiol Chem 172:82.

Ewing, W. R. 1963. *Poultry Nutrition.* 5th ed. Pasadena, Calif.: Ewing.

Fritz, J. C., et al. 1946. J Nutr 31:387–96.

Fuller, H. L. 1964. Vitam Horm 22:659.

Ganguly, J., et al. 1952. Arch Biochem Biophys 30:275–82.

Garvey, W. T., and R. E. Olsen. 1978. J Nutr 108:1078–86.

George, J. C., and D. Jyoti. 1955. J Anim Morphol Physiol 2:2–29.

Gillis, M. B., G. F. Heuser, and L. C. Norris. 1948. J Nutr 35:351–63.

Gillis, M. B., L. C. Norris, and G. F. Heuser. 1954. J Nutr 35:195.

Goodman, H. M. 1969. Proc 1st Pigeon Conference, Cook College, Rutgers University, New Brunswick, N.J., p. 48.

Goodman, H. M., and P. Griminger. 1969. Poult Sci 48:2058–63.

Gries, C. L., and M. L. Scott. 1972a. J Nutr 102:1287–96.

______. 1972b. J Nutr 102:1259–68.

Griminger, P. 1981. Proc 12th Pigeon Conference, Cook College, Rutgers University, New Brunswick, N.J., pp. 14–25.

______. 1983. In *Physiology and Behavior of the Pigeon,* ed. M. Abs, pp. 19–31. London: Academic.

______. 1986. In *Avian Physiology,* 4th ed., ed. P. Sturkie, pp. 328–47. New York: Springer-Verlag.

Griminger, P., and J. L. Gamarsk. 1972. Poult Sci 51:1464.

Griminger, P., and C. G. Scanes. 1986. In *Avian Physiology,* 4th ed., ed. P. Sturkie, pp. 308–27. New York: Springer-Verlag.

Harrap, B. S., and E. F. Wood. 1967. Comp Biochem Physiol 20:449.

Harrow, B. 1946. *Textbook of Biochemistry.* 4th ed. Philadelphia: Saunders.

Hazelwood, R. L. 1986. In *Avian Physiology,* 4th ed., ed. P. Sturkie, p. 325. New York: Springer-Verlag.

Hegsted, D. M., et al. 1941. J Biol Chem 140:191–200.

Hollander, W. F., and O. Riddle. 1946. Poult Sci 25:20–27.

Hooper, J. H., J. L. Halpin, and J. C. Fritz. 1942. Poult Sci 21:472.

Hopkins, D. T., and M. C. Nesheim. 1967. Poult Sci 46:872–81.

Hutcheson, T. B., T. K. Wolfe, and M. S. Kipps. 1936. *The Production of Field Crops.* 2d ed. New York: McGraw-Hill.

Jensen, L. S., et al. 1956. Poult Sci 35:810–16.

Jones, B. D. 1985. *Malassimilation Syndromes.* U Mo Vet, Columbia: University of Missouri Veterinary Hospital.

Jukes, T. H. 1947. Ann Rev Biochem, 16:194.

Jukes, T. H., and F. H. Bird. 1942. Proc Soc Exp Biol Med 49:231.

Justin, J. R. 1985. Personal communication, Department of Agronomy, Cook College, Rutgers University, New Brunswick, N.J.

Keane, K. W., R. A. Collins, and M. B. Gillis. 1956. Poult Sci 35:1216.

Kienholz, E. W., 1961. J Nutr 75:211–21.

Kim, C. S., and C. H. Hill. 1966. Biochem Biophys Res Commun 24:395.

Leach, R. M. 1974. J Nutr 104:684–86.

Leveille, G. A. 1969. Comp Biochem Physiol 28:431.

Levi, W. M. 1941. *The Pigeon.* Columbia, S.C.: Bryan.

Lyons, M., and W. M. Insko. 1937. Ky Agric Exp Stn Bull 371.

Maas, J., L. E. Davis, and C. Hempstead. 1984. Am J Vet Res 45:2347.

Maynard, L. A., and J. K. Loosli. 1962. *Animal Nutrition.* 5th ed. New York: McGraw-Hill.

Michel, J. N., and M. L. Sunde. 1985. Poult Sci 64:669–74.

Mineral Nutrition. 1967. Dawe's Frontiers in Nutrition (Dawe's Laboratories, Chicago) 197:755–56.

Moon, R. D., and H. P. Zeigler. 1979. Physiol Behav 22:1171-82.
Morrison, F. B. 1951. *Feeds and Feeding.* Ithaca, N.Y.: Morrison.
National Academy of Sciences-National Research Council. 1977. *Nutrient Requirements of Poultry.* Washington, D.C.: National Academy Press.
Noguchi, T., A. H. Cantor, and M. L. Scott. 1973. J Nutr 103:1502-11.
Norris, L. C., and M. L. Scott. 1965. In *Diseases of Poultry,* 5th ed., ed. H. E. Biester and L. H. Schwarte, pp. 144-80. Ames: Iowa State University Press.
Phillis, J. W. 1976. *Veterinary Physiology.* Philadelphia: Saunders.
Platt, C. 1951. Poult Sci 30:196-98.
Poley, W. E., A. L. Moxon, and K. W. Franke. 1937. Poult Sci 16:219-25.
Riesen, W. H., et al. 1947. J Biol Chem 167:143.
Robertson, E. I., et al. 1947. Proc Soc Exp Biol Med 64:441-43.
Ross, E., and L. Adamson. 1961. J Nutr 74:329-34.
Russell, W. C., M. W. Taylor, and L. J. Polskin. 1942. J Nutr 24:199.
Sallis, J. D., and E. S. Holdsworth. 1962. Am J Physiol 203:497.
Schaible, P. J. 1970. *Poultry Feeds and Nutrition.* Westport, Conn.: AVI.
Scott, M. L. 1953. Poult Sci 32:670-77.
______. 1962. Nutr Abstr Rev 32:1-8.
Scott, M. L., and G. F. Heuser. 1954. Proc 10th World Poultry Congress 11:255-58.
Scott, M. L., and L. C. Norris. 1965. In *Diseases of Poultry,* 5th ed., ed. H. E. Biester and L. H. Schwarte, pp. 191-95. Ames: Iowa State University Press.
Scott, M. L., et al. 1967. J Nutr 91:573-83.
Scott, M. L., M. C. Nesheim, and R. J. Young. 1969. *Nutrition of the Chicken.* Ithaca, N.Y.: Scott.
Scott, M. L., R. E. Austic, and C. L. Gries. 1978. In *Diseases of Poultry,* 7th ed., ed. M. S. Hofstad, pp. 38-64. Ames: Iowa State University Press.
Scott, M. L., M. C. Nesheim, and R. J. Young. 1982. *Nutrition of the Chicken.* 3d ed. Ithaca, N.Y.: Scott.
Seifried, O. 1930. J Exp Med 52:519-31, 533-38.
Shapiro, D. 1985. Poultry, Jan., pp. 52-55.
Siegmund, O. H., ed. 1979. *The Merck Veterinary Manual.* Rahway, N.J.: Merck.
Singsen, E. P., et al. 1953. Poult Sci 32:924.
Smith, S. 1985. Personal communication, Ag Way, Inc. Ithaca, N.Y.
Sodeman, W. A. 1950. *Pathologic Physiology.* Philadelphia: Saunders.
Standard Brands. 1954. Vitamin D Kinds and Units. Memo. New York: Standard Brands.
Stecher, P. G., ed. 1960. *The Merck Index.* 7th ed. Rahway, N.J.: Merck.
Sturkie, P. D. 1954. *Avian Physiology.* Ithaca, N.Y.: Comstock.
Sunde, M. L., et al. 1950a. Poult Sci 29:220-26.
______. 1950b. Poult Sci 29:696-702.
Thompson, J. N., and M. L. Scott. 1969. J Nutr 97:335-42.
U.S. Department of Agriculture. 1960. Squab Raising. USDA Farmers Bull 684.
Visscher, M. B. 1938. Proc Soc Exp Biol Med 38:323-35.
Vogel, K. 1980. *Die Taube: VEB Deutscher,* p. 317. Berlin: Landwirtschaftsverlag.
White-Stevens, R. 1956. Personal communication, Department of Food Science, Cook College, Rutgers University, New Brunswick, N.J.
Wilgus, H. S., L. C. Norris, and G. F. Heuser. 1937. J Nutr 14:155-67.
Wolter, J., P. P. Boidot, and M. Morice. 1970. Rec Mit Vet 146:1-13.
Yacowitz, H. 1976. Proc 9th Pigeon Conference, pp. 45-49. Cook College, Rutgers University, New Brunswick, N. J.
Yacowitz, H., L. C. Norris, and G. F. Heuser. 1951. J Biol Chem 192:141.
Young, R. J., L. C. Norris, and G. F. Heuser. 1955. J Nutr 55:353-62.
Zeigler, H. P., H. L. Green, and J. Siegel. 1972. Physiol Behav 8:127-34.

7

Neoplastic Growths

Tumors

Tumors, or neoplasms, are uncontrolled, progressive new cell growths that serve no useful purpose. They should not be considered as a single entity but as a multiplicity of diseases with different initiating factors for each new tissue growth. Neoplasms are divided into two groups: those with a known cause, such as RNA or DNA viruses or chemical carcinogens, and those with unknown etiology. The key that stops organized cell growth resides within every cell. The inducing factor, or cause, turns the key back on to initiate neoplasia.

A benign tumor is a localized growth that does not spread in the body. A malignant growth, or cancer, spreads from the tissue of origin to other sites in the body by extension or by blood or lymph. The body of an animal or bird is composed of different types of cells, or building blocks, and each tumor growth is classified on the basis of the cell type from which it is derived (Table 7.1).

The following forms of neoplastic growth have been reported in pigeons. Klumpp and Wagner (1986) observed an adenocarcinoma in the oviduct of a pigeon and a bronchogenic adenocarcinoma in another. Sohkar et al. (1979) identified a basal cell carcinoma in the skin of a pigeon. An embryonal nephroma was diagnosed by Feldman and recorded by Levi (1941). This probably involves undeveloped adrenal "rest" cells in the kidney, but Levi gave no description of the growth. A seminoma, an uncommon malignant embryonal adenocarcinoma arising from sperm cells of the testicle, is rarely reported in other than the dog but have been observed in the collared turtle dove, Amazon green parrot, budgerigar, Jardine warbler, and cockerel (Campbell 1969). Turk et

Table 7.1. Partial list of forms of growth that develop from various cell types

	Growth	
Cell Type	Benign	Malignant
Connective tissues		
Fibrous tissue	Fibroma	Fibrosarcoma
Fat cell	Lipoma	Liposarcoma
Cartilage	Chondroma	Chondrosarcoma
Bone	Osteoma	Osteogenic sarcoma
Nerve sheath	Neurofibroma	Neurogenic sarcoma
Lymphocytic blood tissue	Lymphoma	Lymphosarcoma
Blood vessel	Angioma	Angiosarcoma
Epithelial tissues		
Glandular tissue	Adenoma	Adenocarcinoma
Skin tissue		Basal cell carcinoma
Mucin gland		Mucoid carcinoma

al. (1981) described the microscopic changes in a pigeon as irregular seminiferous tubules distended by spherical to oval seminoma cells and epididymal stroma infiltrated by mononuclear cells and mitotic figures. Klumpp and Wagner (1986) noted a pigeon over 3 yr of age with a cholangiocarcinoma; this growth involved the gallbladder. Rambow et al. (1981) observed a malignant lymphoma, a lymphosarcoma, under the eye of a pigeon in which gross nodular lesions were present in the liver, bone marrow, kidneys, oviduct, intestine, cranial nerves, and pituitary. Chalmers (1986) observed diffuse lymphoid leukosis in a 14-year-old-female pigeon. Luthgen and Valder (1976) also reported finding a lymphosarcoma. Lymphomas of the spleen and liver were described by Klumpp and Wagner (1986). Sra et al. (1981) sectioned a fibrosarcoma, a fibroblastic growth, from a pigeon in India. The first known cancer-producing virus of chickens, Rous sarcoma, has a wide host range that includes pigeons (Purchase and Payne 1984). Several cell types are activated by this virus, including fibroblasts, macrophagelike cells, and mast cells (Haguenau and Beard 1962). Fatty tumors, or lipomas, were pictured on the flank of a pigeon by Levi (1941), and Gandal (1961) resected a capillary hemangioendothelioma, a growth involving the inner lining of capillaries. Levi (1941) also reported this type of growth on the back of a White Carneau. Hasholt and Petrak (1982) noted a multifold increase in granular white cells in pigeons and classified the condition as granulocytic leukemia. Promoter sequences of poxvirus genes may enhance the latent tumorlike growth aspects of the poxvirus. This overgrowth of tissue was noted by Hartig and Frese (1973).

Internal tumors may cause a pendulous abdomen and are usually palpable through the abdominal wall or skin. They may involve any tissue or structure (Fig. 7.1). They cause general loss of condition, lethargy, and loss of appetite. External tumors are easily recognized as a lump or abnormal mass that distorts the features or displaces feathers or tissues (Fig. 7.2). Internal or external growths may gradually encroach on vital structures and interfere with their form or function in such a way as to cause hemorrhage, malfunction, and death.

As a rule some small, benign, circumscribed lesions may be removed surgically, but malignant growths are not generally treatable. Invasive tumor cellular infiltration make it difficult to extract all involved tissues, and unless all tumorous cells are removed, growth continues. Each problem must be evaluated separately by a veterinarian.

Fig. 7.1. Tumor of the kidney.

Cysts

A cyst is a closed, walled sac that is filled with fluid. It is usually lined by epithelial-type cells. Cysts are of several kinds. Retention cysts are the result of duct blockage and are found in the kidney and pancreas. Van Alstine and Trampel (1984) reported polycystic kidneys found in a pigeon. These are fairly common. Cystic tumors are found in the ovary and thyroid and reach considerable size. Developmental cysts result from partial development of the right oviduct (Wolffian cysts) or from residual embryonal kidney structures. Fluid-filled cysts also form from distention of bursae or tendon sheaths, and air-filled cysts develop from ruptured cervical air sacs. The author has encountered each type of cyst in pigeons, with the exception of thyroid cysts. Treatment involves surgery and is generally not indicated.

References

Campbell, J. G. 1969. *Tumours of the Fowl.* Philadelphia: Lippincott, pp. 163–66.

Chalmers, G. A. 1986. Avian Dis 30:241–44.

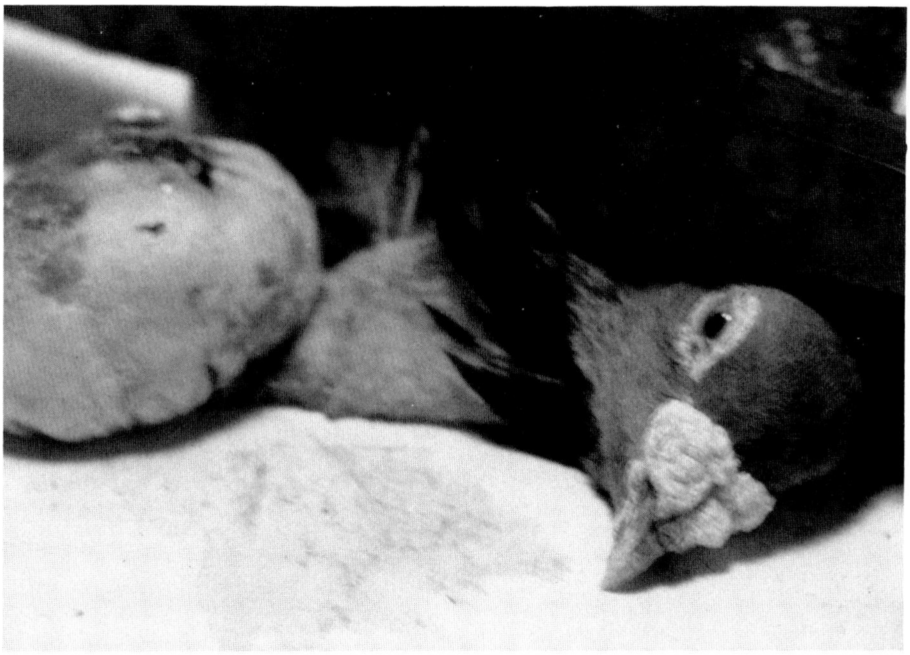

Fig. 7.2. Tumor of the skin.

Gandal, C. P. 1961. Avian Dis 5:250-52.

Haguenau, F., and J. W. Beard. 1962. In *Tumors Induced by Viruses,* ed. A. J. Dalton and F. Haguenau, pp. 1-60. New York: Academic.

Hartig, F., and K. Frese. 1973. Zentralbl Veterinaermed 20:153-60.

Hasholt, J., and M. L. Petrak. 1982. In *Diseases of Cage and Aviary Birds,* 2d ed., ed. M. L. Petrak, pp. 449-57. Philadelphia: Lea & Febiger.

Klumpp, S. A., and W. D. Wagner. 1986. Avian Dis 30:740-50.

Levi, W. M. 1941. *The Pigeon.* Columbia, S.C.: Bryan, pp. 305-307.

Luthgen, W., and W. A. Valder. 1976. Prakt Tieraerztl 57:75-76.

Purchase, H. G., and L. N. Payne. 1984. In *Diseases of Poultry,* 8th ed., ed. M. S. Hofstad et al. Ames: Iowa State University Press.

Rambow, V. J., J. C. Murphy, and J. G. Fox. 1981. J Am Vet Med Assoc 179:1266-68.

Sohkar, S. M., et al. 1979. Avian Pathol 8:69-75.

Sra, I. S., et al. 1981. Indian J Poult Sci 16:428-29.

Turk, J. R., J. Kim, and A. M. Gallina. 1981. Avian Dis 25:752.

Van Alstine, W. G., and D. W. Trampel. 1984. Avian Dis 28:758-64.

8

Toxins and Chemical Poisons

Poisons are substances that, when applied to the body surface, ingested, inhaled, or developed in the body, may cause degenerative changes or disturbances to body tissues and their functions.

Toxins are more or less unstable poisonous substances from living or dead plant or animal sources that by strict definition engender a specific antitoxin response and signs of poisoning after a period in the body. Microbial toxins may be classed as endotoxins that are confined to the cell until the cell disintegrates. They may also be soluble exotoxins excreted by the organism. Bird losses from toxins and poisons result from autointoxication, bacterial toxins, fungal toxins, and chemical poisons.

Autointoxication

Definition. Autointoxication, or self-poisoning, is caused by noneliminated waste products generated within the body during cellular metabolism or that develop from intestinal decomposition.

Cause. In pigeons, uric acid produced by the kidneys accumulates in the body if water is insufficient to flush out the kidneys. The white crystals also precipitate throughout the body if the kidneys are damaged. Birds produce a preponderance of uric acid instead of urea. The livers of chickens and pigeons produce hypoxanthine, which is then oxidized to uric acid in the kidneys of the pigeon by xanthine oxidase (Sturkie 1954).

Intestinal intoxication in pigeons may result from the feeding of high-moisture grains, the ingestion of excess foreign substances and fiber, or the overconsumption of new grit or other substances, particularly when first given. Any of these mechanisms may produce faulty elimination of intestinal contents.

Signs. Toxins absorbed from the intestine and circulated throughout the body result in depression, loss of appetite, weakness, prostration, and death.

Necropsy. In autointoxication, examination of dead specimens reveals congestion of muscle tissue and viscera, dehydration, accumulation of urates in kidney tubules, and/or discolored dark digestive tract contents.

Treatment. Treatment for autointoxication must increase available water and combine this with the removal of the offending material and the cause of the problem. Milk of magnesia, 1/4 tsp/bird per os may be helpful. Sulfathalidine (Merck) at 25–50 mg/bird per os given once may also benefit birds with intestinal disturbances.

References

Sturkie, P. D. 1954. *Avian Physiology.* Ithaca, N.Y.: Comstock, pp. 224–25.

Bacterial Toxin: Botulism

Definition and Synonyms. Botulism is a disease of birds, animals, and humans caused by the ingestion of bacterial exotoxins produced by *Clostridium botulinum.* Food intoxication, limberneck, and western duck sickness are terms applied to this disease.

Cause and Classification. The organism is placed in class Schizomycetes, order Eubacteriales, family Bacillaceae, and genus and species *Clostridium botulinum.*

C. botulinum is divided into types A to E, depending on the nature of the immunologically distinct neuroexotoxins. Bergey's classification presently lists nonproteolytic varieties, which include some strains of type B and known strains of type C, D, and E as *C. botulinum.* Proteolytic members including type A and some strains of type B are listed as *C. parabotulinum.* Most American type B organisms are proteolytic whereas European type B are not (Burrows 1954). The most common toxins observed in birds are types A and C, but types B and E have also been reported in birds. Western duck sickness is caused by type C toxin and limberneck of chickens and pheasants is produced by types A and B (Burrows 1954).

Nature. *C. botulinum* is a large, motile, gram-positive, noncapsulated, sporulating rod with rounded ends that measures 4–8 μm in length. The oval spore usually forms near the end of the rod. It usually occurs singly but may be found in short chains. It is a strict anaerobe that grows best from 30–35°C but will grow and produce toxin at 37.5°C. It grows best at a neutral or slightly alkaline pH (7.5–9). Fermentation occurs in dextrose, maltose, salicin, and levulose. Other sugars are variable. Broth cultures exude a butyric odor. On solid media, irregular, small, flat, grayish colonies develop. Older colonies darken and become opaque or yellowish brown. Gelatin is liquefied, and litmus milk is slowly reduced but not coagulated. Nitrites and indole are not formed (Kelser and Schoening 1948; Burrows 1954; Davis et al. 1973).

Distribution

GEOGRAPHY AND HISTORY. Botulism was first noted in Germany in 1785; it derives its name from the Latin *botulus* (sausage) (Burrows 1954). In Germany, van Ermengem (1896) isolated a spore-bearing, toxin-forming anaerobe from ham and named it *Bacillus botulinus.* It was later shown to be a nonproteolytic type B strain (Burrows 1954). In 1917 Graham et al. in Kentucky identified *C. botulinum* as the cause of forage poisoning in horses.

INCIDENCE. The botulism organism is a widely dispersed saprophyte found in soil throughout the world.

SEASON. The toxin is readily formed when botulinus spores germinate, grow, and multiply under anaerobic conditions. This often occurs in warm late summer in stagnant ponds, which have an absence of oxygen. Under such conditions anaerobic decomposition of plant and animal tissue normally takes place. This process is assisted by *Pseudomonas aeruginosa,* a common soil organism, which contributes to the production of toxin by utilizing oxygen (Quortrup and Sudheimer 1934).

AGE, SEX, AND BREED. The toxin affects birds of any age, but adult flying birds usually have more of an opportunity to ingest toxin.

Neither sex nor breed alters the incidence.

SPECIES. Chickens were first observed by Dickson in 1917 with limberneck. Kalmbach and Gunderson (1934) investigated western duck sickness and listed 69 affected species of birds in 21 families. Herons, geese, ducks, hawks, sandpipers, gulls, and blackbirds were included. The disease was reported by Palmer and Baker (1922) in swans, Theiler (1927) in the ostrich, Coburn and Quortrup (1938) in turkeys, and Vadlamudi et al. (1959) in pheasants.

Bengston (1924) inoculated guinea pigs subcutaneously and determined the minimum lethal (toxin) dose (MLD) to be 0.00012 mg/kg. Kalmbach (1939) reported that the vulture was quite resistant to the toxin, withstanding 300,000 guinea pig MLDs.

Transmission. Botulism toxicity is not infectious. It is not transferred from one bird to another as a contagious disease. Ingested droppings containing the toxin can cause trouble, but this is unlikely. Maggots of the flesh-eating bottle fly *Lucilia caesar* may be present in decomposing carcasses, and toxins formed in the dead birds may be present in and on the maggots. In this case, if maggots are eaten, botulism can develop. Fortunately, pigeons are essentially seed eaters and seldom vary their diet.

Epidemiology

DISTRIBUTION IN BODY. The toxin is absorbed from the gastrointestinal tract and carried by the bloodstream to presynaptic terminals of cholinergic nerves where it interferes with nerve transmission and thus muscle enervation.

ENTRY AND EXIT. The intestinal contents may occasionally contain botulinus organisms, but this may bear no relationship to the disease. The organism often reproduces and produces toxin in the ceca. Since pigeons have no functional ceca, toxin production within the body is limited. Ingested preformed toxin produces the intoxication, with the possible exception of crop-formed toxin.

Feces may contain organisms and toxin, but "fixed" toxin is not released from the body.

PATHOGENICITY. The botulinus organism itself is virtually noninvasive in the live bird, but muscle tissues of the dead bird may be invaded with the production of harmful exotoxin. Of the toxins, type A is the most potent intravenously and type C most toxic orally (Gross 1984). The activity of type C toxin appears to be due to the presence of a specific bacteriophage, CE beta. As a group, botulinus toxins are the most potent known toxins and act by blocking the release of acetylcholine. The crude toxin, treated with trypsin for a limited time, can be potentiated 10 to 1000 times by partial proteolysis (Davis et al. 1973). This suggests that ingested toxin may be potentiated by digestive enzymes of the proventriculus and intestine.

IMMUNITY. Botulism toxins may be detoxified by formaldehyde to form toxoids that may be used to produce active immunization in healthy individuals sometime before exposure. Clinical botulism does not induce immunity. The toxin causes death as antibodies are formed.

Morbidity and Mortality. Only those birds that have ingested botulinus toxin will become sick, but not all of the birds that have ingested toxin will die. Death depends on the amount of toxin ingested.

Signs. Paralysis of the legs, wings, and neck are characteristic of botulism. The neck is often limp; thus the disease is also called limberneck. Complete body paralysis may ensue, and feathers pull out easily.

Necropsy. Gross internal changes are lacking in botulism. Mild enteritis may be observed.

Diagnosis. Paralytic signs, loose feathers, and the lack of gross internal lesions, together with an appropriate history, is usually sufficient for a diagnosis in birds. In questionable cases demonstration of toxin in centrifuged gizzard or intestinal contents of sick birds is necessary.

Differential Diagnosis. Paramyxovirus infection, exotic herpes, and Newcastle disease must be considered.

Biological Properties

STRAINS. Strains differ in toxin-producing ability. Meyer and Gunnison (1929) described 15 subtypes.

SEROLOGY. Clostridial infection can be classified on the basis of the flagellar (H) antigen and the lipopolysaccharide somatic (O) antigen. H antigens are thermolabile, but O antigens are somewhat thermostable (Davis et al. 1973).

TISSUE AFFINITY. The botulinus toxin is neurotoxic and affects nerve tissue.

STAINING. Young cultures are gram-positive and stain with ordinary aniline dyes.

STABILITY. Although the toxins are composed of protein, proventriculus acidity and proteolytic enzymes do not destroy the toxins when ingested (Burrows 1954). Botulinus toxin is relatively heat-labile. It is inactivated in 10 min at 100°C and 30 min at 80°C (Schoenholz and Meyer 1924). The spores are relatively heat-resistant. Steam pressure sterilization destroys spores in 10 min at 120°C, but boiling requires 5 hr (Gross 1984).

CULTURE. Brain-heart infusion agar streaked prior to hardening with a ground liver tissue sample and incubated for 2 days at 37°C in an anaerobic jar is suggested. For further study, mice injected intraperitoneally with serum from sick birds may also be indicated (Smith 1975).

Prevention. Decomposed feed or canned food should be avoided, although canned food that is acid is less likely to cause poisoning. Dead birds should be removed from the loft immediately and live birds kept away from their droppings. In 1957 C. A. Brandly used aluminum hydroxide-adsorbed botulinus antitoxin for protection of pheasants (Levine 1965). Type C toxoid and Freund's incomplete adjuvant were used twice at 3-wk intervals in quail, ducks, and pheasants by Boroff and Reilly (1962).

Treatment. Antitoxin is effective before the toxin has been "fixed" to nerve endings; thus antitoxin administered immediately following entrance of toxin will often prevent intoxication. Type C antitoxin, which is seldom available, or equine polyvalent antitoxin containing types A, B, and E given intravenously or/and intraperitoneally is often helpful. In chickens, crop lavage with weak pink potassium permanganate is beneficial, but pigeons have crop glands that may be burned by the disinfectant. Laxatives are seldom of value in sick birds because the toxin is often absorbed by the time a laxative takes effect. They can be valuable in birds that lack evidence of poisoning. An individual bird may be given 0.5 tsp of a stock solution of epsom salts (1 tsp/pt water).

References

Bengston, I. A. 1924. U S Publ Health Serv Hyg Lab Bull 136.

Boroff, D. A., and J. R. Reilly. 1962. Int Arch Allergy Appl Immunol 3:809–16.

Burrows, W. 1954. *Textbook of Microbiology.* 16th. ed. Philadelphia: Saunders, p. 521.

Coburn, D. R., and E. R. Quortrup. 1938. J Am Vet Med Assoc 93:385.

Davis, B. D., et al. 1973. *Microbiology.* 2d ed. Hagerstown, Md.: Harper & Row, pp. 833–84.

Dickson, E. C. 1917. J Am Vet Med Assoc 50:612.

Graham R., A. L. Brueckner, and Pontius. 1917. Ky Agric Exp Stn Bull 207.

Gross, W. B. 1984. In *Diseases of Poultry,* 8th ed., ed.

M. S. Hofstad et al., pp.257–58. Ames: Iowa State University Press.
Kalmbach, E. R. 1939. J Am Vet Med Assoc 94:187.
Kalmbach, E. R., and M. F. Gunderson. 1934. USDA Tech Bull 411.
Kelser, R. A., and H. W. Schoening. 1948. *A Manual of Veterinary Bacteriology.* Baltimore: Williams & Wilkins, pp. 350–57.
Levine, N. D. 1965. In *Diseases of Poultry,* 5th ed., ed. H. E. Biester and L. H. Schwarte, pp. 456–61. Ames: Iowa State University Press.
Meyer, K., and J. B. Gunnison. 1929. J Infect Dis 45:96, 106, 119, 135.
Palmer, C. C., and H. R. Baker. 1922. Ohio State Univ Vet Alum Q 10:93.
Quortrup, E. R., and R. L. Sudheimer. 1943. J Bacteriol 45:551.
Schoenholz, P., and K. F. Meyer. 1924. J Infect Dis 35:361.
Smith, L. D. S. 1975. In *Isolation and Identification of Avian Pathogens,* ed. S. B. Hitchner et al., pp. 91–92. Ithaca: Arnold.
Theiler, A. 1927. Union S Afr Dept Agric Rep 11–12, Dir Vet Ed Res 2:821.
Vadlamudi, S., V. H. Lee, and R. P. Hanson. 1959. Avian Dis 3:344–50.
van Ermengem. 1896. Centralbl Bakteriol 19.

Fungal Toxins

Definition. Mycotoxicosis is poisoning by the ingestion of fungal toxins in food or feed. These toxins are metabolites released by the growth of toxigenic fungi in or on the feedstuff.

Cause and Classification. Mycotoxins are generally synthesized by fungi in or on grains and other plant materials. Most of the toxins in stored plant material, such as grains and silage, are produced by three genera of fungi: *Aspergillus, Penicillium,* and *Fusarium.* Some of the substances, such as ergot, invade living plants and others invade decaying plant material (Smith 1982). There are over 200 mycotoxins and at least 20 are toxic for humans, animals, and birds.

The most common toxins and some of the fungi of origin are aflatoxin (*Aspergillus* sp.), zearalenone (*Fusarium* sp., *Gibberella zea*), T-2 toxin (*Fusarium tricincium*), Ochratoxin (*Aspergillus* sp., *Penicillium* sp.), Citrinin (*Penicillium citrinum, Penicillium* sp.), and ergot (*Claviceps* sp.).

Other mycotoxins include sterigmatocystin, patulin, roquefortin, penitrem A, fumitremorgen B, roridin A, alpha cyclopiazonic acid, and penicillic acid (Gorst-Allman and Steyn 1978; Smith 1982).

Nature. Mycotoxins are low-molecular-weight, nonantigenic, fungal metabolites that at high levels can produce acute, even fatal, disease and at lower levels can be carcinogenic and mutagenic. These toxicants represent a diverse group of chemical compounds. The structure of six aflatoxins have been determined and all conform to the coumarin-lactone- and bifuran-type structure (Smith 1969). Other bifuran compounds (e.g., furazolidone) have played a major role in the control of gram-negative bacteria in birds and animals. It is strange that one group of somewhat similar compounds has been so poisonous while the other has been so beneficial.

Distribution

GEOGRAPHY AND HISTORY. Mold toxicosis was first recognized in Russian horses in 1930 (Creek 1962). This stimulated research in Maryland during World War II to identify the toxic fungi. In 1961, Blount in Great Britain reported the death of turkey poults resulting from Brazilian ground nut meal or peanut meal. Since then numerous reports have appeared in literature.

INCIDENCE. Hesseltine (1975) indicated that 2–3% of the corn crop may be lost each year from mold damage. Fungi are universally distributed. Every feed sample may contain mold spores. Failure to find spores in feed generally represents inadequate examination. The presence of mold does not mean that the feed is unfit for bird or animal consumption, but heavily mold-contaminated grain or feed should be avoided because it is more likely to have toxigenic fungi. It has been reported that 85% of grade 3 corn contains toxin (Bennett 1974). This grade may be used for alcohol production.

AGE, SEX, AND BREED. Young birds appear more susceptible to mycotoxins, and mortality has reached 30% in chickens (Wilkes 1959), but there is no consistent pattern. Sex and breed factors do not alter the incidence.

Transmission. Mycotoxins are not transferred from one bird to another. They are not infectious.

Epidemiology

DISTRIBUTION IN BODY. Mycotoxins spread throughout the body.

ENTRY AND EXIT. Mycotoxins can enter the body by ingestion, inhalation, or direct contact. Most toxins leave the body in the urine and feces. Hen's eggs do not bear detectable levels (Wilkes 1959). Cow's milk excretes about 1–4% of the ingested toxin (Masri et al. 1969). It is conceivable then that pigeon crop milk could also carry the toxin.

PATHOGENICITY. Molds damage the texture, flavor, odor, color, and chemical composition of feedstuffs. They cause loss of germination, dark germs, mustiness, and heating of stored grains. Mycotoxins persist long after molds cease to grow and cause harm if ingested.

Wogan and Newberne (1967) found that 18 of 22 male cancer-sensitive rats developed liver cancer after being fed 1.0 ppm aflatoxin B_1 continuously for 35–41 wk. This establishes aflatoxin B_1 as one of the most potent carcinogens known. They can even interfere with native resistance mechanisms and immunogenic reactions, leaving the bird or animal more susceptible to disease (Pier et al. 1980). It depresses protein synthesis, binds with cellular DNA, and inhibits the formation of RNA. Fat transport and synthesis are altered in chickens, permitting the accumulation of liver fat. Molds may tie up selenium or utilize it depriving the bird of needed selenium for lipase production. Without sufficient lipase the lipid-bile salt micelles cannot be formed and this interferes with the absorption of fatty acids and fat-soluble vitamins (Wilkes 1959). Warden (Vitamin D 1969) reported an increased demand for lysine when turkeys were fed moldy meal. Evidence was also presented to show that arginine was destroyed by aflatoxin.

SUSCEPTIBILITY. Very small amounts of toxin in parts per billion (ppb) can cause health problems. The harm created varies with the type of mycotoxin and the dose. Birds also differ in their tolerance for these toxins. Ducks show evidence of aflatoxin intoxication at 30 ppb in feed, turkeys tolerate up to 250 ppb, pheasants 500 ppb, chicks 1250 ppb, and laying hens 2500 ppb (Wilkes 1959). Pigeons were not included in the report. Stoloff (1972) indicated 20 ppb were considered toxic for humans, and Bennett (1974) compared this to a peanut in a 10-lb sample. This level falls within the 10–100 ppm that cause poisoning in animals, reported by Wogan (1972). The absence of reported cases in pigeons is probably due to several factors. For the most part, loft pigeons consume whole grains of good quality; second, they appear to have considerable inate resistance; and last, there is a general lack of recognition of the problem.

Signs. Signs of mycotoxicosis vary. Low-level toxin consumption seldom presents dramatic evidence. On the other hand, acute signs include increased vascular fragility and blotch or pinpoint hemorrhages of the skin and underlying tissues. Birds lack appetite and are depressed. Other signs include diarrhea, unthriftiness, paleness, poor hatchability, tremors, incoordination, and death.

Necropsy. Internally, multiple hemorrhages and fatty infiltration of the liver may be present following 1 ppb ingested aflatoxin. Kidney and liver necrosis, altered gizzard lining, regression of the bursa, diarrhea, prolonged blood clotting time, and lymphoid depression occurs with 0.6 ppm ochratoxin. T-2 toxin causes severe mouth ulcers, intestinal inflammation, liver hematomas, and prolonged blood clotting time in broilers at 4 ppm (Pier et al. 1980). Similar lesions may occur in pigeons. Ergot may cause gangrene of toes and digestive disturbances (Wyatt et al. 1972).

Micropathology. In chronic mycotoxicosis the blood cellular pattern is altered and liver and kidney changes are present. The bone marrow shows depletion of red blood centers, hypoplasia of white blood cells, and replacement of lymphocytic foci with fat cells (Forgacs et al. 1962).

Diagnosis. In mycotoxicosis, the cause of the trouble is not immediately evident. Visual examination of the feed may not indicate fungus growth or the presence of a toxin. Laboratory methods must be employed. Bennett (1974) outlined methods of extraction and identification of toxins.

Differential Diagnosis. Signs of plant toxins can easily be confused with drug or chemical poisoning.

Factors Contributing to Toxicity. Mycotoxins can develop before or during harvest. Storage fungi are able to grow and produce toxin on stored feedstuffs with a moisture level above 12%. Higher moisture levels of 20–21% wet weight are required for some field fungi and most types of advanced decay fungi. Insect damage is one of the factors that permits fungus growth on living plants. Warm temperatures encourage growth. Fungi grow slowly at 40–50°F and more rapidly at 80–90°F. Longer storage time also increases the possibility of mold growth. Grain can also take on atmospheric moisture with prolonged storage. This may leave grains with higher than 12% moisture. In humid tropical countries, environmental conditions will always make it difficult to maintain properly dried grain or ground feed. Present-day field harvest methods require that shelled corn be bin-dried to 12% moisture within 24 hr of harvest. The wound at the tip of the corn kernel created by field shelling permits fungus entrance unless the grain is quickly dried. *Aspergillus flavus* will grow in starchy cereal seeds, such as wheat, barley, rice, corn, and sorghum when the moisture content is above 13%; in soybeans, above 12%; and in flaxseed, above 10% (Christensen and Kaufmann 1972). M. Zuber observed that even though *A. flavus* grows on soybeans it fails to produce toxin (Weinberg 1975). As a general rule, if the relative humidity is above 50% and the temperature is 50–100°F, *A. flavus* will grow and produce toxin (Thomas 1971). In addition, Bottomly et al. (1950) demonstrated a logrithmic increase in molds in corn as the relative humidity was raised from 75% to 100%.

Biological Properties

STABILITY. Mycotoxins are very stable and many toxins withstand considerable heat. For example, milk may contain toxin when cows are fed silage or corn contaminated with aflatoxin, but pasteurization will not significantly reduce the toxicity (Allcroft and Carnaghan 1962).

Prevention. Suggestions for mycotoxin prevention have met with varied degrees of success. The farmer or grain grower should harvest only mature, low-moisture grain. Newly harvested grain should be dried to 12% moisture within 24 hr. Storage should protect against weather and provide a relative humidity of under 50%. Efforts should also be made to avoid mechanical damage and to control insect and rodent damage.

Grain and feedstuffs can be treated to reduce fungus growth. During storage, fungicides such as 0.1% propionic acid (Warden 1969) or 1–10 lb propionate/t feed may be applied (Wilkes 1951; Thomas 1971). If grain is stored outside and becomes wet during rainy weather, Thomas (1971) suggests spraying the surface of the pile with a solution of 1 lb calcium propionate in 1½ gal water. Propionic acid is hazardous to handle and corrosive, but it is effective as a volatile, metabolizable, fungistatic antigermination compound when used in conjunction with drying procedures (Smith 1982). Thiabendazole (Merck) at 45 mg/lb grain has been helpful as a fungistat (Siegmund 1979). In addition, it has been shown that 500 ppm 8-hydroxyquinoline added to moldy broiler mash suppressed fungus growth and toxin formation (Forgacs et al. 1963). A newer fungicide that permits corn storage at 17% moisture is Mertect 34OF (Merck); 1 oz/33 bu corn applied in 3 qt water is effective for at least a year (Smith 1987). Another product from the Asian neem tree also appears to have value.

The pigeon fancier and the feed supplier should insist on dry, clean, bright-colored, undamaged whole grain for birds and livestock. If feedstuffs are free from toxigenic molds, there can be no mycotoxin formation. Ground grain, mash, or pellets may contain mycotoxin only if an ingredient contains toxigenic fungi. Pelleting temperature will not destroy the toxin. Where advanced agricultural procedures are used, high levels of mycotoxins are rarely found in grain.

Removal of mycotoxin is difficult, expensive, and not always effective. Defatted meals can be detoxified, but the process does not work well on whole grain. Researchers at the USDA Peoria station developed an experimental ammoniation procedure which was highly effective (Weinberg 1975). It has not been used commercially.

Fortunately pigeons usually discriminate between good and off-colored seed if they are fed a grain diet. Feed provided in clean, dry hoppers and not on the ground also reduces the chances of mycotoxin ingestion.

Treatment. Treatment procedures should first remove the cause or source. Visibly wet or moldy feed should be discontinued, and only dry, wholesome grain or pellets should be provided in clean, dry hoppers.

Antibiotic treatment is usually of little value and may in fact increase the problem. Antibiotics are produced by fungi to fight bacteria. If bacterial populations are suppressed by antibiotics, fungi often grow better, and this is not desirable. Other medications, such as fat-soluble vitamins and selenium, may be increased and the feed fat increased to 15%. Proteins should also be increased to 30% and the salt content of feed lowered (this includes pigeon grit, which contains salt). These are temporary measures. Other treatments are not effective.

References

Allcroft, R., and R. B. Carnaghan. 1962. Vet Rec 74:863–64.

Bennett, G. 1974. 46th Northeast Conference on Avian Diseases, Pennsylvania State University, University Park.

Blount, W. P. 1961. Br Turkey Fed 9:52.

Bottomly, R. A., C. M. Christensen, and W. F. Giddes. 1950. Cereal Chem 27:271–96.

Christensen, C., and H. H. Kaufmann. 1972. Farm Tech Agri-fieldman 28:22–29.

Creek, R. D. 1962. Poult Livest Comment, E. I. Du Pont, 19:2.

Forgacs, J., et al. 1962. Avian Dis 6:363–80.

Forgacs, J., H. Koch, and R. H. White-Stevens. 1963. Avian Dis 7:56.

Gorst-Allman, C. P., and P. S. Steyn. 1978. Chromatography 175:325.

Hesseltine, C. W. 1975. Sci News 107:13–15.

Masri, M. S., V. C. Garcia, and J. R. Page. 1969. Vet Rec, Feb 8, pp. 146–47.

Pier, A. C., J. L. Richard, and S. J. Cysewiski. 1980. J Am Vet Med Assoc 176:719.

Siegmund, O. H., ed. 1979. *The Merck Veterinary Manual.* 5th ed. Rahway, N.J.: Merck.

Smith, D. 1987. Farm J, Sept, pp. 38–39.

Smith, J. E. 1982. World Poult Sci 38:201–12.

Smith, K. J. 1969. Proc 29th Meeting of the American Feed Manufacturers Association Nutrition Council pp. 11–16.

Stoloff, L. 1972. Farm Tech Agri-fieldman 28:60–63.
Thomas, R. D. 1971. Anim Nutr Health, March, p. 5.
Vitamin D. 1969. Dawe's Frontiers of Nutrition (Dawe's Laboratories, Chicago) 216:831.
Weinberg, J. 1975. Sci News 107:13–15.
Wilkes, C. G. 1951. Merck Tech Serv 980:274.
______. 1959. Merck Tech Serv 290:1172.
Wogan, G. N. 1972. Farm Tech Agri-fieldman 28:7a.
Wogan, G. N., and P. M. Newberne. 1967. Cancer Res 27:2370–76.
Wyatt, R. D., et al. 1972. Avian Dis 16:1123–30.

Chemical Poisons

Introduction. This section identifies poison types and causes so that poisoning can be avoided. Where possible one or more basic suppliers of chemical agents have been included to permit a source for further reference. Space does not permit a detailed description of each poison and its effects. Common signs of poisoning are depression, dehydration, excessive thirst, cyanosis, weakness, convulsions, paralysis, and death. Diagnosis requires a chemical evaluation of each case and a determination of the type of poison.

Poisoning occurs in pigeons as a result of improper use of chemicals or the incidental ingestion of poisons, such as the following:

1. Chemotherapeutic agents used in or on pigeons, such as vitamins, coccidiostats, wormers, bird insecticides, and growth stimulants.
2. Chemical compounds (Buck 1982) used about the premises and loft, such as disinfectants, fungicides and insecticides, fumigants, rat poisons, heavy metals, and herbicides.
3. Plant poisons that may be accidentally ingested, such as the phytotoxins present in seeds, roots, stems, leaves, and flowers of poisonous plants.

Drug products have been widely used to control various conditions, diseases, and pests. Unfortunately pigeons have been poisoned because owners have not followed label directions and have given medications without cause. The amount given or used must be weighed or measured. The idea that more is better has killed birds. Owners should treat their birds only when a specific level and kind of treatment is indicated by a reliable diagnosis. Birds must not be medicated without reason. Excessive medication is harmful. When drugs enter the body they find their way to the liver and finally to the kidneys for excretion. Cell changes or damage may occur throughout the body. For this reason and because therapeutic drug levels often approach toxic levels, label instructions must always be followed.

Chemotherapeutic Agents

VITAMIN PREPARATIONS AND FEED SUPPLEMENTS. When these are fed in excess they can be just as toxic as other chemical poisons.

COCCIDIOSTATS. Coccidiostats such as the sulfonamides are therapeutic agents used for the control of coccidiosis when fecal examinations reveal treatable levels of coccidiosis. Sulfamerazine (SM) and sulfaquinoxaline (SQ) are the drugs most often used. SQ is much more toxic than SM but both can kill pigeons. The dose and dosage schedule must be strictly followed. Other coccidiostats have been used on pigeons and these include Nitrofurazone NFZ (Rhodia), Zoalene (Dow), Monensin (Elanco, discontinued), and Amprolium (Merck). Each may be toxic if used too long or at too high a level. Paresis and staggering may be observed in pigeons when excess medication is given.

VERMIFUGES. Vermifuges are substances given for the removal of worms. They should be specific for the type of worm present. Tapeworm medications are distinct from roundworm or capillary worm drugs. Again, fecal examinations enable the identification of parasite eggs released from the intestine. This permits the use of the proper wormer. Hygromycin B (Lilly) is one that cannot safely be given to pigeons (Kocan 1972).

INSECTICIDE POWDER. An insecticide powder, such as dog or cat flea powder, is used on pigeons once a week for the control of lice, and louse flies can be poisonous if ingested or inhaled. Pyrethrin (Fairfield American, Prentiss Drug), Sevin (Union Carbide), or Malathion (American Cyanamide) are the active ingredients in many flea and louse powders. Sodium fluoride powder, which is often used for louse control, is quite irritating and toxic and Co-ral 0.25% (Chemagro) is another body insecticide, used for the control of feather mites, that can be toxic for pigeons. Gloves and a mask should used in applying any powder.

GROWTH STIMULANTS. These have little use in pigeon feed except in squab operations. Arsenic trioxide or 3-nitro-4-hydroxyphenylarsonic acid have been used in poultry feed as growth stimulants, but if the feed is not properly mixed or an error in dosage occurs, poisoning often results.

Chemical Compounds

DISINFECTANTS. These are valuable chemical compounds used in premise disease control. Care must be taken to use them correctly. Pigeon fanciers like to put disinfectants such as potassium permanganate in the bath water. If the bath water is ingested or the eyes become wet, harm occurs. The bath pan may be disinfected with potassium permanganate at 0.1 g/gal or with 2 tbsp sodium hypochlorite 5.25% (Clorox, Clorox Co.)/pt water. Following disinfection, the bath pan should be rinsed. The bath water itself should remain pure water.

Disinfectants used in the loft may be applied to feed or water containers, but these must be washed clean before use. Products such as phenols, carbolineum, or cresol should not be applied to perches or nest boxes unless sufficient drying time is permitted prior to replacing the birds. The fumes are too toxic and burns on the feet can occur. Lye applied with a broom to the building walls and floor is also caustic and must be washed off. Again, the person applying the chemicals must be cautious. Lye must never be sprayed, and boots, rubber gloves, and eye protection are required. Quaternary ammonium compounds (quats) form another group of premise disinfectants. They produce contact cellular necrosis similar to T-2 mycotoxin (Reuber et al. 1970). They should be carefully painted or applied at 200–400 ppm in a very coarse spray. Inhaled spray droplets can cause serious pseudomembranous tracheal lesions. Attendants should not use quats or other disinfectants on birds or in their water or feed. If individual birds are to be treated, a veterinarian should be consulted.

PREMISE FUNGICIDES. Premise fungicides (Sina 1987) are chemical compounds such as copper sulfate 0.2%, folpet (Phaltan, Chevron), anilazine (Dyrene, Chevron), and benomyl (Benlate, Du Pont) which have been widely used. Each is toxic if ingested or inhaled. Grains used for seed are often treated with organic seed protectants, such as thiram (Arasan, Du Pont, discontinued) or captan (Orthocide, Chevron). Such grains should be dyed to identify them because these treated poisonous grains have been mixed on occasion with other feed grains.

PREMISE INSECTICIDES. Premise insecticides (Sina 1987) are chemical compounds, including tars and oils, chlorinated hydrocarbons, organophosphates, carbamates, and pyrethrins. Coal tar, creosote, and crankcase drain oil each serve to plug the spiracles of mites as they crawl through the liquid on the perch.

Chlorinated hydrocarbons (Sina 1987) include DDT (dichlorodiphenyltrichloroethane) (Neocid, Ciba-Geigy), endosulfan (Endosan, Hoechst AG; Thiodan, FMC), chlordecone (Kepone, Allied, discontinued), aldrin (Aldrex, Shell), chlordane (Termide, Sandoz) for termites, dieldrin (Dieldrex, Shell), lindane (Isotox, Los Angeles Chem.), toxaphene (Strobane T90, Agro-Quimicas), methoxychlor (Marlate, Kincaid), polychlorinated biphenols (PCBs), heptachlor (Sandoz). Dieldrin, a seed protectant, has killed pigeons (Blakley 1982).

Organophosphate compounds (Sina 1987) are body parasite insecticides that act by acetylcholinesterase inhibition: diazinon (Spectracide, Ciba-Geigy); Malathion (American Cyanamid; Calmathion, Rhone-Poulenc); dimethoate (Cygon, American Cyanamid); chlorpyrifos (Dursban, Dow) for fire ants, ticks, and fleas; parathion (Bladan, Bayer AG). In addition, dyfonate (N2790, Stauffer Chem.) is a soil parasite insecticide; phosmet (Imidan, Stauffer Chem.) is an animal-body insecticide. Disulfoton (Disyston, Bayer A G) and ronnel (Ectoral, Dow, now discontinued) are used to control ticks. Dichlorvos (Vapona, Shell, now discontinued) has been used by pigeon fanciers to control louse flies, but it is ineffective in the form of pest strips.

Carbamates (Sina 1987), another group of chemical insecticides, inhibit cholinesterase: methomyl (Lannate, Du Pont); propoxur (Baygon, Bayer AG), a roach bait; bendiocarb (Ficam, FBC Ltd.); aldicarb (Temik, Union Carbide); carbofuran (Furadan, FMC); and carbaryl (Sevin, Union Carbide; Dicarbam, BASF).

Pyrethrins and their synthesized products are commonly used insecticides. If applied on birds in excess of label directions they can cause harm.

Other insecticides (Sina 1987) include Amdro and Rotenone. Amdro (American Cyanamid), a hydramethylnon chemical, is used for roach and fire ant control. When used outside and without following directions, the product can be a problem for pigeons. Rotenone (Rotacide, Fairfield), a widely used insecticide, can produce poisoning if used excessively.

FUMIGANTS. These are chemical compounds, such as methyl bromide and ethylene dibromide, that are commercial insecticides. They have been used in grain storage warehouses to kill grain mites and beetles. Residual gas in nonaerated grain has infrequently resulted in poultry and human poisoning. Another fumigant, formaldehyde, which is used in hatcheries, is unlikely to be a problem for the pigeon fancier. Sulfur dioxide gas generated by burning sulfur candles has disinfectant action under closed loft conditions, but the gas can be fatal for both humans and birds. Proper aeration is therefore necessary before birds are returned. Black Leaf 40 vapors, which contain nicotine sulfate and are used for the night control of lice and mites on birds, may be applied with an oil can but never sprayed on perches. The vapors arising from the liquid can be extremely poisonous for people and pigeons in poorly ventilated lofts. Naphthalene or paradichlorobenzene mothballs have been added to nests in an effort to protect pigeons from lice, mites, and louse flies. Unless the white mothballs are enclosed in a small cloth bag, birds may be poisoned when they ingest them as the balls get smaller. Vapona strips have also been tried in lofts to reduce louse fly populations. If the loft is closed, the vapors are toxic for the birds and attendants. If windows are open, the vapors are not effective. Carbon monoxide gas is seldom a cause of poisoning in pigeons; only where faulty or improperly ventilated stoves are used as space heaters does this occur.

RAT POISONS AND MOUSE BAITS. These compounds (Buck 1982) must be carefully distributed in restricted containers to prevent ingestion by pigeons and other animals. Red squill (Rodine) is one of the oldest and best rat poisons, but it is difficult to obtain. It should be mixed in feed without hand contact and left out only at night. Newer rodenticides include alpha naphthylthiourea (ANTU), sodium monofluoracetate, zinc phosphide, and the anticoagulants warfarin, brodifacoum, diphacinone, chlorphacinone, and indanedione.

Thallium, another of the presently used professional rat poisons, results in goutlike signs and acute death in birds. Poisons that have been used for a long time in poison baits are arsenic and strychnine. Sodium arsenate and lead arsenate have also been used in ant or roach baits.

METAL POISONS. These include lead, mercury, and phosphorus. Lead shot or BB's mistaken for seeds may induce lead poisoning in pigeons. Other sources of lead include lead arsenate, lead acetate, and paint oxides and carbonates. These are seldom cause for trouble in pigeons except where pigeons are raised near smelters. Mercurial ointment and organophosphorus compounds have been used as insecticides and have caused poisoning in pigeons.

HERBICIDES. Herbicides (Sina 1987) such as esters of propionic acid and 2,4-dichlorophenoxyacetic acid (2,4D) are produced by Bayer AG, BASF, and Rhone-Poulenc. The triazines, dipyridyls, glyphosphates, and arsenicals all offer the possibility of contact or ingested toxicity. Recently sprayed wet lawns may provide foot exposure. Numerous pigeon poisoning cases have incriminated herbicides, but factual evidence is not presented.

OTHER DRUGS. *Borates,* which include boric acid and sodium borate, are often used for roach control. Uncovered baits could cause poisoning.

Sodium bicarbonate has been given to birds when sour crops were suspected. This drug readily plugs bird kidneys and causes death.

Fertilizers containing sodium or potassium nitrate can also affect kidneys as well as the liver and heart.

Sodium chloride is often a problem in pigeons. Grit may contain in excess of 0.5% salt. Watery droppings increase, but where water is available poisoning seldom occurs.

Plant poisons or phytotoxins provide another means of poisoning for pigeons. Soil selenium may become incorporated in plant grains. Where high soil levels are present, toxic grain levels in excess of 0.5 ppm may occur. Corn cockle seeds (*Argostemma githago*) are highly toxic but very unpalatable, so poisoning is unlikely from this source. Corn cockle weed seed is accidentally harvested with wheat, however, and can be mixed with pigeon grains. Crotolaria seed (*Crotolaria spectabilis*) is harvested with corn or soybeans and produces dropsy and liver failure in birds. Only 10 seeds containing the toxic alkaloid monocrotalin are needed to cause abdominal distention. Toxic coffee bean seeds (*Cassia occidentalis* and *C. obtusifolia*) can also be harvested with corn or soybeans. Rapeseed may accidentally be mixed with pigeon grains and cause liver damage. Sesbania seeds (*Daubentonia longifolia*) and coyotillo (*Karwinskia humboldtiana*), which are grown in Mexico; cacao bean waste (*Theobroma cacao*); castor beans (*Ricinus communis*); and (*Glottidium vesicarium*) seeds are all toxic, but the possibility of pigeon poisoning from these plants is limited. Vetch (*Vicia species*), a poisonous plant grown largely in Idaho and Oregon, may accidentally contaminate pigeon grains. The seed contains a neurotoxic amino acid (Ressler 1962). Ergot, masses of sclerotia from the fungus *Claviceps purpurea,* grows in the seed-bearing heads of rye and other small grains. It produces a hard, black mass containing at least 40 harmful alkaloids. Rye is seldom included in pigeon grain, so this poison is not encountered. Fortunately pigeons have a discriminating taste ability and discard undesirable seeds unless they are quite hungry. This largely limits phytotoxin poisoning.

References

Blakley, B. R. 1982. Can Vet J 23:267–68.

Buck, W. B. 1982. Animal Poison Control Center Rep 1981–82, University of Illinois, Urbana, pp. 1-28.

Kocan, R. M. 1972. Avian Dis 16:714–17.

Ressler, C. 1962. J Biol Chem 237:733–35.

Reuber, H. W., T. A. Rude, and T. A. Jorgenson. 1970. Avian Dis 14:211–18.

Sina, C. 1987. *Chemicals Handbook.* Willoughby, Ohio: Meister.

IV

PARASITIC DISEASES

9

Internal Parasite Infestations

Internal parasites of pigeons include multicellular roundworms, tapeworms and flukes, and unicellular protozoa. This chapter presents the innate nature of each group of parasites and the methods for their prevention and control. Control measures involve good management practices, which include the prevention of fecal contamination of feed and water and the reduction of worm eggs and intermediate hosts in and about the loft and premise.

WORM INFESTATIONS

Worms are small, limbless, elongated invertebrates, usually with a soft, naked body.

Roundworms

Definition. Roundworms, or nematodes, are common multicellular, usually elongated, cylindrical, unsegmented worms. The adults, found in many parts of the body, are among the most important parasites of pigeons.

Classification. The parasites are placed in phylum Nemhelminthes, class Nematoda. Yamaguti (1961) listed 25 families of roundworms in 9 orders as avian parasites. Members of at least 11 families are reported to infest pigeons. These family classifications may change with time. Table 9.1 lists those roundworms reported to affect pigeons.

Nature

LIFE CYCLE. The life cycle of pigeon roundworms may or may not require an invertebrate intermediate host, such as an earthworm, grasshopper, pillbug, or cockroach. Some species of worm eggs passed in pigeon droppings must be ingested by an intermediate host for further development. In this the indirect cycle, the pigeon becomes infected only upon ingestion of infective larvae in an intermediate host. Some roundworms, however, require no intermediate host and the simple ingestion by pigeons of embryonated eggs or larvae in the droppings serves to complete the direct cycle. With proper moisture and temperature, worm eggs embryonate or develop to the infective stage on the ground. The time required for development depends somewhat on the species of roundworm. Within the pigeon, copulation between male and female worms eventually takes place, with the production of eggs. The developmental stage when eggs are deposited varies with the species. Some eggs are unsegmented; others may contain partly developed larvae that hatch in a few days.

Table 9.1. Roundworms reported in pigeons

Family, Genus, and Species	Reported by	Location	Affected Species[a]	Pigeon Incidence[b]	Intermediate Hosts	Body Location
Ascarididae						
Ascaridia columbae[c]	Gmelin (1790)	Indonesia, United States[m]	Do	+ + +	None	Lungs, intestine
A. lineata	Schneider (1866)		C, D, Go, Pa, T	+		Intestine
A. gadrii	Khanum (1976)	Norway		+		Intestine
A. razia	Akhtar (1937)	India		+		Intestine
A. perspicillum	Rudolphi (1803)	Europe, Japan, Hawaii, India	R	+		Intestine
Heterakidae						
Heterakis gallinarum	Schrank (1788)	United States	C, D, Gf, Go, Gr, T, Pa, Ph, Q	+		Intestine
Trichostrongylidae						
Trichostrongylus tenuis[c]	Mehlis (1846)	United States	C, D, Gf, Go, Pa, Ph, T, Q	+ +	None	Intestine
Ornithostrongylus indicus	Deshmukh (1969)	India		+		Intestine
O. oruei	Vassiliades (1969)	Senegal		+		Intestine
O. quadriradiatus[c]	Stevenson (1904)	Cuba, Israel, New Jersey, Washington, D.C., Taiwan, New York[d]	C, D, G, O	+ +	None	Intestine
Amidostomatidae						
Amidostomum anseris[c]	Zeder (1800)	New York, Delaware, Pennsylvania, Washington, D.C.	D, Go	+ +	None	Gizzard
A. skrajabini	Boulenger (1926)	United States	C[n],D	+	None	Gizzard
Epomidiostomum uncinatum[c]	Lundahl (1841)		C[n], D, Go	+	None	Gizzard
Acuariidae						
Dispharynx nastua[c]	Rudolphi (1819)	California, New York, Great Lakes	C, Gf, Gr, Pa, P, Q, T, Ph	+ +	Pillbug, sowbug	Proventriculus
Tetrameridae						
Tetrameres americana[c]	Cram (1927)	California[e,f], France[g], Italy[h], Morocco[i], Portugal[j]	C, D, Gr, Q, T	+ +	Grasshopper, cockroach, pillbug, sowbug	Proventriculus
T. fissispina[c]	Diesing (1861)	Europe, South America, Africa, India[g], Australia, Kansas, Oklahoma[k]	C, D, Gf, Go, Q, T	+ +	Grasshopper, cockroach, earthworm	Proventriculus
Microtetrameres helix[c]	Cram (1927)	North America, Russia		+	Grasshopper	Proventriculus
Thelaziidae						
Oxyspirura mansoni[c]	Cobbold (1879)	United States	C, D, Gf, Gr, P, Pf, Q, T	+	Cockroach	Conjuctiva
Dipetalonematidae						
Paroncocerca alii	Deshmukh (1969)	India		+		Crop

Table 9.1. (*Continued*)

Family, Genus, and Species	Reported by	Location	Affected Species[a]	Pigeon Incidence[b]	Intermediate Hosts	Body Location
Filariidae						
Pelecitus mazzanti	Railliet (1895)	Italy		+		Neck subcutaneous
P. clava	Wedl (1856)	Europe, Asia, Russia, Australia		+		Neck subcutaneous
Microfilaria columbae	Baylis (1939)	Calcutta		+		
Syngamidae						
Syngamus trachea[c]	Montagu (1871)	Germany[l], United States	C, D[n], Gf, Go Ph, Q, T, Pf		Cockroach Earthworm or none	Trachea
Trichuridae						
Capillaria columbae[c]	Rudolphi (1819)	Indonesia, United States[m]	C, Do, T	+ +	None	Intestine
C. caudinflata[c]	Molin (1858)		C, D, Gf, Go, Pa, Ph, Q, T	+ + +	Earthworm	Intestine
C. contorta[c]	Ceplin (1839)	Florida, New York, Virginia	C, D, Gf, Gr, Pa, Ph, Q, T	+ +	None or earthworm	Crop
C. dujardini	Travassos (1915)	Denmark, England	C, Pa, Ph	+		Intestine
C. obsignata[c]	Madsen (1945)	Hertfordshire, United States	C, Gf, Go, Q, T	+ +	None	Intestine
C. philippinensis	Cross (1983)	Philippines		+		Intestine

Source: Pilitt (1986) recorded all species to infest pigeons.
Note: Blanks indicate a lack of information.
[a]C = chickens, Do = doves, D = ducks, Gf = guinea fowl, Go = geese, Gr = grouse, P = passerines, Pa = partridges, Pf = peafowl, Ph = pheasants, Q = quail, R = rats, T = turkeys.
[b]+ = incidental, + + = occasional, + + + = more common.
[c]Wehr (1952, 1959, 1965); Ruff (1984).
[d]Leibovitz (1962).
[e]Mathey et al. (1956).
[f]Raggi and Baker (1957).
[g]Bechade and Bechade (1953).
[h]Romboli (1942).
[i]Santucci et al. (1953).
[j]Alves da Cruz and de Costa (1954).
[k]Ewing et al. (1967).
[l]Gerlach and Heller (1972).
[m]Wehr and Hwang (1964).
[n]Experimental.

FORM. The roundworm is elongated and cylindrical, with a tapered head and tail. The tough outer cuticle may bear *alae* (longitudinal folds) that are confined to the anterior or posterior ends. The latter are prominent on the tail of the male. Roundworms have an alimentary tract with a mouth, esophagus, intestine, rectum, and cloacal or anal opening. They usually have separate sexes and usually reproduce by eggs, which eventually reach the outside. The nervous system consists of a nerve ring around the esophagus. A pair of amphid sensory receptors appear on the anterior end and a pair of phasmids may be situated behind the anus. There is no circulatory system, and respiration occurs through the cuticle covering the worm. The excretory system usually consists of a cylindrical cavity and an excretory pore (Chandler 1949).

Distribution

GEOGRAPHY. Roundworms are found wherever pigeons are raised.

INCIDENCE. Roundworm incidence varies with the species and geographic location.

SEASON. For those species requiring an intermediate host, warm weather favors roundworm transmission.

AGE, SEX, AND BREED. Young birds appear to have less resistance and are more seriously affected, but all ages can become infected. *Capillaria* sp. appear more often in older birds.

Roundworm incidence does not appear to be altered by sex, breed, or strain of bird.

SPECIES. Each roundworm species infests specific definitive hosts (Table 9.1).

CARRIER. Mildly roundworm-infested pigeons often serve as subclinical carriers.

INTERMEDIATE HOSTS. Intermediate hosts are essential for the transmission of several roundworms. *Tetrameres* (Cram 1931 a, b; Young 1981) requires either a pill bug (*Armadillidium vulgare*), sowbug (*Porcellio scaber*), cockroach (*Blatella germanica*), or grasshopper (*Melanoplus femurrubrum* or *M. differentialis*). *Dispharynx* (Cram 1931a) requires pillbugs or sowbugs. *Oxyspirura* (Wehr 1952, 1959, 1965) requires a cockroach *Pycnoscelus (Leucophaea) surinamensis*. *Capillaria* (Wehr 1936) may involve the earthworm (*Eisenia foetidus* or *Allolobophora caliginosa*); this depends on the species of threadworm. *Syngamus* may or may not include earthworms and cockroaches (Hwang 1961).

Transmission.

Direct transmission, the ingestion of embryonated eggs, introduces the parasite to the host. With indirect transmission, an invertebrate intermediate host must be involved.

Epidemiology

DISTRIBUTION IN BODY. Specific species of pigeon roundworms locate themselves in the crop, esophagus, proventriculus, gizzard, intestine, liver, lungs, eye, neck, and trachea.

ENTRY AND EXIT. Eggs or larvae are ingested by the pigeon, and eggs leave in the droppings.

PATHOGENICITY. Infection is often a factor of exposure. The ingestion of a single embryonated egg may not cause infestation. Also, both male and female worms are essential for reproduction.

IMMUNITY. Ascarid resistance appears to be related to the amount of mucus produced by goblet cells of the intestine and to the intake of vitamin A.

Incubation Period.

The period of time from the ingestion of an infectious roundworm egg or an intermediate host until eggs are produced by the parasite varies with the worm species. Less than 2 wk or up to 1 mo or more may be required.

Course.

Worms do not normally terminate their existence within the host's body without the use of a vermifuge; thus the course is of indefinite duration unless treatment is given.

Morbidity.

The number of roundworm-infected birds in a loft depends on the exposure. If sanitation is poor and parasites are introduced, the egg numbers multiple rapidly, increasing the infection rate.

Mortality.

Ornithostrongylus quadriradiatus and *Dispharynx nastua* appear to be the most serious roundworms of pigeons, but other species are also important.

Signs.

Heavily roundworm-infected birds lose appetite and become unthrifty, thin, anemic, and poorly feathered. They are reluctant to move or fly, and when forced to move they move slowly and with an unsteady gait. If forced to fly they are easily exhausted. On the other hand, a few worms may go unnoticed. Tracheal worms, which are uncommon, produce only mild respiratory difficulty in experimentally infected pigeons (Wehr 1959).

Necropsy.

Numerous *Capillaria,* (threadworms) produce marked thickening and inflammation of either the crop or intestinal mucosa, depending on the species present. Ascarids and other intestinal worms are usually accompanied by copious amounts of catarrhal mucus in the intestine. *Ascaridia columbae* second-stage larvae often migrate to the liver and lungs but do not develop (Wehr and Hwang 1964). Such larvae produce little damage until they die. Stomach or gizzard worms are embedded in the respective tissues. Their presence creates considerable inflammation and necrosis. Blackened hemorrhagic areas of the gizzard lining may become loosened and often slough in places. Dark, round *Tetrameres* may be so numerous in the glands of the wall of the proventriculus as to virtually block the stomach lumen. *Dispharynx* worms also bury their heads in the mucosa of the proventriculus with the formation of ulcers and extensive thickening. *O. quadriradiatus* is a serious bloodsucker in the small intestine and frequently causes hemorrhage and catarrhal enteritis. Eye worms usually cause pasty eyelids with the accumulation of inflammatory cheesy exudate. The nictitating membrane swells and is frequently moved in an effort to remove the irritation under the conjunctiva. The eyes themselves become reddened and watery in early stages.

Micropathology. Parasitism usually increases the numbers of eosinophiles in a blood smear. Localized tissue destruction with hemorrhage may occur wherever parasites are found.

Diagnosis. The presence of roundworm eggs in the droppings can be established by a centrifuged fecal sample mixed and suspended in a concentrated sugar solution. The type of egg or worm must be identified to establish a diagnosis.

Biological Properties. Worm eggs are quite stable. They can overwinter in the soil. Earthworms, as mechanical hosts, can ingest worm eggs and carry them below the freezing level, protecting them until spring.

Prevention. Breeding birds and their progeny should never be housed in a dirty loft. Routine cleaning and sanitation is essential for the health of all birds. Droppings should be removed routinely and buried under 2 ft of soil or removed from the premises. Good sanitation interrupts the breeding potential of intermediate hosts that reproduce in manure and debris. Low-growing shade also harbors earthworms, grasshoppers, pillbugs, sowbugs, and other insects.

A 2-inch-mesh porch floor or exercise yard may be used to exclude the birds from dirt or soil where worm eggs accumulate. Only methyl bromide fumigation has been successful in killing worms and eggs in the soil (Clapham 1950). Annual removal of the soil in exercise yards and its replacement with fresh sand will accomplish the same result.

In addition to sanitation, insecticides should be employed to curtail intermediate hosts. Carbaryl, (Sevin, Union Carbide) or Malathion (American Cyanamid) 1% may be sprayed weekly in and about the premises. Care must be taken to avoid inhalation of any insecticide. Chloropyrifos (Dursban, Dow) 0.5% spray is an effective 3-wk residual premises insecticide. It must not be applied on the birds.

Control. Roundworm control measures encompass prevention and treatment methods. Treatment should be provided only when an examination of droppings reveals the presence of worm eggs. When treatment is indicated, the dose must be accurate and should be measured. Routine use of any vermifuge without good reason is harmful to birds.

The actual expulsion of worms by a worm treatment does not necessarily kill worms or their eggs. Therefore, unless droppings are removed after treatment, it may simply spread worms to the other birds. Worm eggs are difficult to destroy, but they can be killed with lye at the rate of 1 lb/5 gal water. Lye must never be sprayed, and extreme caution must be used in its application. It must not be used when birds are in the loft, as it will burn feet and kill birds. Rubber boots and gloves are suggested for anyone working with lye. Fresh water should be used to flush away the lye 24 hr after its application.

Treatment. The type and method of treatment used to remove roundworms depends on the type of parasite. Common intestinal worms may be removed with piperazine hydrochloride or citrate (anhydrous) tablets at the rate of about 60 mg/lb body weight. It should not be given to birds under 3 wk of age and it need not be repeated unless fecal exams reveal continued infestation. Whitney (1957) gave pigeons 8 g/gal water for 3 days. In chickens piperazine has been used in the feed at a rate of 0.1–0.4% for 1 day or in the water at 0.1–0.2% and repeated at 30-day intervals where necessary. Vianello and Vicenzoni (1955) gave piperazine citrate to chickens at 300–400 mg/kg body weight with success.

Capillaria may be removed with L-levamisole hydrochloride (tetramisole, Tramisol, American Cyanamid), at the rate of 1 g/gal water for 1 day only. Ripercol, DL-tetramisole, sold in Europe, has about half the potency of L-levamisole (Alford 1986). It should not be repeated under 3 wk and should not be given to birds raised for human food within 7 days of slaughter. Higher levels may kill or harm pigeons. Schock and Cooper (1978) recommended levamisole at the rate of 33 mg/kg body weight. Janssen Laboratories sells Spartakon, a 20-mg Ripercol tablet for pigeons.

The dose is one tablet per adult bird. One-half tablet should be sufficient for small birds. Thienpoint et al. (1966) and Bruynooghe et al. (1968) removed *C. obsignata* from chickens with DL-tetramisole at 40 mg/kg body weight. Later Kates et al. (1969) used L-tetramisole at 30 mg/kg body weight in turkeys for *C. obsignata.*

Coumaphos (Meldane, Chemagro) at 40 ppm in feed and Hygromycin B (Elanco) were approved for control of capillarids in chickens, but Hygromycin B must not be used in pigeons. Coumaphos was reported to have activity against *C. obsignata* by Eleazer (1969). Hendricks (1962, 1963) indicated that methyridine was effective against *C. obsignata* in chickens. Norton and Joyner (1965) injected methyridine 100–150 mg/kg subcutaneously (SC) with good results against larvae and adults. Wehr et al. (1967) reported the SC injection of a 5% aqueous solution of the same drug in the leg of a pigeon for the control of *C. obsignata.* Injections of 25–45 mg/bird were 99–100% effective, but great care must be taken to avoid skin contamination, which causes tissue necrosis. Clarke (1962) used individual doses of 25 and 50 mg haloxon/kg body weight in chickens for capillary worms. Norton and Joyner (1965) gave 50–60 mg/kg but found haloxon to be less effective against larvae.

Panigrahy et al. (1982) suggested levamisole at the rate of 33 mg/kg body weight for the control of *Tetrameres.* Young (1981) used 5 mg mebendazole/bird morning and night for 3 days with success. Gizzard worms should also respond to this dosage level.

Ruff (1984) quotes Enigk and Dey-Hazra (1971a, b) in the treatment of *Trichostrongylus tenuis.* Cambendazole (Merck, discontinued) at 30 mg/kg, pyrantel tartrate (Strongid T, Pfizer) at 50 mg/kg, thiabendazole (TBZ, Merck) at 75 mg/kg, and citarin at 40 mg/kg were effective. Also mebendazole (Telmin, Pitman-Moore) at 10 mg/kg for 3 days was very effective (Enigk et al. 1973). Telmin, which is not cleared for poultry in the United States, has been used at 400 g/t for 14 days. *D. nasuta* was not completely controlled by mebendazole (Ruff 1984).

Amidostomum anseris adults and larvae were controlled in geese by cambendazole at 60 mg/kg (Enigk and Dey-Hazra 1971a) and Enigk et al. (1973) found mebendazole at 10 mg/kg for 3 days very effective.

Eye worms may be removed manually from beneath the nictitating membrane of the eyelid. A local anesthetic should be employed as suggested by Saunders (1929).

Syngamus worm adults and larvae, which are not common in pigeons, can be treated with one dose of thiabendazole at 0.3–1.5 g/kg body weight, which is the dose for chickens and turkeys (Horton-Smith et al. 1963). Third-stage larvae are unaffected. The turkey feed dosage level is 0.5% (Wehr and Hwang 1964), and this is 36.7 g/100 lb feed using a 62% sheep wormer (Melancon 1986). Barium antimonyl tartrate dust has been effective in treating pheasant tracheal worms. Exposure to the dust kills the worms in the trachea. If worms are killed too fast, the dead worms may plug the trachea and kill the bird. The quantity of dust and the duration of exposure in a closed container must be controlled.

For *O. quadriradiatus* in pigeons Leibovitz (1962) used a 75-mg suspension of thiabendazole for 4 days.

References

Alford, B. 1986. Personal communication, American Cyanamid.

Alves da Cruz, and A. Rodrigues de Costa. 1954. Dir Jeral Serv Pecu B Pecu 22:3.

Bechade, R., and A. M. Bechade. 1953. Rec Med Vet 129:645–50.

Bruynooghe, D., D. Thienpoint, and O. F. Vanparijs. 1968. Vet Rec 82:701–6.

Chandler, A. C. 1949. *Introduction to Parasitology.* 8th ed. New York: Wiley.

Clapham, P. A. 1950. J Helminthol 24:137.

Clarke, M. L. 1962. Vet Rec 74:1431–32.

Cram, E. B. 1931a. USDA Tech Bull 227:1–27.

______. 1931b. J Parasitol 18:52.

Eleazer, T. H. 1969. Avian Dis 14:228–30.

Enigk, K., and A. Dey-Hazra. 1971a. Deutsch Tieraerztl Wochenschr 78:178–81.

______. 1971b. Berl Muench Tieraerztl Wochenschr 84:11–14.

Enigk, K., A. Dey-Hazra, and J. Batke. 1973. Avian Pathol 2:67–74.

Ewing, S. A., J. L. West, and A. L. Malle. 1967. Avian Dis 11:407–12.

Gerlach, H., and G. Heller. 1972. Kleinrierpraxis 17:247.

Hendericks, J. 1962. Tijdschr Diergeneeskd 87:314–22.

______. 1963. Tijdschr Diergeneeskd 88:418–24.

Horton-Smith, C., P. L. Long, and J. G. Rowell. 1963. Br Poult Sci 4:217–24.

Hwang, J. C. 1961. J Parasitol 47:20.

Kates, K. C., M. L. Colgalazier, and F. D. Enzie. 1969. Trans Am Microsc Soc 88:142–48.

Leibovitz, L. 1962. Avian Dis 6:380.

Mathey, W. J., H. E. Adler, and P. J. Siddle. 1956. Am J Vet Res 17:521.

Melancon, J. 1986. Personal communication, Merck Research, Minneapolis, Minn.

Norton, C. C., and L. P. Joyner. 1965. J Comp Pathol 75:137–45.

Panigrahy, B., et al. 1982. J Am Vet Med Assoc 181:384–86.

Raggi, L. G., and N. F. Baker. 1957. Avian Dis 1:227–34.

Romboli, B. 1942. Nuova Vet 20:228.
Ruff, M. D. 1984. *Diseases of Poultry,* 8th ed., ed. M. S. Hofstad et al., pp. 614-48. Ames: Iowa State University Press.
Santucci, J., R. Sendral, and J. Haag. 1953. Rev Med Vet 104:335.
Saunders, D. A. 1929. Fla Agric Exp Stn Bull 206:565.
Schock, R. C., and R. Cooper. 1978. Mod Vet Pract 59:439-43.
Thienpoint, D., et al. 1966. Nature 209:1084-86.
Vianello, G., and V. Vicenzoni. 1955. Clin Vet 78:365.
Wehr, E. E. 1936. N Am Vet 17:18-20.
______. 1952. In *Diseases of Poultry,* 3d ed., ed. H. E. Biester and L. H. Schwarte, pp. 835-85. Ames: Iowa State University Press.
______. 1959. In *Diseases of Poultry,* 4th ed., ed. H. E. Biester, pp. 741-81. Ames: Iowa State University Press.
______. 1965. In *Diseases of Poultry,* 5th ed., ed. H. E. Biester and L. H. Schwarte, pp. 965-1005. Ames: Iowa State University Press.
Wehr, E. E., and J. C. Hwang. 1964. J Parasitol 50:131-37.
______. 1967. Avian Dis 11:44-48.
Wehr, E. E., et al. 1967. Avian Dis 11:322-26.
Whitney, L. F. 1957. Vet Bull 27:475.
Yamaguti, S. 1961. *Systema Helminthum.* Vol 3. New York: Interscience.
Young, R. A. 1981. Vet Med Small Anim Clin 76:426-27.

Tapeworms

Definition. Tapeworms, or cestodes, are long, flat, white, usually ribbonlike, segmented worms that vary greatly in size. The adults are primarily found in the host's intestines.

Classification. Pigeon tapeworms are placed in phylum Platyhelminthes, class Cestoidea, subclass Cestodaria (no scolex) with order Pseudophyllidea and subclass Eucestoda (with scolex) with order Cyclophyllidea. The majority of pigeon genera are in the second order. Table 9.2 gives the families, genera, and species reported to infest pigeons.

Nature

LIFE CYCLE. The life cycle of every *Eucestoda* pigeon tapeworm in which the cycle is known requires an invertebrate as an intermediate host (IH) (Wehr 1965). To complete the cycle, tapeworm segments or tapeworm eggs must be eaten by such invertebrates as beetles, ants, flies, grasshoppers, slugs, earthworms, and snails. Each tapeworm requires specific IHs for transmission, the kind of tapeworm that a pigeon gets is thus directly related to the IHs eaten by the pigeon. The tapeworm segments or eggs are passed in the droppings of the infected pigeon and available for ingestion by the invertebrate IH.

Spherical oncosphere (from *onkos*–hook) embryos develop within the tapeworm egg (Fig. 9.1). *Eucestoda* spp. develop embryos that bear three pairs of clawlike hooks. Following ingestion by a proper arthropod IH these eggs hatch in the digestive tract of the IH, forming small, round cysticercoid larvae (Fig. 9.2). These penetrate the intestinal wall and migrate to a suitable location where they form bladderlike cysts. The process is completed in about 2 wk. Cysts remain viable and infective in the IH for many months. When the infective IH is eaten by the pigeon, each cystic tapeworm cysticercus is released and the invaginated scolex with its hooks is thrust out to attach in the intestine of the pigeon. Segment growth then commences.

The life cycle varies with the species. *Diphyllobothrium mansoni* is the only member of the Pseudophyllidea order reported to infest pigeons. Dogs and cats commonly pass the tapeworm eggs in the feces and these develop in 18 days. Freely swimming ciliated coracidia are released from the eggs to be ingested by a cyclops, a freshwater crustacean, in which procercoid larvae develop. When pigeons, fowl, pigs, rats, fish, frogs, or snakes drink the infested cyclops in water, plerocercoid larvae form in the IH. These may be distributed throughout the body. Dogs or cats eat the pigeon or other IH and develop the intestinal tapeworms. The pigeon does not develop the adult tapeworm (Hegner et al. 1929).

Form. *Eucestoda* tapeworms are composed of a head, neck, and variable number of segments. The number of segments varies with the species. The *scolex* (head) is an organ of attachment. It bears four cup-shaped suckers that encircle the *rostellum* (a retractile crown). Rostellar hooks may or may not be present and spines may arm the suckers. The size, shape, and number varies with the species. The neck is the short, narrow, unsegmented section joining the scolex with the segmented proglottids. In most species it is from this area that new growth or *strobila* (a chain of segments) develops. Maturing segments contain male and or female reproductive organs. The female system includes an ovary, oviduct, uterus, shell gland, yolk glands, vagina, and fertilization canal. *Cyclophyllidea* spp. have no uterine pore and eggs remain within the uterus until the segment is shed. The older terminal gravid segments break off and are shed in the droppings of infested birds. Male structures include testes, vas efferentia, vas deferens, cirrus, genital pore, and seminal receptacle.

Table 9.2. Tapeworms reported in pigeons

Order, Family, Genus, and Species[a]	Reported by	Location	Affected Species[b]	Intermediate Host
Order Cyclophyllidea (with scolex)				
Davaineidae				
Raillietina (R) nagpurensis	Sprehn (1932)	India		
R. (R) joyeuxi	Lopez-Neyra (1943)	Spain		
R. kantipura	Sharma (1943)	Nepal		
R. (R) korkei	Southwell (1939)	Boma		
R. microcantha		Europe, Africa		
	Fuhrmann (1908)	Russia, Granada		
R. nripendra	Sharma (1943)	Nepal		
R. taiwanensis	Yamaguti (1935)	Taiwan		
R. (sic) insignis	McCulloch (1939)	Nigeria		
R. (skrajabinia) bonini	Megnin (1899)	Germany		
R. clerci	Fuhrmann (1932)	Africa, Asia		
R. echinobothrida +[c]	Megnin (1881)	Worldwide	C, T	Land snail, ant
R. (polychalix) johri	Ortlepp (1938)	India		
R. volzi	Fuhrmann (1905)	Sumatra		
R. weissi valliclusa	Joyeux (1934)	France		
R. (R) torquata	Fuhrmann (1932)	India, Burma, Nepal		
R. tokyoensis	Sawada (1960)	Tokyo		
R. (R) tetragona[c]	Molin (1858)	Hertfordshire	C, T, Gf, Q, Pf	Ant, housefly
R. (R) quadritesticulata	Sprehn (1932)	India		
R. (R) spiralis	Singh (1960)	India		
R. (R) laminata	Hilmy (1936)	Burma		
Cotugnia fleari	Mughal (1967)			
C. magna	Burt (1940)	Ceylon		
C. aurangabodensis	Shinde (1969)	India		
C. polyacantha	Fuhrmann (1909)	Cairo		
Davainea crassula	Clerc (1910)	England, Urals		
D. microcantha	Skrjabin (1914)			
Hymenolepididae				
Hymenolepis columbae	Zeder (1800)	Europe, Africa		
H. clausa	Hughes (1940)	Oklahoma		
H. macracanthos	Sharma (1943)	Nepal		
H. rugosus	Clerc (1906)	Urals, Nepal		
H. serrata	Fuhrmann (1932)			
H. sphenocephala	Rudolphi (1810)	Europe		
Mayhewia nebraskensis	Rolan (1969)	Nebraska		
Anoplocephalidae				
Aporina delafondi +[d,e]	Railliet (1892)	Iowa, Texas, Pennsylvania, District of Columbia		
Nepalesia joodhari	Sharma (1943)	Nepal		
Dilepididae				
Choanotaenia infundibulum	Block (1779)		C, T	Housefly, beetle, termites, grasshopper
Parachoanotaenia columbi	Borgarenko (1976)			
Amoebotaenia cuneata	Linstow (1872)	Egypt, Rangoon	C, T	Earthworm
Taeniidae				
Coenurus sp.	Johri (1954)	India		
Taenia cordifer	Bosc (1802)	India		
Order Pseudophyllidea (no scolex)				
Diphyllobothriidae				
Diphyllobothrium mansoni	Joyeux (1934)	United States, Japan, Africa China	Dog, cat	Cyclops
Uncertain classification				
Epomidiostomum uncinatum				
Columbia allahabodi	Srivastara (1966)	India		
Bothreocephalus columbarum	Bertkaw (1896)	Burma		

Source: Pilitt (1986) recorded all species to infest pigeons.
Note: Blanks indicate a lack of information.
[a]+ = primary pigeon species.
[b]C = chickens, D = ducks, Gf = guinea fowl, Go = geese, Pf = peafowl, Q = quail, T = turkeys.
[c]Graham (1937).
[d]Becklund (1964).
[e]Wehr 1965.

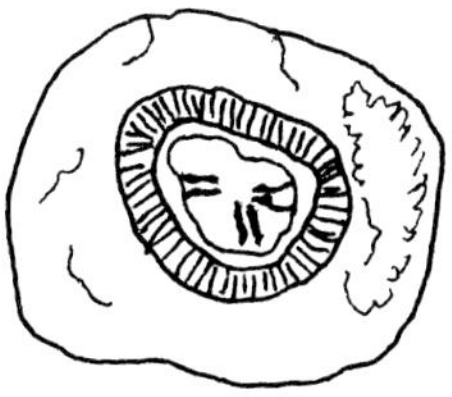

Fig. 9.1. Tapeworm egg. The egg has a thick wall and three sets of hooks.

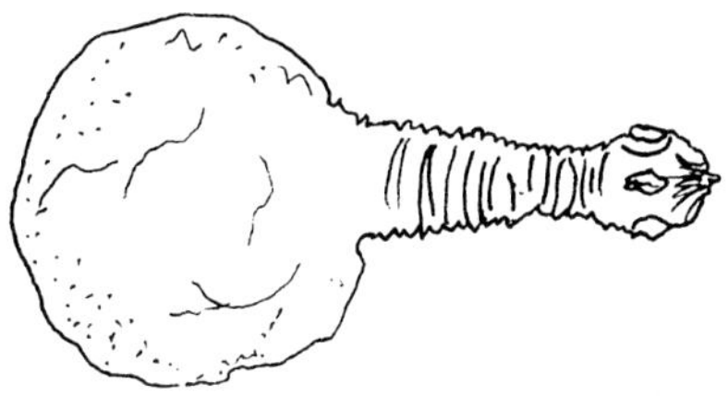

Fig. 9.2. Tapeworm evaginated cystericoid.

Tapeworms do not have a digestive tract or alimentary canal. Food is absorbed through the body wall. The excretory system consists of a pair of lateral longitudinal tubes that are connected in each segment to flame cells and a transverse tube near the posterior margin of the proglottid. Nerve trunks and musculature connect the segments.

Bird *Cestodaria* do not form a chain of distinct segments. They have a head with two *bothria* (lateral sucking grooves) but no true scolex. A set of reproductive organs, male and female, share a genital opening. Eggs are discharged from the uterine pore to the intestinal lumen of the host.

See Chandler (1949) for a discussion of tapeworm form.

Distribution

GEOGRAPHY. Pigeon tapeworms are found wherever pigeons are raised.

INCIDENCE. *Eucestoda* tapeworms are uncommon where the IHs are well controlled.

SEASON. Tapeworm transmission occurs more readily in warm weather and in climates or areas where IHs or cyclops are prevalent.

AGE, SEX, AND BREED. All ages can be infested with tapeworms, but older chickens develop some resistance and tolerate moderate infestation without outward signs. This may also be true of pigeons. Both sexes and all breeds become infested equally.

SPECIES. Each *Eucestoda* tapeworm species infests specific definitive hosts, including animals and birds.

CARRIERS. Pigeons may remain infested for prolonged periods without clinical evidence of infestation.

TRANSMISSION. An IH is essential for completion of the life cycle.

Epidemiology

DISTRIBUTION IN BODY. Adult *Eucestoda* tapeworms are found in the lumen of the intestine and digestive tract. The pigeon does not develop *Cestodaria* tapeworms. *Plerocercoid* larvae may develop in the flesh of the pigeon. If a dog or cat eats the infested pigeon, the dog or cat gets the tapeworm. The pigeon is an IH.

ENTRY AND EXIT. *Eucestoda* tapeworm eggs are passed in the pigeon feces but are largely confined within the gravid tapeworm segments in *Cyclophyllidea* species. The free tapeworm eggs or gravid segments are ingested by the IH along with its food. In *Eucestoda* spp., larval cysticercoid tapeworms enter the pigeon in the infested IH as the IH is ingested. *Pseudophyllidea* spp., described previously, require a procercoid in a cyclops and pleocercoid in an IH for transmission to occur.

PATHOGENICITY. Pigeons usually become infested with *Eucestoda* by the ingestion of an IH that contains a cysticercoid. The release of a few such cysts produces little or no gross change in the intestine, but severity of symptoms appears to be directly related to numbers of parasites since tapeworms do not suck blood.

IMMUNITY. Resistance to tapeworm infection can be demonstrated as age increases.

Incubation. In *Eucestoda,* the cysticercoid is released when the IH is ingested by the pigeon. The procercoid in *Pseudophyllidea* is released in the intestine shortly after ingestion of the cyclops by the pigeon. The development interval may be considered the incubation period.

Course. Tapeworm scolices seldom become dislodged without chemical vermifuge action. The bodies of the worms may break off, but the scolex remains in the intestinal wall where it grows additional proglottids; thus the course is prolonged.

Morbidity and Mortality. Sickness is evident only when large numbers of worms are present, and birds seldom die from tapeworm infestation alone.

Signs. Heavily tapeworm-infested birds may become unthrifty and light in weight. Squabs may be retarded and pale.

Necropsy. Large numbers of *Eucestoda* tapeworms produce intestinal irritation and inflammation with catarrhal enteritis. Some atrophy of the muscles may also be noted.

Diagnosis. The gross examination of fresh droppings is an unreliable method of diagnosis unless proglottids are found. Worm segments and eggs may not be observed because they are intermittently present in the droppings, eggs are too small to see without magnification, and eggs may be enclosed in the proglottid case, which may not be observed. A necropsy may reveal the segmented worm in the intestine. Certainly this diagnostic method is reliable. It also enables the identification of *R. echinobothrida,* the head of which produces up to pea-sized nodules in the wall of the intestine and has been confused with tuberculosis. It is one of the longest tapeworms (34 cm long by 4 mm wide) (Reid 1984).

Biological Properties. Eggs are quite stable and withstand most chemical treatments, but drying eventually reduces viable numbers. Self-sterilization takes place over a period of time in stored manure (Wehr 1965).

Prevention. Pigeons raised under natural conditions easily come in contact with intermediate hosts, many of which may serve to transmit a single tapeworm species. To prevent the ingestion of an IH is virtually impossible. The aim of prevention is to reduce the numbers of IHs to which the pigeon is exposed. Insecticides applied in or on the manure and about the loft are helpful. Shade or protection produced by low-growing weeds and brush and by accumulated trash and debris should be removed. The area below a loft building should be enclosed to exclude pigeons. The floor of the flight area should be wired to permit droppings to pass through but prevent pigeon contact with the IH in the manure. Body waste should be removed routinely and either buried deeply or carefully removed in covered containers.

Treatment. Although not a permanent solution to the problem, starvation of birds for 24 hr increases peristaltic activity and shears tapeworm body segments. It does not remove heads or scolices; thus tapeworm segments will continue to grow and the worms will remain.

Information pertaining to the chemical removal of pigeon tapeworms has not been recently reported and drugs suggested for poultry have not been entirely satisfactory. Butynorate (dibutyltin dilaurate, Tinostat, Salisbury) has USDA approval for use on poultry but does not have clearance for pigeons.

Other products that may be tried experimentally include niclosamide (Yomesan, Chemagro) (Boisvenue and Hendrix 1965), mebendazole (Telmin, Pitman-Moore) (Matta 1980), bunamidine hydrochloride (Scolaban, Wellcome), and praziquantel (Droncit, Miles). Manufacturer's directions must be evaluated by a veterinarian to determine possible experimental oral treatment levels.

For the control of intermediate hosts, spraying or dusting with an insecticide is recommended. Carbaryl (Sevin, Union Carbide) 3–5% dust or Malathion (American Cyanamid) 4% dust, or 1% spray with either product is effective. Caution must be employed in the application of any insecticide, avoiding skin and eye contact and lung inhalation.

References

Becklund, W. W. 1964. Am J Vet Res 25:1380–1416.

Boisvenue, R. J., and J. C. Hendrix. 1965. J Parasitol 51:519–22.

Chandler, A. C. 1949. *Introduction to Parasitology.* 8th ed. New York: Wiley.

Graham, R., et al. 1937. Ill Agric Exp Stn Circ 469, p. 50.

Hegner, R., F. M. Root, and D. L. Augustine. 1929. *Animal Parasitology.* New York: Appleton-Century, pp. 305–7.

Matta, S. C. 1980. Indian J Poult Sci 15:207–10.

Pilitt, P. A. 1986. Personal communication, USDA Agricultural Research Service.

Reid, W. M. 1984. In *Diseases of Poultry,* 8th ed., ed. M. S. Hofstad, pp. 649–67. Ames: Iowa State University Press.

Wehr, E. E. 1965. In *Diseases of Poultry,* 5th ed., ed. H. E. Biester and L. H. Schwarte, pp. 965–1005. Ames: Iowa State University Press.

Flukes

Definition. Flukes, or trematodes, of pigeons are short, flat, unsegmented, leaflike, occasionally cylindrical worms that are parasitic internally and bear adhesive suckers.

Classification. The parasite is placed in phylum Platyhelminthes, class Trematoda, order Digenea, and two suborders, Gasterostomata and Prosostomata (Dawes 1946). Table 9.3 lists the families, genera, and species reported to affect pigeons.

Nature

LIFE CYCLE. The life cycle of digenetic flukes, which are parasites of pigeons, involves two or more asexual generations and an alternation of hosts. Thousands of eggs pass from infected pigeons in the droppings. A ciliated embryo (miracidium) develops within each egg, and these hatch in water or in the intestines of a particular mollusk or snail, which is the intermediate host. Freely swimming miracidia must quickly find and enter a specific snail or die.

Once within the snail the miracidium transforms into an irregular, saclike mother sporocyst. After 1 or 2 wk of development, the mother sporocyst in the snail produces one or more generations of rediae by asexual reproduction.

After further growth each rediae produces freely swimming cercariae. In some species of flukes the mother sporocyst produces daughter sporocysts, which also produce cercariae. One miracidia may result in thousands of cercariae.

Cercariae become metacercariae when they encyst. Most cercariae have tails, which they lose when they encyst. Encystment occurs within a second intermediate host or on submerged objects in the water. After a period of encystment the metacercaria becomes infective for the final definitive host, which, in this case, is the pigeon. In order to complete their development, larval cercariae must reach the definitive host, such as the pigeon, either directly or indirectly before sexual reproduction with the production of eggs can occur (Kingston 1984).

FORM. The flattened pouchlike body is secured in the host by an oral sucker and in most flukes by a second ventral sucker. The digestive and excretory systems are branched to carry food to all parts of the body and to carry waste products out. The mouth is located at the bottom of an oral sucker, usually at the anterior end of the body. The muscular pharynx near the mouth branches into two blind cecal sacs with no anus. The excretory system has a number of branched tubules extending posteriorly that collect fluid from flame cells. The fluid from two collecting tubules empty into a bladder and thence to the excretory pore at the posterior end of the fluke. The nervous system consists of a small ganglion near the pharynx and a few nerve branches. The reproductive process is specialized and complex. All pigeon flukes, except those in the family Schistosomatidae, are hermaphroditic, that is, they have both male and female reproductive systems. Female glands include a vitelline gland and duct, an ootype surrounded by Mehlis' gland, a short oviduct leading to the ovary, an oviduct, and a Laurer's canal or vestigial vagina, which may be connected to the seminal receptacle. The male system has two or more testes, a vas efferens and vas deferens, and a cirrus and pouch. Mature fertilized eggs, zygotes, move along the uterus to the genital sinus from the ootype where fertilization occurs. The male duct joins the female system at the genital pore. Fig. 9.3 presents an outline of the anatomy of a fluke.

See Chandler (1949) for a discussion of fluke form.

Distribution

GEOGRAPHY. Flukes are found in pigeons throughout the world. Krause (1925) collected 5000 *E. revoltum* from 8 pigeons in Rostock and Van Heelsbergen (1927) in Holland counted 1550 in 1 pigeon. Bolle (1925) and Zunker (1925) also observed heavy infections. Beaver (1937) in Illinois experimentally infected pigeons with the same species.

INCIDENCE. Uncommon in the United States, flukes occur where marshy areas and the proper snail, tadpole, or fish are present. *Echinostoma, Hypoderaeum,* and *Tanaisia* species are among the most common.

SEASON. Warm weather is conducive to the completion of the fluke life cycle.

AGE, SEX, AND BREED. Any age, sex, or breed can be affected by flukes.

Transmission. Parasitism is totally dependent on the presence of water and the appropriate first intermediate molluscan host. These are usually gastropod snails. In some flukes second intermediate hosts are required, which include insects, snails, tadpoles, and fish.

Epidemiology

DISTRIBUTION IN BODY. Pigeon flukes have been found in the intestine, liver, portal veins, pancreas, bile duct, kidney tubules, and eyes.

ENTRY AND EXIT. Flukes gain entrance by mouth following ingestion by the pigeon of encysted metacercariae. Cysts may be formed on any solid object or in a second intermediate host.

Eggs passed by mature flukes in the droppings of pigeons embryonate under favorable environmental conditions of warmth and moisture to become infectious ciliated embryos. In some instances when the egg is laid it contains a fully formed miracidium, which may be released when it reaches water. In other fluke species it is released when ingested by a snail host.

PATHOGENICITY. The flukes are much less important than roundworms, but when numerous flukes are involved, serious losses can occur. The damage produced is proportional to the number of flukes within an individual bird, partly because flukes suck blood.

Each species of fluke must find specific types of intermediate hosts and specific definitive hosts to complete the cycle. Each stage requires time for development before it becomes infectious.

Incubation. Each species requires time for each stage. The complete fluke life cycle may take several weeks or months.

Course. The course is indefinite when only a few flukes are present, but large numbers of flukes can cause death within a few days.

Table 9.3 Flukes reported in pigeons

Family, Genus, and Species	Reported by	Location	Affected Species[a]	Pigeon Incidence[b]	Intermediate Hosts	Body Location
Echinostomatidae						
Echinoparyphium aconiatum	Dietz (1909)					Intestine
E. flexum	Malek (1977)	United States				Intestine
E. paraulum	Dietz (1909)	Germany, Russia				
E. recurvatum[c]	Linstow (1873)	Europe, Asia Africa, North America	C, D	+ +	Snail Tadpole	Intestine
E. schulzi	Matevosian (1938)	Bushkiria				
Echinostoma erraticum	Lutz (1924)	Brazil				
E. exile	Lutz (1924)	Brazil				
E. nudicatudatum	Nasir (1960)					
E. revolutum[c,d,e]	Froelich (1802)	Nepal, Europe, Philippines	C, D, Go, M, T, S	+ +	Snail, fish, *planaria*	Intestine, cloaca
Hypoderaeum conoideum[c,d,e,f]	Bittner (1927)	Europe, Japan, Siberia	C, D, Go,	+ + +	Snail, tadpole	Intestine
Brachylaemidae						
Postharmostomum gallinum[c,e]	Witenberg (1923)	Europe, Asia, Africa, Hawaii, Puerto Rico	C, Gf, T	+ +	Land snail	Cecal area
P. commutatum	Dekmans (1929)	Puerto Rico				
Glaphyrostomum indicum	Mukherjee (1964)	India				
Brachylaema degiustii	Nasir (1966)					Intestine
B. fuscatum[c]	Rudolphi (1819)	Europe, Asia Africa, North America, England	C, D	+ +		Esophagus Intestine
B. nicolli		Kirghizia				Intestine
B. mazzantii	Travassos (1927)	Brazil				
Heterophyidae						
Cryptocotyle concavum[c]	Creplin (1825)	Europe, North America	C, D, M, T	+ +	Snail, fish	Intestine
Apophallus muhlingi	Jagerskiold (1899)	Europe	Dog, cat			
Pygidiopsis genata	Ciurea (1934)	China, Israel	Dog, cat			Intestine
	Denmark, Egypt					
Rossicotrema donicum	Ciurea (1934)	Europe, North America				Intestine
Metagonimus romanicus	Ciurea (1934)					
Distomum mesotomum	Stossich (1898)	Europe				
Strigeidae						
Apatemon gracilis[c]	Sprehn (1932)	Europe		+		
A. bdellocystis	Lutz (1933)					
Strigea strigis	Sudarikov (1960)					
Cotylurus cornutus[c]	Rudolphi (1808)			+		Intestine
Tetracotyl sphaerula	Sudarikov (1962)	Japan, Siberia				
Diplostomidae						
Posthodiplostomum minimum	MacCallum (1921)					
Clinostomidae						
Clinostomum attenuatum[c]	Cart (1913)	Nebraska		+		

Table 9.3 (*continued*)

Family, Genus, and Species	Reported by	Location	Affected Species[a]	Pigeon Incidence[b]	Intermediate Hosts	Body Location
Philophthalmidae						
Philophthalmus columbae	Karyakarte (1968)				Snail	Eye
Eucotylidae						
Tanaisia bragai[c,e]	dos Santos (1934)	Brazil, United States, Puerto Rico, Philippines	C, T	+ + +	Land snail	Kidney
T. domestica	Nasir (1972)	Venezuela				Kidney
Opisthorchiidae						
Amphimerus elongatus[c]	Gower (1938)	Calcutta	C, D	+ +	Snail, fish	Liver, pancreas
Schistosomatidae						
Bilharziella polonica[c]	Kowalewski		D	+ +	Snail	Circulation, liver
Austrobilharzia penneri	Short (1961)	Florida				Liver
A. terrigalensis	Bearup (1956)					Liver

Source: Pilitt (1986) recorded all species to infest pigeons.
Note: Blanks indicate a lack of information.
[a]C = chickens, D = ducks, Gf = guinea fowl, Go = geese, M = mammals, S = swans, T = turkeys.
[b]+ = incidental, + + = occasional, + + + = more common.
[c]Kingston (1984).
[d]McDonald (1969).
[e]Price (1965).
[f]Stunkart and Dunihue (1931).

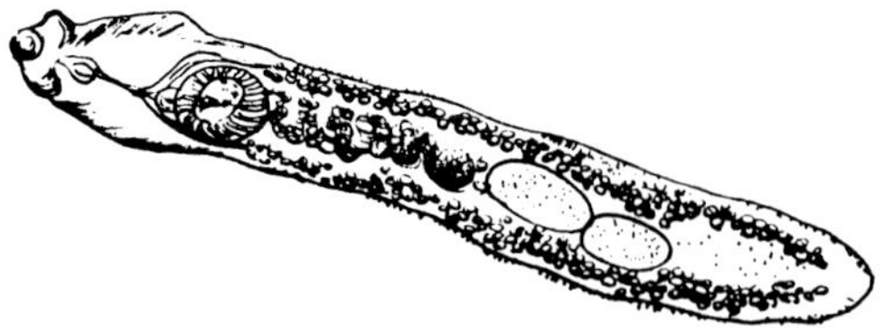

Fig. 9.3. Fluke, *Echinostoma revoltum.*

Morbidity and Mortality. The number of sick or dead birds largely depends on the numbers and species of flukes in the birds. *T. bragai*-infected pigeons were observed by Maldonado and Hoffman (1941) and no ill effects were noted. Kingston (1984) reported that the damage from *A. elongatus,* which causes pancreatic duct blockage, largely depends on the number of flukes present.

Signs. In light fluke infestations specific evidence is seldom noted. However, parasitism may cause hemorrhagic diarrhea, anemia, loss of weight, weakness, and death. Pectoral muscle atrophy and liver damage may also occur.

Necropsy. Findings depend on the type and location of the parasites. Intestinal inflammation, hemorrhage, and congestion may be noted with intestinal flukes. Distention of the kidneys and ureters may be observed when flukes invade these locations.

Diagnosis. Fluke infestations are usually detected at necropsy, but routine fecal flotations may be necessary to establish the presence of worm eggs in partly decomposed birds.

Biological Properties. Few ciliated embryos within eggs survive if they do not reach water. Freely swimming miracidia also die in a few hours unless a proper snail is found. The cercariae are also susceptible to drying and chemical parasiticides. Snails require water and their numbers are dissipated by drying.

Prevention and Control. Flukes require at least one snail in the life cycle. Thus control measures must be directed toward elimination of this intermediate host. Pigeon lofts should not be located near marshy areas. Where wet areas are nearby, drainage is necessary to prevent snail reproduction.

Powdered copper sulfate may be used at 20–100 ppm on marshy areas. One part in 500,000 parts of water in streams will kill most snails, but it will also kill fish. Livestock are not harmed by this level (Kingston 1984). Land snails may be destroyed by contact poisons such as metaldehyde. Potassium aluminum sulfate has also been effective (Malek and Cheng 1974).

Treatment. Oviduct flukes in chickens have been treated with 1–1.7 ml carbon tetrachloride on 3 consecutive days (Schmid 1930). This drug is very dangerous for people and is not recommended. Giddings (1988) used praziquantel with success in a toucan at 10 mg/kg body weight for 14 days. Other newer parasiticides should also be tried.

References

Beaver, P. C. 1937. Ill Biol Monogr 15:96.

Bolle, W. 1925. Dtsch Tieraerztl Wochenschr 33:529.

Chandler, A. C. 1949. *Introduction to Parasitology.* 8th ed. New York: Wiley.

Dawes, B. 1946. *The Trematoda.* Cambridge, Engl.: Cambridge University Press.

Giddings, R. F. 1988. J Am Vet Med Assoc 193:1555–56.

Kingston, N. 1984. In *Diseases of Poultry,* 8th ed., ed. M. S. Hofstad, et al., pp. 668–90. Ames: Iowa State University Press.

Krause, C. 1925. Berl Muench Tieraerztl Wochenschr 41:262–63.

Maldonado, J. F., and W. A. Hoffman. 1941. J Parasitol 27:91.

Malek, E. A., and T. C. Cheng. 1974. *Medical and Economic Malacology.* New York: Academic.

McDonald, M. E. 1969. Bur Sport Fish Wildl Spec Sci Rep Wildl 126.

Price, E. W. 1965. In *Diseases of Poultry,* 5th ed., ed. H. E. Biester and L. H. Schwarte, pp. 1035–55. Ames: Iowa State University Press.

Schmid, F. 1930. Tieraerztl Rundsch 36:313.

Stunkart, H. W., and F. W. Dunihue. 1931. Biol Bull 60:179–86.

Van Heelsbergen, T. 1927. Tijdschr Diergeneeskd 54:414–16.

Zunker, M. 1925. Berl Muench Tieraerztl Wochenschr 41:483.

PROTOZOAN DISEASES

Single-celled protozoa form an important group of parasites in the animal kingdom. Nine such parasitic diseases are recognized in pigeons. These protozoa are separated into three classes: Lobosasida, which have pseudopodia; Zoomastigophorasida, which have flagella; and Sporozoasida, which reproduce by sporulation and have no organs of movement.

Amoebiasis

Definition. Amoebas are irregularly shaped, single-celled, motile, primitive protozoan parasites that are essentially confined to the intestine of birds, animals, and humans and often produce an inflammation of the intestine with diarrhea.

Classification. Amoeba are placed in the phylum Sarcomastigophora, class Lobosasida, order Amoebidorida, and family Amoebidae. The genus is *Acanthamoeba* (Volkonsky 1931) and the species *polyphaga* (Pusch Karew 1913), reported by Kadlec (1978) and Levine (1985).

Life Cycle. Reproduction occurs without sex, by binary fission. The amoeba becomes round, condenses, and forms a cyst wall. The nucleus divides followed by the cytoplasm. The number of nuclei formed in reproduction depends on the genus (Chandler 1949).

Form. The small amoebic trophozoites have no cilia and no flagella. At room or body temperature they move and ingest food by means of pseudopodia, which do not branch or anastimose. *A. polyphaga* forms tapering hyaline projections called acanthopods (Kadlec 1978). The cytoplasm contains food vacuoles and a nucleus. Delicate, protective, double-walled cysts may enable amoebas to withstand conditions outside the body. Fig. 9.4 outlines the form of the parasite.

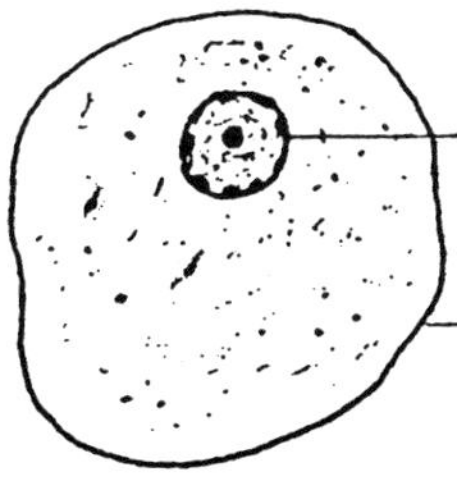

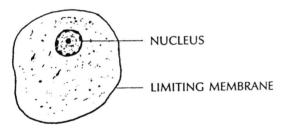

Fig. 9.4. Amoeba. The limiting membrane tends to flow and enables movement.

Distribution. Amoebas appear to be widespread. Some are pathogenic; others are not. The feeding and drinking habits of pigeons naturally introduce the parasites to pigeons, and they are probably more common, particularly in warm weather, than reports would indicate. Amoebic cysts were reported in pigeons by Jaskoski and Plank (1967) in Chicago, but the species was not determined. Kadlec (1978) noted *A. polyphaga* in Czechoslovakia in pigeons, cows, bulls, rabbits, and turkeys. Infected birds serve as the reservoir for other bird species.

Transmission. The parasite is transmitted to pigeons in feces-contaminated feed or water. Kadlec (1978) indicated that the parasites are confined to the intestine of pigeons and are passed in the droppings. Levine (1973) indicated that cysts can remain alive in moist feces for 28 days, which increases chances for transmission.

Signs and Necropsy. Evidence of infection may not be observed since mortality seldom occurs. During necropsy excessive mucus may be found in the lumen of the intestine resulting from an inflammation of the mucosa.

Diagnosis. To identify the parasite, it is necessary to prepare wet, warm intestinal smears in saline on warmed slides. Phase contrast microscopy also aids in diagnosis. Flotation in zinc sulfate to concentrate the cysts helps in any study of the parasite. Lugol's iodine diluted 1:5 is effective in staining the organism, but haematoxylin is a good differential stain after Schaudinn's fixation (Levine 1973).

Control. Naturally, traffic in live birds encourages spread of the disease. It is thus important to isolate and treat infected birds.

Treatment. Furazolidone (NF180, Hess & Clark) at the rate of 2 lb/t feed for 7 days is suggested. Lyght (1966) lists human agents of choice as carbarsone, chiniofon, chloroquine, glycobiarsol, iodochlorhydroxyquin, emetine, and paromomycin. Unfortunately, many human products can not be given to pigeons. When NF180 is not available, human agents should be tried only with the advice of a veterinarian.

References

Chandler, A. C. 1949. *Introduction to Parasitology.* 8th ed. New York: Wiley, pp. 88–115.

Jaskoski, B. J., and J. D. Plank. 1967. Avian Dis 11:342–44.

Kadlec, V. 1978. J Protozool 25:235–37.

Levine, N. D. 1973. *Protozoan Parasites of Domestic Animals and of Man.* 2d ed. Minneapolis: Burgess, pp. 129–55.

______. 1985. *Veterinary Protozoology.* Ames: Iowa State University Press.

Lyght, C. E., ed. 1966. *The Merck Manual.* 11th ed. Rahway, N.J.: Merck, p. 1614.

Coccidiosis

Definition and Synonyms. Coccidiosis is an acute or chronic, infectious, transmissible intestinal parasitic disease of pigeons, other birds, and animals, caused by an intracellular, rather host-specific protozoan parasite. It is characterized by inflammation, invasion, dislocation, and destruction of intestinal epithelial cells.

The term *cocci* also implies coccidiosis.

Cause and Classification. The organism is placed in phylum Apicomplexa, class Sporozoasida, order Eucoccidiorida, family Eimeriidae, and genera and species *Eimeria labbeana* (Pinto 1928) (*E. pfeifferi, E. columbarum* Nieschulz 1925, 1935) reported by Levine (1985). Other species include *E. sphenocercae* (Ray 1952) and *E. columbae* (Mitra and Gupta 1937), *E. tropicalis* (Mulhotra and Ray 1961), and *Wenyonella columbae* (Holdar and Ray-Choudhury 1974), the last three reported by Levine (1985). Each of these species has been found in India.

Nature

LIFE CYCLE. Pigeon coccidia complete their life cycle in the epithelial cells lining the small intestine. Both asexual and sexual stages occur there. The asexual cycle, schizogony, begins after the oocyst is shed in the feces of the pigeon. The unsegmented oocysts undergo development, sporulation or maturation, outside the host (Fig. 9.5). This takes 3 days for *E. labbeana* (Boughton and Volk 1938) and 5 days for *E. sphenocercae* (Ray 1952). According to Boughton (1934, 1937) 80% of the oocysts appear in the feces between 9 A.M. and 3 P.M. During this process, oocysts of all *Eimeria* species form four rounded sporocysts inside each egglike oocyst, and these in turn form two elongated curved sporozoites (Lund and Farr 1965). Levine (1985) reported that the *W. columbae* oocyst forms four sporocysts with four sporozoites per sporulated sporocyst, which differs from *Eimeria.* Oocyst production in the pigeon may last several days, and during this period many oocysts may be released in the feces. Under favorable conditions of moisture and temperature, oocysts may remain infective for several months.

When viable sporulated *Eimeria* oocysts are ingested by the pigeon, eight freely swimming sporozoites escape to initiate the development of schizonts in individual epithelial cells of the small intestine. Tyzzer (1929), in his study of *E. tenella* of chickens, estimated that about 900 merozoites were released from each first generation schizont. However, Levine (1973) indicates that only 15–20 merozoites develop in *E. labbeana.* These enter other epithelial cells to form second-stage schizonts that produce only 16 merozoites each; Nieschulz (1935) believed these to be extracellular. Eventually merozoites appear with the ability to form male and female gametocytes. This begins the sexual cycle, or sporogony.

Fig. 9.5. Sporulated oocyst. Sporocysts within indicate that the oocyst is infectious.

The male microgametocyte within an individual cell matures with the production of many spermlike flagellated microgametes. The female macrogametocyte within another cell in the wall of the intestine forms one macrogamete. When the macrogamete is fertilized by the freely swimming microgamete, a zygote develops. The zygote becomes the oocyst, which is released when the epithelial cell of the intestinal wall ruptures.

FORM. Each state of the coccidia has a different form and size. The oocyst itself consists of a compound wall, an outer resistant capsule, and an inner thin envelope surrounding the rounded mass of nucleated protoplasm and jellylike material (Labbe 1896; Henry 1932; Becker 1952). According to Boughton and Volk (1938) oocysts of *E. labbeana* are slightly elliptical, 15–26 μm by 14–24 μm, with an average of 20 μm by 18 μm. There is no micropyle and no residual body. Oocysts of *E. labbeana* produced early in the prepatent period may be smaller than those later in the period (Reid et al. 1984). The sporocyst of *E. labbeana* is oval with a plug at the more pointed pole. Each sporocyst contains a residual body. Two elongated sporozoites escape from each sporocyst. Yakimoff and Timofeieff (1939) studied and reported on this phase of the cycle of *E. labbeana.* The rounded trophozoite or schizont in the individual cells of the small intestine contains many sickle-shaped merozoites after completion of asexual nuclear divisions within the schizont (Srivastava 1966). The micro- and macrogametocytes are rounded bodies conforming to the size and shape of the distended individual epithelial cells that they occupy.

The oocyst of *E. sphenocercae* was first reported in India by Ray (1952). It is kidney-shaped and has a protruding micropyle. The size is 25.–17.5 μm by 15–12.5 μm with an average of 19.18 μm by 12.6 μm. The sporulation time is 5–6 days. Ray (1952) also gave the size of the sporocysts as 18.75–17.5 μm by 13.75–12.5 μm. These were broadly oval and contained residual bodies. Levine (1973) described the oocyst of *E. columbae* as subspherical, measuring 16 by 14 μm. It has a thin wall and a residual body. The sporulation time is 4–5 days.

SEROLOGY. Coccidiosis does not engender a circulating antibody response that can be readily measured by serological tests. McDermott and Stauber (1954) demonstrated agglutination of merozoites of *E. tenella* and agglutinins in the sera of experimentally infected chickens. Pierce et al. (1962) and Rose and Long (1962) also detected serum precipitins in chickens. It is possible that similar serology may be observed in pigeons.

Distribution and Incidence. Coccidiosis is very common throughout the world wherever pigeons are raised. All pigeons get the infection and persistent low levels of exposure may be expected and desired. In one study, Schwabe (1979) in New Jersey reviewed cases between 1975 and 1979 and found 76.8% of 3116 New Jersey specimens had one or more oocysts and 66.6% of 2354 out-of-state specimens also contained oocysts.

SEASON, AGE, SEX, AND BREED. Spring and summer warmth increases sporulation and thus exposure to infected oocysts, but the disease can occur at any time. Young birds are quite susceptible to coccidiosis and often get overwhelming infections because they have not received previous exposure. Older birds develop a degree of localized immunity from prior exposure and thus seldom get serious fatal infection. There is no sex or breed difference.

SPECIES. Stabler and Holt (1962) reported *E. labbeana* in ringdoves and turtledoves. Nieschulz (1935) tried to infect chickens with *E. labbeana* without success. Most *Eimeria* species of coccidia exhibit marked host specificity. Chickens, turkeys, ducks, and geese each have their own species. Animals also have their own species, and these do not interchange with those of birds.

CARRIERS. Pigeons may serve as inapparent carriers and shedders of coccidia for a period of time before and after they show signs of illness. Schwabe (1979) reported that oocysts were shed by pigeons for up to 5–6 wk after clinical evidence of infection. Pigeons are not continuous carriers after they develop immunity to the infection. As immunity dissipates they may be reinfected. Rehm and Hasslinger (1969) reported reinfection in 5 wk following treatment. Pigeons may also be mechanical carriers of oocysts on their feet and in their digestive tracts. This means that they may mechanically carry the oocysts of another bird species.

Transmission. Sporulated oocysts must be ingested for infection to become established. Contaminated feed and water are often the source of infection, but because pigeons habitually eat off the ground or floor where oocysts accumulate, transmission occurs naturally. Mechanical transmission also occurs by flies, shoes of attendants and visitors, feed sacks, pigeon baskets, free-flying birds, and by dogs, cats, and rats.

Epidemiology

DISTRIBUTION IN BODY. In the pigeon, the parasites are entirely confined to the digestive tract and the schizonts and sexual gametocytes are confined to the intestinal wall.

ENTRY AND EXIT. Ingestion of sporulated oocysts may establish infection. Nonsporulated oocysts will not. Experimental merozoite infection of the crop has been demonstrated in chicks (Levine 1940). Perhaps this may occur in pigeons as well. Nonsporulated, noninfectious oocysts are shed in the feces of infected birds.

PATHOGENICITY. Pigeons must ingest a specific number of oocysts per species of coccidia for clinical infection to establish. Each species of oocyst has its own natural ability to produce clinical disease; one species may require 1000 oocysts and another 2000 before clinical coccidiosis develops. The continued ingestion of a few oocysts may cause mild infection but not clinical disease. Exposure to low-level infection does not harm the bird and is necessary in maintaining immunity.

IMMUNITY. Birds may casually develop a degree of protective immunity to coccidiosis by repeatedly picking up small numbers of infective sporulated oocysts. Constant exposure to infection maintains and reinforces the immune response. If a management practice excludes ingestion of oocysts, immunity is lost. When birds are placed on wire and are prevented access to their droppings, they may become susceptible to serious coccidiosis. It must be remembered that localized tissue immunity does not provide cross-protection between species. Each species of coccidia establishes its own immune reaction.

Incubation. From the time of ingestion of coccidia oocysts and the release of sporozoites until the shedding of oocysts, a period of 6 days is required for *E. labbeana* (Levine 1973).

Course. The course of coccidiosis is to a large extent related to the age of the birds, the magnitude of exposure, the immunity, the stress imposed before and during infection, and the treatment given.

Morbidity and Mortality. When loft infection occurs, most of the birds will be sick until the disease is either controlled by medication or immunity interrupts further development. Mortality seldom occurs in adult birds, but young squabs can die.

Signs. A sudden decrease in either water or feed consumption may be the first signs of coccidiosis, but adult birds seldom show clinical evidence of infection. General unthriftiness, loss of appetite, weakness, dehydration, liquid feces, loss of weight, lack of desire to move or fly, and lack of alertness and the tendency to remain with the eyes closed may be signs of infection. In young birds, diarrhea, dehydration, and emaciation often lead to death.

Necropsy. Swelling and thickening of the intestinal wall with mild to extensive inflammation and engorgement of blood vessels (so-called "blush") may be noted along the small intestine. In chronic infections the intestine may evert (roll out on itself) when opened, indicating that the inner wall is thickened. The intestine often becomes filled with grayish white mucus, which is usually devoid of blood. Oocysts are not produced in great numbers; thus the degree of infection may be hard to estimate.

Diagnosis. A glass rod may be used to collect floating oocysts or other parasite forms on the surface of a centrifuged sample. They can be concentrated on the top of the solution in a test tube by centrifuging 1 tbsp concentrated sucrose solution together with ½ tsp feces. The centrifuge should be operated at low speed for 15 min. A drop or two collected from the fluid surface and placed on a glass slide may be examined under low-power light microscopy. This gives a relative indication of infection. It is essential to arbitrarily grade the level of infection by counting the numbers of oocysts in about 10 microscopic fields. A combined total of less than 5 oocysts means +1; less than 10, +2; less than 20, +3; and more than 20, +4. A reading of +1 is not considered treatable and is of no consequence. It could mean recycled oocysts or oocysts from some other source, such as free-flying birds or from animals. Oocyst species are not easily distinguished by routine procedures. In any diagnosis the history is important in evaluating a problem.

Differential Diagnosis. In the absence of oocysts or schizonts in a centrifuged sample, enteritis must be explained on the basis of other diseases. *Salmonella,* capillary worms, and *E. coli* each may cause diarrhea.

Biological Properties

STRAINS. Coccidial strain differences do occur, and drug-resistant strains may develop.

STABILITY. Within the intestine, various reproducing stages of coccidia are susceptible to coccidiostats (McDougald 1982). The oocyst is very resistant to disinfectants, including 5% formalin, 5% copper sulfate, 10% sulfuric acid, 5% potassium hydroxide, 5% potassium iodide, and peracetic acid (Reid et al. 1984).

Sporozoites and merozoites do not survive long outside the body. Oocysts, on the other hand, are very resistant and will survive for a year in soil. Drying, direct sunlight, heat, lack of oxygen, and bacterial and fungal activity are detrimental (Lund and Farr 1965). Long (1974) has shown that chicken oocysts will survive and grow in chicken embryos.

Prevention. Prevention is not the aim. Every bird must get coccidiosis and be continually exposed in order to establish and maintain an immune response. The aim is to control the amount that birds get by the use of drugs designed to suppress the parasites.

Control. Routine fecal examinations establish the level of oocyst buildup in individual pens or groups of birds. High levels of exposure are reflected in varying degrees of clinical disease. By preventing excessive exposure, serious disease and death can be avoided.

After a reliable diagnosis is established, there are several ways to control the disease. The first is good loft management. Avoidance of crowding and provision of reasonably clean, dry floors or litter and screened floor porches is important. Good ventilation is needed, as are clean feed hoppers and waterpans. Hoppers and waterpans should exclude droppings, and they must not be filled too full, to avoid spillage and billing of feed. Shed-type roof construction instead of A-type or modified-A-type prevents wet floor areas from moisture condensation, as trapped warm air in an A roof holds moisture that later falls as it cools. Moisture is necessary for oocyst sporulation; by keeping the floor fairly dry the number of infective oocysts is controlled.

The second control method entails the use of chemical coccidiostats, which limit the extent of infection. Coccidiostats do not necessarily suppress all coccidiosis stages within the bird, but they interfere with some stage of their development. Nitrofurazone (NFZ Soluble 4.59%, Rhodia, Hess & Clark) is presently used in drinking water in mild +2 levels of coccidiosis at the rate of 2 level tbsp/2½ gal water for 5 days. The mixture should be prepared fresh daily. The treatment may be repeated if necessary, but fecal samples should be reexamined to determine if satisfactory results have been obtained. Resistant strains may require further treatment.

Treatment. When heavy +3 or +4 levels are

noted, sodium sulfamethazine (Sulmet 12.5%, American Cyanamid) may be used in drinking water at 1 oz (30 ml or 2 tbsp)/gal for 2 days, followed by 4 days off, on 1 day, off 4 days, and on 1 day. If the temperature is above 85°F, half days should be substituted for full days. The schedule should not be repeated without 1 wk off sulfa medication.

Sulfaquinoxaline (SQ 3.2%, Merck) should be discreetly used on +4 levels of coccidiosis in which the birds are obviously sick or dying. The therapeutic dosage rate and schedule for the 3.2% solution of sulfaquinoxaline is 0.025% or 1 oz/gal drinking water 2 days on, 3 days off, 2 on, 3 off, and 2 on. Temperature restrictions also apply, and the schedule of treatment must not be repeated within a week. Sulfaquinoxaline solution, as well as all other coccidiostats or drugs, may be toxic if used beyond recommended levels. Treatment must therefore be employed with caution regardless of the medication, following the directions on the bottle and the advice of a veterinarian.

Scupin (1966) and Rehm and Hasslenger (1969) each reported on the use of amprolium 0.018% and ethopabate 0.0114% for 1 wk. Amprolium 9.6% (Merck) may be used in the water at the rate of 8 fl oz/50 gal (0.012%) for fairly severe outbreaks or 4 fl oz/50 gal (0.006%) for milder +2 or +3 coccidiosis levels. The solution may be used on days off during sulfa treatments if infection levels require further treatment. Amprolium is also suggested for continuous 5- to 7-day treatments in place of NFZ. It is nontoxic if repeated or continued at the lower level.

Sulfadimethoxine (Agribon, Hoffmann-LaRoche) has been used effectively in drinking water at 0.025% for turkeys and 0.05% for chickens for 6 days. Triple-sulfa combinations are less toxic when compared to single sulfa drugs given at the same dosage levels. Other coccidiostats have also been used with success, including Appertex (Janssen). Unfortunately most drugs are not cleared for pigeon use because of the cost of marketing approvals. Whenever drugs are used, regardless of clearance, the directions of the manufacturer should always be followed with reasonable judgment.

References

Becker, E. R. 1952. In *Diseases of Poultry,* 3d ed., ed. H. E. Biester and L. H. Schwarte, pp. 943–77. Ames: Iowa State University Press.

Boughton, D. C. 1934. J Parasitol 20:329.

______. 1937. J Parasitol 23:291–93.

Boughton, D. C., and J. J. Volk. 1938. Bird Banding 9:139–53.

Henry, D. P. 1932. Univ Calif Pub Zool 37:269–78.

Labbe, A. 1896. Arch Zool Exp 4:517–654.

Levine, N. D. 1973. *Protozoan Parasites of Domestic Animals and of Man,* 2d ed. Minneapolis: Burgess, p. 295.

______. 1985. *Veterinary Protozoology.* Ames: Iowa State University Press.

Levine, P. P. 1940. J Parasitol 26:237.

Long, P. L. 1974. Avian Pathol 3:255–68.

Lund, E. E., and M. M. Farr. 1965. In *Diseases of Poultry,* 5th ed., ed. H. E. Biester and L. H. Schwarte, pp. 1056–96. Ames: Iowa State University Press.

McDermott, J. J., and L. A. Stauber. 1954. J Parasitol 40:23.

McDougald, L. R. 1982. In *Biology of Coccidia,* ed. P. L. Long, p. 373. Baltimore: University Park Press.

Nieschulz, O. 1925. Arch Protistenkd 5:479–94.

______. 1935. Centralbl Bakteriol 134:390–93.

Pierce, A. E., P. L. Long, and C. Horton-Smith. 1962. Immunology 5:129.

Pinto, C. 1928. C R Soc Biol 98:1571.

Ray, D. K. 1952. J Parasitol 38:546–47.

Rehm, H., and M. A. Hasslinger. 1969. Kleintierkund Prax 14:98–99.

Reid, W. M., P. L. Long, and L. R. McDonald. 1984. In *Diseases of Poultry,* 8th ed., ed. M. S. Hofstad, pp. 692–717. Ames: Iowa State University Press.

Rose, M. E., and P. L. Long. 1962. Immunology 5:79.

Schwabe, O. 1979. Proc Pigeon Health Disease Conf, Cook College, Rutgers University, New Brunswick, N.J.

Scupin, E. 1966. Dtsch Tieraerztl Wochenschr 73:35–37.

Srivastava, H. K. 1966. Indian J Vet Sci 36:221–26.

Stabler, R. M., and P. A. Holt. 1962. Proc Helminth Soc 29:76.

Tyzzer, E. E. 1929. Am J Hyg 10:1-115.

Yakimoff, W., and P. Timofeieff. 1939. Bull Soc Pathol Exot 32:284–85.

Hexamitiasis

Definition and Synonyms. *Hexamita* is an uncommon flagellated protozoan parasite of pigeons that confines itself to the intestinal tract. The condition is also called Hexamita infection and *Octomitus columbae* infection.

Cause and Classification. The organism is placed in phylum Sarcomastigophora, class Zoomastigophorasida, order Diplomonadorida, family Monocercomonadidae, and genera and species *Spironucleus columbae* (Noller and Buttgereit 1923) (syn. *Hexamita columbae* and *Octomitus columbae*), reported by Levine (1985).

Nature. Multiplication of *Hexamita* is by longitudinal binary fission. The piriform body is bilaterally symmetrical with two anterior nuclei. The parasites have six anterior flagella and two posterior ones. Three arise on each side anterior to the two nuclei. In addition, one arises on each side near the nuclei and these pass posteriorly through the body to emerge near the posterior end (Levine 1985).

Distribution

GEOGRAPHY. The pigeon *Hexamita* parasites are not common, but they have been identified in Germany (Noller and Buttgereit 1923) and in California (Hinshaw and McNeil 1941; and McNeil and Hinshaw 1941).

SEASON. Warm temperature favors the survival of the organism and thus its transmission.

AGE, SEX, AND BREED. Hexamitiasis occurs primarily in young birds, but there is no evidence that sex or breed alters the incidence.

CARRIERS. Adult birds that have survived earlier *Hexamita* attacks may serve as carriers.

Transmission. The ingestion of fresh, live Hexamita in the droppings in contaminated feed or water serves to transmit infection.

Epidemiology

DISTRIBUTION IN BODY. *Hexamita* is found in the digestive tract from the gizzard to the vent.

ENTRY AND EXIT. *Hexamita* is ingested and is passed in the droppings.

PATHOGENICITY. Only a few *Hexamita* organisms are required to establish infection. They multiply so rapidly that birds become sick almost overnight.

IMMUNITY. It appears that older birds are less susceptible to *Hexamita,* which means that a degree of immunity is established or that they become more resistant with age.

Incubation. In turkeys, *H. meleagridis* produces signs in 4–7 days after ingestion of the parasites. Pigeons probably develop signs in the same length of time. The course is variable.

Morbidity and Mortality. Most of the flock will become infected at the same time; thus most of the birds will be sick after the initial onset. Losses can develop soon after signs appear.

Signs and Necropsy. Foamy, watery diarrhea, dehydration, loss of weight, weakness, unkempt feathers, and death may be observed. Catarrhal inflammation of the intestine with a lack of tone, foamy, mucus-coated feces and congestion are typical internal findings.

Diagnosis. A diagnosis is established in a pigeon on finding in a fresh fecal smear a fast-swimming, single-celled parasite that fails to stay within the microscopic field. One must constantly follow the parasite across the slide. In the live bird, gently scraping the cloacal area may yield parasites. Trichomonads spiral and stay within the microscopic field.

Biological Properties

STAINING. Smears may be fixed with Carnoy's fixative and stained with Giemsa for study. Schaudinn's fluid is a good fixative and Honigberg's staining technique is reliable.

STABILITY. The parasites are easily destroyed by drying and disinfectants.

Prevention. Good sanitation, clean feed hoppers and waterpans, and exclusion of other birds should reduce transmission.

Control and Treatment. Control involves good sanitary management and chemotherapy. Reid et al. (1984) suggest the use of butynorate (0.0375%) (Tinostat, Salsbury), chlortetracycline (0.0055%) (Aureomycin, American Cyanamid), and furazolidone (0.022%) (NF180, Hess & Clark), alone or with oxytetracycline (0.044%) (Terramycin, Pfizer).

References

Hinshaw, W. R., and E. McNeil. 1941. Am J Vet Res 2:453–58.

Levine, N. D. 1985. *Veterinary Protozoology.* Ames: Iowa State University Press.

McNeil, E., and W. R. Hinshaw. 1941. J Parasitol 27:185.

Noller, W., and F. Buttgereit. 1923. Zentralbl Bakteriol 75:239.

Reid, W. M., P. L. Long, and L. R. McDougald. 1984. In *Diseases of Poultry,* 8th ed., ed. M. S. Hofstad et al., pp. 725–26. Ames: Iowa State University Press.

Leucocytozoon Infection

Definition. *Leucocytozoon* infection is a rare protozoan blood disease of pigeons similar to *Plasmodium* infection but transmitted by blackflies or midges.

Classification. The parasite is placed in phylum Apicomplexa, class Sporozoasida, order Eucoccidiorida, family Plasmodiidae, and genus *Leucocytozoon* (Levine et al. 1980; Levine 1985). Jansen (1952) reported *L. marchouxi* (Mathis and Leger 1910) (syn. *L. turtur*) in pigeons and doves in South Africa.

Nature

LIFE CYCLE. The cycle of this species is unknown (Levine 1973), but it is probably similar to *Plasmodium* with the exception that schizogony is confined to pigeon tissue cells and endothelial cells of various organs. It is known that only male and female gametocytes are found in pigeon white blood cells, which become grossly distorted and spindle-shaped. Schizonts do not appear in the bloodstream.

The *Simulium* blackflies or *Culicoides* midges probably serve as vectors and receive the gametocytes when they bite. In the flies, microgametes, escape the microgametocytes and fertilize the macrogametes, forming zygotes. Ookinetes form from the zygotes in the stomach wall of the fly. These form oocysts, which release a few sporozoites. Sporozoites are injected into the pigeon by the fly bite. Schizonts form in the pigeon tissue cells and are subdivided into cytomeres with release of merozoites. The merozoites enter white blood cells to form male and female gametocytes or gamonts (Levine 1985).

FORM. Macrogametocytes (female) are rounded or elliptical sacs. They stain dark blue with a reddish nucleus with Giemsa. Microgametocytes (male) stain pale blue and have a diffuse, large, pale pink nucleus with Giemsa. The host cell cytoplasm rarely forms a narrow border around the sac of the male parasite, but it may be observed around some of the macrogametocytes. Cytoplasm usually appears in the "horns" of the gamonts (Levine 1985). The nucleus of white cells has several lobes, which may be separated; thus the host cell nucleus may be compressed to one side or appear split in some cells.

Distribution

GEOGRAPHY. The only report in pigeons was from South Africa, but the *L. marchouxi* species occurs worldwide (Levine 1973).

INCIDENCE. The disease is rare in pigeons, but this parasite species is fairly common in wild mourning doves (Levine 1973).

SEASON, AGE, AND BREED. Blackflies and midges are prevalent during warm weather and warm climates. Either may transmit the blood parasite, which persists in the host throughout the year. Sex and breed do not appear related to the incidence of disease.

SPECIES. Pigeons and doves are affected. Levine and Kantor (1959) noted *Leucocytozoon* infection in 17 species of columborid birds, including turtledoves and mourning doves.

CARRIERS. Infected doves are the apparent natural host and reservoir.

Transmission. The parasites cannot be transmitted by blood inoculation because only gametocytes appear in blood. *Simulium* blackflies or *Culicoides* midges serve as vectors.

Epidemiology

DISTRIBUTION IN BODY. The gametocytes appear in the bloodstream, and schizonts and cytomeres are localized in organ tissue cells. In a waterfowl species Chernin (1952) noted that gametocytes disappear from the blood and reappear in the spring. This must be kept in mind when evaluating treatment trials.

ENTRY AND EXIT. The bite of an infected blackfly or midge introduces sporozoites to the pigeon, and noninfected flies in turn receive gametocytes from infected pigeons.

PATHOGENICITY. Fourteen-day-old infected nestlings observed by Levine (1954) showed no signs of illness, but once a bird is infected it appears to remain infected.

Incubation. Sporozoites appear in the salivary glands of the fly within a few days after biting an infected pigeon.

Signs and Necropsy. The organism does not appear to be a serious pathogen for pigeons. Thus losses are unlikely, but general weakness, and inactivity may be observed. An enlarged spleen, paleness, and watery blood may be present. Pigment granules (hemozoin) are not formed from the breakdown of hemoglobin.

Diagnosis. Stained blood smears or tissue sections enable a diagnosis. *Leucocytozoon* parasites stain readily with Wright's, Giemsa, and Romanowsky stains after methyl alcohol fixation. Distorted spindle-shaped red blood cells with characteristic displaced nuclei depict infection. Tissue sections reveal endothelial parasitism and schizonts.

Stability. *Leucocytozoon* parasites do not survive outside the pigeon or fly vector.

Prevention. Pigeons should be raised away from water where blackflies and midges breed.

Treatment. No treatment has been reported effective for *Leucocytozoon* infection in pigeons. Plasmochin, atabrin, and quinine have been tried in other species without success. Pyrimethamine, at 1 ppm, and sulfadimethoxine, at 10 ppm, given together are considered a preventative for *L. caulleryi* in chickens by Fallis et al. (1974). Clopidol (Coyden, Dow) in feed, at 0.025%, controlled but did not eliminate *L. smithi* in turkeys, as reported by Siccardi et al. (1974). The drug has been continuously fed chickens for coccidiosis at 0.0125%.

References

Chernin, E. 1952. Am J Hyg 56:101–18.

Fallis, A. M., S. S. Desser, and R. A. Khan. 1974. Adv Parasitol 12:1–67.

Jansen, B. C. 1952. J Vet Res Onderstepoort 25:3–4.

Levine, N. D. 1954. J Protozool 1:140–143.

______. 1973. *Protozoon Parasites of Domestic Animals and of Man.* 2d ed. Minneapolis: Burgess, pp. 277–85.

______. 1985. *Veterinary Protozoology,* Ames: Iowa State University Press.

Levine, N. D., and S. Kantor. 1959. Wildl Dis 1:1–38.

Levine, N. D., et al. 1980. J Protozool 27:37–58.

Siccardi, F. J., H. O. Rutherford, and W. T. Derieux. 1974. Avian Dis 18:21–32.

Malaria

Definition. Malaria is an uncommon protozoan parasitic disease of pigeons caused by a single-celled *Plasmodium* parasite. It is characterized by red blood cell destruction, asexual replication in tissue cells, and male and female gametocytes in erythrocytes. It is also called avian malaria and *Plasmodium* infection.

Cause and Classification. The parasite is placed in phylum Apicomplexa, class Sporozoasida, order Eucoccidiorida, family Plasmodiidae, and genus and species *Plasmodium relictum* (Grassi-Tletti 1891), reported by Levine (1973, 1985).

Nature

LIFE CYCLE. The pigeon serves as the intermediate host in which nonsexual *Plasmodium* replications (schizogony) take place in tissue cells, followed by merozoite invasion of erythrocytes with the formation of sexually immature male and female gametocytes. Culicine mosquitoes of the *Aedes, Culex,* and *Culiseta* genera, but rarely *Anopheles,* presumably serve as the definitive host in which maturation of gametes, fertilization, and sporogony take place.

Sporozoites produced by the stomach of the mosquito are stored in its salivary glands. The bite of an infected mosquito injects these sporozoites into the pigeon. When injected they do not initially enter the red blood cells of the pigeon. Instead they enter lymphoid and macrophage cells and reticuloendothelial cells of the liver, lung, spleen, brain, and kidney (James and Tate 1938). First-generation schizonts, called cryptozoites, yield freely swimming merozoites, which reenter similar cells and replicate in second-generation schizonts, called metacryptozoites. After two or more exoerythrocytic schizogonic cycles, merozoites enter the bloodstream, but in some species the exoerythrocytic forms may multiply indefinitely. Red cells are invaded eventually and male and female gametocytes are formed. The initial stage of erythrocyte invasion is called the trophozoite (Huff and Coulston 1944, 1946).

In some species when erythrocytic parasites are injected instead of sporozoites, these may enter tissues as exoerythrocytic forms. Huff and Coulston (1946) have called these phanerozoites. They do not contain pigment granules.

The mosquito receives the undeveloped gametocytes as it bites the infected pigeon. Male and female gametes mature within the gametocytes. Fertilization of the female gamete occurs with the formation of a zygote. Each ookinete from a zygote penetrates the stomach wall and develops into an oocyst, which produces many sporozoites. These are released to the salivary glands of the mosquito.

FORM. The parasite as first observed in a red cell has the appearance of a signet ring. The center is vacuolated and is bounded by a thin ring of light cytoplasm. A rather large nucleus dominates one side of the ring. As the schizonts develop and enlarge, numerous chromatin clusters appear, individually confined within one or more thin, irregular, saclike membranes. Male and female gametocytes occupy most or all of the red cell cytoplasm and are lightly stippled throughout.

Distribution

GEOGRAPHY. Malaria is worldwide. Becker et al. (1956) reported the disease in pigeons in Iowa. Pelaez et al. (1951) noted infection in Mexico.

INCIDENCE. Malaria is rare in pigeons in the United States.

SEASON, AGE, SEX, AND BREED. The infection may be diagnosed at any time of the year, but transmission occurs in warm climates and during warm weather when mosquitoes are present. Birds of any age, sex, and breed are susceptible.

SPECIES. There are probably several species in the genus *Plasmodium* that affect pigeons, but only *P. relictum* (*P. praecox*) is confirmed. Sergent and Sergent (1904) observed the organism in Algeria. Coatney (1938) in Nebraska studied a strain of *P. relictum* that was also infectious for canaries and chickens. Mathey (1955) noted the species in California, and Becker et al. (1956) reported the organism in Iowa. Manwell (1933) showed that *P. gallinaceum,* which is pathogenic for chickens, does not affect pigeons.

CARRIERS. Infected birds serve as nonsymptomatic carriers in early stages of infection.

Transmission. The bite of a mosquito transfers infection. *Plasmodium* can be transmitted from one pigeon to another by injections of blood from an infected pigeon. This is because the asexual schizont stages occur in red blood cells.

Epidemiology

DISTRIBUTION IN BODY. The parasites may appear in fluids and cells throughout the body.

ENTRY AND EXIT. The blood parasites enter the body of the pigeon during the bite of an infected mosquito and leave the same way.

PATHOGENICITY. The number of sporozoites required to establish infection in a pigeon will vary with the individual bird, the parasite species, and other imposed stress factors. Some species of the parasite do not establish infection. Huff and Coulston (1946), McGhee

(1970), and Soni and Cox (1975) have studied the immunogenic factors preventing infection.

As the parasites grow and break out of the blood and tissue cells, cellular destruction occurs, resulting in anemia and the release of blood pigment and waste products. Actually this destruction is more serious for the pigeon but less so for the mourning dove and canary (Levine 1985).

Incubation. The time that elapses between the mosquito bite and the clinical onset of infection will largely parallel infectivity. The erythrocytic stage occurs about 10 days after the infective bite (Chandler 1949).

Course. Since treatment for malaria is unsuccessful, the disease continues unabated. The extent of the disability and the death rate depends on the numbers and species of parasites, as well as other stress factors.

Morbidity and Mortality. The *Plasmodium* infection rate within a loft depends to a large extent on the numbers of infected mosquitoes biting the birds. Mortality of 90% has been reported in fowl (Springer 1984).

Signs and Necropsy. In malaria, severe anemia may be evident, together with weakness and mortality. Naturally the lack of oxygen-carrying capacity of the blood may cause some respiratory distress. Pale skin and breast tissues may be encountered as the carcass is opened. Generally birds with malaria will present enlarged pale spleens and kidneys.

Micropathology. When second-generation merozoites invade red blood cells prior to the formation of gametocytes they form the shape of a vacuolated signet ring. With Romanowsky's stain the cytoplasm stains blue and the nucleus red. The parasites ingest hemoglobin of the host red blood cells and appear as light brown hemozoin granules when stained with Wright's stain.

Diagnosis. Air-dried blood smears stained with Wright's stain and examined under the oil objective of a light microscope enable a diagnosis of malaria. Many red cells will be observed to contain thin-walled sacs within the cytoplasm that are irregular in size and shape and contain stippled chromatic aggregations.

Biological Properties

TISSUE AFFINITY. In the pigeon,sporozoites invade macrophage, lymphoid, and reticuloendothelial cells and later merozoites enter erythrocytes (James and Tate 1938). In the mosquito the parasites may be found in the stomach and salivary glands.

STAINING. Romanowsky's, Wright's, and Giemsa's stains are effective in staining the parasites.

STABILITY. The parasite does not survive outside of host birds or specific mosquitoes.

CULTURE. The malarial parasites can be grown in chick embryos and in tissue culture.

Prevention and Control. All additions to a loft should be blood-tested to ensure freedom from malarial infection. Mosquitoes must be controlled by draining all water containing trash and other wet breeding areas. A mosquito-screened loft with two screen doors forming an entrance air lock is helpful in mosquito-prevalent areas.

Treatment. There is no medication presently available for treatment or prevention of malaria.

References

Becker, E. R., W. F. Hollander, and W. H. Pattillo. 1956. J Parasitol 42:474.

Chandler, A. C. 1949. *Introduction to Parasitology.* 8th ed. New York: Wiley, pp. 183–215.

Coatney, G. R. 1938. Am J Hyg 27:380.

Huff, C. G., and F. Coulston. 1944. J Infect Dis 75:231–49.

______. 1946. J Infect Dis 78:99–117.

James, S. P., and P. Tate. 1938. Parasitology 30:128.

Levine, N. D. 1973. *Protozoon Parasites of Domestic Animals and of Man.* Minneapolis: Burgess, pp. 259–74.

______. 1985. *Veterinary Protozoology.* Ames: Iowa State University Press.

Manwell, D. R. 1933. Am J Trop Med 13:97.

Mathey, W. J. 1955. Vet Med 50:318.

McGhee, R. B. 1970. In *Immunity to Parasitic Animals,* vol. 2, ed. D. J. Jackson et al. New York: Century Croft.

Pelaez, D. A., et al. 1951. Rev Palud Med Trop 3:59.

Sergent, E., and E. Sergent. 1904. C R Soc Biol 56:132.

Soni, J. L., and H. W. Cox. 1975. Am J Trop Med Hyg 24:206–13.

Springer, W. T. 1984. In *Diseases of Poultry,* 8th ed., ed. M. S. Hofstad et al., pp.730–31. Ames: Iowa State University Press.

Haemoproteus Infection

Definition and Synonyms. *Haemoproteus* infection is common protozoan blood disease of pigeons that is characterized by asexual replication in tissue cells and male and female gametocytes in red blood cells. It is also called pigeon malaria.

Cause and Classification. The parasite belongs in the phylum Apicomplexa, class Sporozoasida, order Eucoccidiorida, family Plasmodiidae (Springer 1984), and genus and species *Haemoproteus columbae* (Kruse 1890), reported by Levine (1985). Another pigeon and dove organism, *H. sacharovi* (Novy and Mac-

Neal 1904) has been reported by Levine and Kantor (1959) and Becker et al. (1956).

Nature

LIFE CYCLE. Sporozoites are injected into the pigeon during the bite of an infected hippoboscid louse fly. Sporozoites are baby swimming parasites from the saliva of the fly. This is the start of the nonsexual cycle called schizogony. The parasites enter cells found in blood vessels of the lungs, liver, spleen, and bone marrow, called reticuloendothelial cells. When infected, they form unpigmented, round masses, called schizonts. Each of these undergoes multiple division to form 15–20 small unpigmented cytomere bodies, each bearing a single nucleus. Cytomeres grow and divide and form multinucleated cytomeres. When the swollen cell can no longer contain the cytomeres, they are then released into the blood capillaries. From the cytomeres, enormous numbers of merozoites are formed in a matter of 4 wk. These tiny, swimming, single-celled parasites may reenter blood vessel cells to repeat the process or enter red blood cells to form male and female gametocytes (Olsen 1962). Wenyon (1926), however, states that cytomeres may be bypassed and schizonts may produce merozoites directly.

The sexual part of the cycle (Adie 1924) takes place in the louse fly or dipteran vector. As the fly sucks blood from an infected bird, it receives male and female gametocytes. In the midgut of the fly, the microgametocyte forms six to eight microgametes or sperm, which swim to the macrogametes or eggs and fertilize them. The fertilized egg is the zygote, which forms an ookinete and later becomes the oocyst in the wall of the fly's stomach. This takes 10–12 days after fly infection. From the oocyst, numerous sporozoites are formed. These migrate to the salivary glands and are injected into the bloodstream of the pigeon as the fly bites.

FORMS. The gametocytes are halter-shaped bodies in the erythrocytes that partially enclose the nucleus of the host cell without displacing it (Fig. 9.6). According to Olsen (1962), the microgametocytes are 11.9–15.3 μm long by 2.5–4.3 μm wide with a large oval nucleus 2.8 by 5.8. The nucleus stains light blue and the cytoplasm pale blue with Wright's stain. The macrogametocytes are 13.7 μm and stain darker than microgametocytes. The cytoplasm of the cells containing the parasites stains normally. True brownish malarial pigments are a constant finding in the cytoplasm of red cells containing mature gametocytes. Pigment granules from ruptured red cells may also be found clumped in the blood serum.

Fig. 9.6. *Haemoproteus* red blood cell microgametocyte. The brownish blood pigment hemosidin within the RBC identifies the parasites.

Distribution

GEOGRAPHY. The disease is worldwide in distribution and has been reported in various states of the United States (Drake and Jones 1930; Huff 1932, 1942; Coatney 1935; Herman 1938, 1944; Wetmore 1941; Herman and Glading 1942; Wood and Herman 1943; Stubbs 1948; Becker et al. 1956; Hanson et al. 1957; Becker 1959; Levine and Kantor 1959; Couch et al. 1961; Levine 1961), in India (Acton and Knowles 1914; Alcock 1914; Narang and Bhatnagar 1969), in England (Baker 1957), in Brazil (Giovannoni 1946), and in Honolulu (Alicata 1939, 1947; Kartman 1949). It's safe to say that wherever pigeons are raised, the disease may be encountered.

INCIDENCE. *Haemoproteus* infection is present in about one-third of the flocks and in about 10–15% of the birds in the eastern United States.

SEASON. Once pigeons are infected with *Haemoproteus,* the parasites remain in the bird all year. There is no observed seasonal variation.

AGE, SEX, AND BREED. Birds of any age may be infected. There is no evident sex or breed difference.

SPECIES. *Haemoproteus* is not only common in pigeons (Coatney 1933, 1936; Coatney and Hickman 1952) but has been reported in mourning doves (Becker 1959), turtledoves (Levine 1961), western mourning doves, western white-winged doves, and band-tailed pigeons (Wood and Herman 1943). Levine and Kantor (1959) tabulated reports of *Haemoproteus* in 45 species belonging to 19 genera of columbiform birds (pigeons and doves) and all of these were probably *H. columbae,* the common pigeon type of the organism.

CARRIER. The *Haemoproteus*-infected bird remains a carrier for life unless cured by chemical treatment.

Transmission. Pigeons become infected when bitten by hippoboscid flies (louse flies) or by other dipteran vectors. In addition to *Pseudolynchia canariensis,* two other louse flies, *Microlynchia pusilla* in South America and *Ornithomyia avicularia* in England, are suspected of being vectors (Baker 1957). In addition, Fallis and Wood (1957) discovered that an orthorraphous insect, a biting midge, *Culicoides* sp., is a suitable intermediate host and transmitting agent of *H. nettionis,* which is another form of *Haemoproteus* in ducks. This suggests midge transmission in pigeons. Furthermore, Hanson et al. (1957) pointed out that louse flies are extremely rare on mourning doves, but *H. columbae* is common in them.

Lastra and Coatney (1950) showed that it is possible to transfer the infection by direct injection of considerable quantities of blood or the transplantation of tissues from donor pigeons taken early in the course of the infection. Olsen (1962) also indicated that birds may be infected by intraperitoneal injections of emulsions of lungs containing merozoites or intestines and salivary glands of infected louse flies. Olsen was unable to produce infection by the injection of red blood cells alone. This was because only gametocytes are found in red blood cells and these must enter the louse fly before transmission occurs.

Epidemiology

DISTRIBUTION IN BODY. *Haemoproteus* first invades the lungs and then the liver. The liver is the reservoir for the tiny swimming merozoites that may be observed in the serum of infected birds. Gametocytes develop in the red blood cells.

ENTRY AND EXIT. The bite of the fly permits entrance and exit for *Haemoproteus* parasites.

PATHOGENICITY. It undoubtedly takes more than a pair of *Haemoproteus* organisms, male and female gametocytes, to establish infection in most birds. Body defenses remove foreign pathogens from the body.

IMMUNITY. Some degree of immunity is present because few birds die from infection.

Incubation. Distribution of parasites to the blood from the liver may be observed within a week following removal of medication to suppress blood infection. Initial release of merozoites following infection requires about 1 month.

Course. The course is indefinite. It may persist until the death of the bird or recovery following medication.

Morbidity and Mortality. A small percentage of flocks will develop 90% infection, but very few birds will actually die. Many are culled from flocks because of debility.

Signs and Necropsy. Most birds present no external evidence of infection from *H. columbae.* On long races infected birds often return late or fail to return. Feathers of most affected birds lack the normal gloss or sheen. There is no diarrhea and eyes remain bright and alert. A necropsy usually reveals an enlarged spleen and an off-color liver. Farmer (1965) reported gizzard enlargement following infection with *H. sacharovi.*

Micropathology. Serial sections of liver tissue may reveal cytomeres amid areas of necrosis and hemorrhage. Every infected bird will have some liver damage.

Diagnosis. The presence of *Haemoproteus* gametocytes inside the cytoplasm of red blood cells is diagnostic. Air-dried blood smears stained with Wright's or Giemsa stain following fixation with methyl alcohol must be examined under oil. Suspicious purplish-brown blood pigments may be found in the serum outside the cells. Further examination of the slide may reveal these same pigments in red cells and these granules identify the gametocytes. It is usually very difficult to discern the nucleus of the gametocyte because the pigment granules stipple the sac. Fifty microscopic fields must be examined on each blood smear before a negative diagnosis is rendered. Even this number permits a margin of error.

A wet, unstained hanging drop of blood under a coverslip enables examination for tiny, motile swimming merozoites. The slide must be warmed to maintain motility. Early blood infection may thus be detected before red cells are invaded.

Blood smears may be prepared by securing a tiny drop of blood from the wing vein on the undersurface of the wing joint. The vein is punctured with a sharp clean needle and a tiny drop of blood collected on one end of a clean slide. The smear is formed by permitting the drop of blood to adhere to the lower end of a second slide held at a 45-degree angle. The blood adheres to the upper slide as it moves forward (pulling backward will crush cells). The air-dried slide is then identified with the leg band number of the bird. A red wax pencil should be used for identifying slides. Other materials are obscured by the stain. Slides may be mailed to the laboratory for staining and examination.

Differential Diagnosis. An enlarged spleen may be caused by paratyphoid. A culture must be made to exclude this disease. *Leucocytozoon, Plasmodium,* and *Trypanosomes* may also cause an enlarged spleen. These can be identified by blood smears.

Biological Properties

STRAINS. Various *Haemoproteus* strains are probably present but have not been described.

STABILITY. *Haemoproteus* parasites do not survive outside the vector or host.

Control and Prevention. A *Haemoproteus*-control program must be employed to eradicate the disease. First, a blood test is made on 10% of the flock. If infection is discovered, all of the birds are blood-tested and positive birds segregated in a mosquito-screened pen. Infected birds are culled carefully and inferior birds removed from the loft. The best infected birds are saved for breeding after the pigeons are dusted with dog or cat flea powder to kill the louse flies. The progeny will remain free of the disease if no louse flies are present. Infected birds are treated and a clean flock established.

In clean lofts, all purchased birds or strays should be blood-tested before they are admitted to the loft. Also, all birds returning from races or shows should be dusted to kill flies before they mix with other birds. The fly undergoes a rapid change from larval to the pupal resting state. These pupae are often overlooked on the loft floor and in flight baskets, training trucks, and nest boxes. Cleaning must be conducted at least every 20 days to remove pupae before the flies emerge (Bishopp 1929).

Treatment. In pigeons, Coatney (1935) has shown that Atabrine and Pamaquin suppressed the gametocytes but not the schizonts. In studies conducted by the author (Tudor et al. 1980), similar results were found with quinacrine hydrochloride (Atabrine, Winthrop) and chloroquine phosphate (Aralen, Winthrop). Gametocytes in the red blood cells and merozoites in the serum disappeared from the blood in 4–6 weeks. When treatment was discontinued active liver schizont infection served to replace the merozoites and gametocytes in the circulating blood. Continued daily medication over a period of 3 mo with 18 mg Atabrine is successful in completely destroying schizont and gametocyte infection. The drug acts to kill swimming merozoites and sporozoites first; thus red cell penetration and infection ceases. The routine death, degeneration, and replacement of old red blood cells accounts for much of the disposal of infected red blood cells. Red blood cells normally survive about 4–6 wk before they are worn out and removed from the circulation. The drug provides a means of curing infected valuable birds, but tablets must be hand-fed, one tablet each day. The drug is poisonous above this dosage level and will kill birds; below this level it is not effective. Water or feed medication will not guarantee an effective nontoxic dose.

Preliminary tests were conducted by the author on ivermectin (Eqvalan, Merck) at 20 mg/ml. Eleven *Haemoproteus*-positive pigeons from two lofts were treated (Table 9.4). The drug was diluted 1 ml/100 ml water and applied by the bird owners with Q-tips to the legs once a week starting 11/21/88 and ending 6 wk later on 12/27/88.

Of blood samples collected from all birds on 12/27/88, five birds, including 02, were definitely positive. The rest were negative or suspicious. Treatments were continued on positive birds on 2/20/89, but apparently because of an outdated drug and continued liver infection, all birds except two were found to be positive by 5/9/89 as indicated by blood tests on 2/20–21/

Table 9.4. Results of blood tests on *Haemoproteus*-positive pigeons treated with Eqvalan

	Loft 1 Birds					Loft 2 Birds					
Band number	10	119	02	118	120	629	1560	638	1261	1259	1020
11/21/88	+	+	+	+	+	+	+	S	+	+	+
12/27/88	–	+	+	–	+	–	+	S	–	–	+
1/9/89[a,b]		–	+		–						
1/13/89						–	–	–	–	–	S
1/16/89	–		+	–							
2/20/89[c]						+	+	–	–	+	–
2/21/89	+	–	+	–	S						
3/20/89	+		+		–						
4/3/89	+	+	+	–	–						
4/9/89						–		+	–	–	+
5/5/89	+	–	+		+						
5/9/89						–	+	+	–	–	+
8/20/89	–	–	+	–							
8/22/89							–		–		

Note: S = suspicious.
[a]End treatment period for all birds.
[b]Blank spaces mean the birds were not available.
[c]Treatment continued on positive (+) birds.

89, 3/20/89, 4/3/89, 4/9/89, 5/5/89, and 5/9/89. However, three of the positive birds reverted to a negative status during this period.

A new supply of Eqvalan water-soluble liquid for horses (10 mg/ml) was then obtained. It was given orally undiluted 10 drops twice a week to bird 02 by the author for 6 wk starting 7/3/89 and ending 8/11/89. The 10 drops were delivered by a 20-gauge needle. The dose was approximately 10 times the dose for a horse on a weight basis. Bird 02 remained highly positive on the basis of Wright-stained blood smears. This strain of *Haemoproteus* appears unaffected by the drug as indicated by an increase in infected red cells in spite of treatment. Blood tests 8/20/89 on four available birds untreated since 5/9/89 found three to be negative. Two other birds tested negative on 8/22/89. A total of seven of the initial eleven were then negative on 8/22/89, suggesting a partial cure.

Obviously caution must be employed in the use of the drug. Rubber gloves should be used to prevent adsorption by the hands of the fancier and the drug should be refrigerated to reduce contamination by microorganisms. The assistance of a veterinarian is essential in establishing the effectiveness of any drug employed.

References

Acton, W. H., and R. Knowles. 1914. India J Med Res 1:663-90.
Adie, H. 1924. Bull Soc Pathol Exot 17:605-13.
Alcock, A. W. 1914. Nature 93:584.
Alicata, J. E. 1939. Hawaii Agric Exp Stn Rept, 1938, pp. 79-82.
______. 1947. Pacific Sci 1:69-84.
Baker, J. R. 1957. J Protozool 4:204-8.
Becker, E. R. 1959. In *Diseases of Poultry,* 4th ed., ed. H. E. Biester and L. H. Schwarte, pp. 888-92. Ames: Iowa State University Press.
Becker, E. R., W. F. Hollander, and W. H. Pattillo. 1956. J Parasitol 42:474-78.
Bishopp, J. C. 1929. Am Pigeon J 18:419-20.
Coatney, G. R. 1933. Am J Hyg 18:133-60.
______. 1935. Am J Hyg 21:249-59.
______. 1936. J Parasitol 22:88-105.
Coatney, G. R., and B. B. Hickman. 1952. J Parasitol 38:12.
Couch, A. B., B. Grabstald, and K. J. Kimbrough. 1961. J Wildl Manage 25:440-42.
Drake, C. J., and R. M. Jones. 1930. J Sci Iowa State Coll 4:253.
Fallis, A. M., and D. M. Wood. 1957. Can J Zool 35:425-35.
Farmer, J. N. 1965. Proc Iowa Acad Sci 71:537-42.
Giovannoni, M. 1946. Argent Biol Tecnol 1:19-24.
Hanson, H. C., et al. 1957. J Parasitol 43:186-93.
Herman, C. M. 1938. Trans Am Microsc Soc 57:132-41.
______. 1944. Bird Banding 15:89-112.
Herman, C. M., and B. Glading. 1942. Calif Fish Game 28:150.
Huff, C. G. 1932. Am J Hyg 16:618-23.
______. 1942. J Infect Dis 71:18-22.
Kartman, L. 1949. Pac Sci 3:127-32.
Lastra, I., and G. R. Coatney. 1950. J Natl Malaria Soc 9:151.
Levine, N. D. 1961. *Protozoan Parasites of Domestic Animals and of Man.* Minneapolis: p. 271. Minneapolis: Burgess, p. 271.
______. 1985. *Veterinary Protozoology.* Ames: Iowa State University Press.
Levine, N. D., and S. Kantor. 1959. Wildl Dis 1:1-38.
Narang, J. R., and P. K. Bhatnagar. 1969. Punjab Vet 8:10-11.
Olsen, O. W. 1962. *Animal Parasites: Their Biology and Life Cycle.* Minneapolis: Burgess, p. 67.
Springer, W. T. 1984. In *Diseases of Poultry,* 8th ed., ed. M. S. Hofstad et al., pp. 730-31. Ames: Iowa State University Press.
Stubbs, E. L. 1948. Univ Pa Bull 48, Vet Ext Q 111.
Tudor, D. C., et al. 1980. Proc Northeast Avian Disease Conference, Cornell University, Ithaca, N.Y.
Wenyon, C. M. 1926. *Protozoology.* Vol 2. London: Bailliere, Tindall & Cox, p. 888.
Wetmore, P. W. 1941. J Parasitol 27:379.
Wood, S. F., and C. M. Herman. 1943. J Parasitol 29:187-96.

Toxoplasmosis

Definition and Synonym. Toxoplasmosis is a prevalent acute or chronic, transmissible intracellular, coccidian protozoan parasitic disease that affects humans, mammals, and birds. Pigeons serve as intermediate hosts and the cat as the final host. Evidence of the disease depends on the number and localization of the parasites.

Toxoplasma infection is another name for this disease.

Cause and Classification. The parasite belongs in the phylum Apicomplexa, class Sporozoasida, order Eucoccidiorida, family Sarcocystidae, genus and species *Toxoplasma gondii* (Levine 1985).

Nature

LIFE CYCLE. *Toxoplasma* parasites complete their life cycle in members of the cat family only. Both asexual and sexual stages occur in the cat following the ingestion of sporulated oocysts, bradyzoite cysts, or tachyzoites. Infected animals and birds may harbor cysts and free-tissue trophozoites. Only following ingestion of bradyzoites in uncooked flesh are the five rapid multiplication stages (types A–E) identified in cats. This is a special type of schizogony called endodyogeny (Frenkel 1973b). In most other animals, humans, and birds, including pigeons, the ingestion of sporulated oocysts and/or cysts containing bradyzoites results in asexual cycles without completion of the life cycle, which precludes the production of oocysts.

The sexual cycle occurs only in the final host, the cat. Four forms of the parasite (schizonts, merozoites, gametocytes, and oocysts) occur only in the epithelial cells lining the cat's small intestine. The cat as the definitive host voids undeveloped, noninfective oocysts in the feces. Sporulation or maturation occurs outside the host and takes 2–4 days at 75°F. During the process two sporocysts form inside each oocyst, and these in turn each form four elongated, curved sporozoites. Oocyst production in the cat lasts approximately 10–20 days, and during this time enormous numbers of oocysts may be released in the feces. Under favorable conditions of moisture and temperature, oocysts will remain infective for several months. Oocyst shedding is noted in cats 3–5 days after the ingestion of cysts containing bradyzoites, 5–10 days or longer following tachyzoite ingestion, and 20–24 days or longer after oocyst ingestion. This time lapse until oocysts appear is called the prepatent period (Jones 1973). When sporulated oocysts are ingested by the cat, eight sporozoites are released to initiate the development of schizonts, which produce merozoites. Eventually merozoites appear with the ability to form male and female gametocytes. As gametocytes mature with the formation of micro- and macrogametes, fertilization of the female takes place. The resulting zygote becomes the oocyst, which is released as the host intestinal cell ruptures.

Some sporozoites, released from oocysts in cats do not enter the sexual cycle in the intestine but are carried to various parts of the body by the circulation and become replicating trophozoites, as occurs in other animals (Jones 1973; Frenkel 1973b).

Unlike most other coccidia, the asexual cycle in intermediate hosts includes two extra intestinal forms of the parasite. These include free proliferative trophozoites or rapidly replicating tachyzoites and chronic resting-state cysts enclosing slowly replicating encysted organisms or bradyzoites.

Ingestion of sporulated oocysts releases sporozoites, and initiates the proliferative or multiplication phase in the intermediate host. These trophozoites are distributed in body tissues by the bloodstream and survive only a short time outside the body (Jones 1973). They are chiefly found in brain and striated muscle but also appear in reticuloendothelial cells of the lungs and spleen, leukocytes, cardiac muscle, and retinal tissues. As immunity in the intermediate host increases, multiplication shifts to intracellular, persistent cysts (Olsen 1974), which become packed with hundreds of naked bradyzoites. When the host cell dies or cysts deteriorate, bradyzoites may be released to infect other cells. Bradyzoites then revert to proliferating tachyzoites (Jones 1973).

The pigeon that consumes *Toxoplasma* oocysts becomes an intermediate host carrying asexual free-tissue trophozoites and cystic stages in its body. It does not produce oocysts.

FORM. Each stage of the parasite has a different form and size. The trophozoite stage of the parasite is a minute crescent or banana-shaped body 4–8 μm long and 2–4 μm in diameter (Siegmund 1979). Trophozoites are pointed on the anterior end and ovoid or rounded posteriorly (Manwell and Drobeck 1953; Scheffield and Melton 1968).

The round to oval cysts vary in size from 50 to 150 μm in diameter and have a thick resilient wall. These contain bradyzoites (Siegmund 1979). The rounded schizonts in the small intestine contain bipolar merozoites. Microgametocytes also appear in the cells of the intestine and contain 12–32 male microgametes. Nearby macrogametocytes containing macrogametes when fertilized form the zygote oocysts shed in the feces (Springer 1984). The oocyst measures 9–11 μm by 11–14 μm (Olsen 1974) and resembles *Isospora bigemina* coccidiosis oocysts (Hutchinson et al. 1971). The two sporocysts within the oocyst are about 6–8 μm in length. They each form four sporozoites 2 by 8 μm (Olsen 1974).

SEROLOGY. In pigeons Miller et al. (1972) found that after feeding 1000 or 2000 oocysts the blood dye titer rose to >128 in some birds. Following intraperitoneal inoculation Jacobs and Melton (1952) observed a dye titer of 1:256 or more that lasted over a year. They further noted that at 18 mo some birds showed 1:4 whereas other titers remained high. Further studies were made by Camargo et al. (1978), who used enzyme linked immunosorbent assays as a means of studying toxoplasmosis serological patterns.

Distribution

GEOGRAPHY. The disease is worldwide. The parasites were first found in the gundi, a North African rodent (*Ctenodactylus gundi*), by Nicolle and Manceaux (1909).

INCIDENCE. The parasite is not common in pigeons and is not generally recognized clinically; a percentage of birds, however, are infected. Most human cases are subclinical, but serologic surveys confirm a high prevalence of infection in humans and animals. Frenkel (1973a) has further indicated that 3000 babies are born with the disease each year. According to Lyght (1966) up to 59% of apparently healthy adults yield positive skin tests. In animals serologic studies have shown that pigs and sheep have high prevalence rates. Lower rates occur in dogs and cats and those are followed by horses and cattle (Siegmund 1979). Dubey et al. (1970) made a separate study of cats in the Kansas City area and found 45% had antibodies.

SEASON, AGE, SEX, AND BREED. *Toxoplasma* infection can occur at any time. Natural resistance is greater in mature individuals (Olsen 1974). There is no reported difference between the sexes or breeds.

SPECIES. Over 50 wild and domestic animals and birds have been reported infected by *Toxoplasma*. Pigeon infection was reported by Carini (1911), Jacobs and Melton (1952, 1966), Miller et al. (1972), and Wenyon (1926).

CARRIERS. Animal and birds may serve as asymptomatic *Toxoplasma* carriers.

Transmission. There are three known infective stages of *Toxoplasma*: bradyzoites in cysts of animal or bird flesh, free tachyzoites in flesh, and sporozoites in oocysts shed in cat feces. Transmission of these may occur by ingestion of uncooked meat and sporulated oocysts, congenital or transplacental infection, and by blood transfusions (Jones 1973). People as intermediate hosts may possibly become infected by accidental ingestion of infectious sporulated cat oocysts. In a few instances the organism appears to have been inhaled, but the ingestion of undercooked rabbit or pork appears to be the primary source (Lyght 1966). Free-tissue trophozoites may be found in blood and in excretions and secretions of infected animals and birds. This stage of the parasite may thus be transferred in milk from humans, cows, goats, sheep, pigs, dogs, cats, rabbits, guinea pigs, and mice (Frenkel 1973a). Presumably this stage could likewise be transferred in the crop milk of the pigeon. In humans, free trophozoites may cross the placenta to the fetus during the last 5–6 mo of pregnancy (Jones 1973). In pigeons, congenital transmission has not been noted but infected chicken eggs have been reported (Sparapini 1950). Iannuzzi and Renieri (1971), however, concluded that the parasite did not survive in unembryonated eggs.

Cats become infected from eating the intermediate host such as infected mice or birds or by eating sporulated oocysts from a cat. In turn, the cat produces oocysts, which serve to infect the intermediate host such as an animal or a pigeon. The pigeon will not likely infect humans unless infected pigeons are eaten in an undercooked condition. In mice, the eggs of *Toxocara cati,* a roundworm of cats, transmitted toxoplasmosis to mice (Jacobs and Melton 1966). In addition, Dubey and Sharnma (1980) considered goat semen as a means of transmission. These findings may open still other avenues of transmission for the pigeon.

Epidemiology

DISTRIBUTION IN BODY. In pigeons, toxoplasmosis cysts may be present in more than one organ or tissue. Miller et al. (1972) found parasites in brain, spleen, liver, lung, heart, kidney, and muscle but no oocysts were shed. Jacobs and Melton (1952) found parasites in the serous exudate of the eyes of pigeons. In chickens, Jacobs and Melton (1966) reported ovary and oviduct infection, in addition to infection in brain, muscle, kidney, gizzard, and intestine. Also in chickens, Bickford and Saunders (1966) found brain, heart, testicle, and pancreas infection after intramuscular inoculation.

ENTRY AND EXIT. In humans, *Toxoplasma* infection may possibly develop as a result of oocyst ingestion or in a few instances by inhalation. Most human cases have been contracted from ingestion of cysts or trophozoites (Lyght 1966). Adult pigeons become infected by oocyst ingestion (Miller et al. 1972). Since adult pigeons feed their squabs, theoretically an infected pigeon can transfer tachyzoites to squabs in crop milk.

In humans, infective trophozoites may be passed in human milk.A newborn infant may likewise carry the parasite. In pigeons the parasite can possibly leave the body in eggs or crop milk.

IMMUNITY. Antibodies definitely limit circulating blood *Toxoplasma* parasites (Jacobs and Melton 1952). Less virulent organisms encyst more readily as immunity increases.

Incubation. Clinical signs of toxoplasmosis may be apparent in 24 hr.

Course. Jacobs and Melton (1952) found *Toxoplasma* parasites in the blood in pigeons for at least 1 yr after inoculation. Some birds had infection for over 18 mo. This suggests a prolonged indefinite course.

Morbidity and Mortality. The number of sick birds in a flock or loft depends on the numbers of *Toxoplasma* oocysts to which the birds are exposed. In a chicken flock, 50% of the birds died following a natural infection (Lund 1972). Similar losses in pigeons may be expected with virulent strains.

Signs. Clinical signs of toxoplasmosis include

fever, loss of weight, inappetence, general debility, incoordination, blindness, nervous signs, and death.

Necropsy. Gross findings of toxoplasmosis often include liver and spleen enlargement, lung congestion, and heart and brain involvement. Jacobs and Melton (1952) also reported eye infection following intraperitoneal inoculation of trophozoites in pigeons. Miller et al. (1972) recovered the parasite from many organ lesions after feeding cat oocysts.

Micropathology. Cysts may be identified in brain or eye tissue sections. The meninges of the brain often show inflammatory changes, with the accumulation of lymphocytes and plasma cells about areas of degeneration and necrosis and about blood vessels.

Diagnosis. The dye test in pigeons is quite effective (Springer 1984). According to Bickford and Sanders (1966), the Sabin-Feldman dye test was of little value in chickens because antibody levels were not detectable. The indirect fluorescent antibody test has been used to detect early immunoglobulins in cats and the indirect hemagglutination test and the complement fixation test also detect the same antibodies (McKinney 1973). The enzyme linked immunosorbent assay test, a sensitive serological test, has been used by Camargo et al. (1978). Voller et al. (1976) reported success with the use of a microplate enzyme immunoassay.

Differential Diagnosis. The diagnosis of toxoplasmosis is often difficult. In pigeons, exotic Newcastle disease, paramyxovirus and herpesvirus infections, and salmonellosis must be considered.

Biological Properties

STRAINS. A single species is recognized, but organisms vary in their degree of virulence for pigeons. Jacobs and Melton (1952) noted a greater degree of parasitemia in pigeons when birds were inoculated with the virulent RH strain.

STAINING. The trophozoites stain well with Giemsa (Siegmund 1979). Romanowsky stain is reported by Wenyon (1926) as effective. Wright's stain may also be used for impression smears.

STABILITY. Oocysts in general withstand most chemical disinfectants. They will survive several months on the ground. Yilmoz and Hopkins (1972) determined that shaded oocysts were infective for over a year. Strong iodine and ammonia solutions will inactivate oocysts, and they will be destroyed by dry heat at 70°C and by boiling water (Siegmund 1979).

CULTURE. Inoculation of mice with suspensions of brain, lung, or spleen by intraperitoneal or intracranial routes is preferred (Frenkel 1981). Chicken embryos may be inoculated in the chorioallantoic cavity. Death occurs in 7–10 days postinoculation. Yellowish white lesions may be noted on the chorioallantoic membrane. Cell cultures have also been used for isolation purposes.

Prevention. To prevent the disease in pigeons, cats and cat feces must be excluded from the loft environment and pigeons fed only in clean hoppers. House cats should be fed only dry, canned, or cooked food and prevented from hunting animals or birds. To prevent the disease in humans, meat should be thoroughly cooked before it is eaten. Hands should be washed after handling raw meat. Changing cat litter pans every 2 days if the animal hunts will enable the attendant to avoid contact with sporulated oocysts. Cat feces and litter should be disposed of in a plastic bag and buried under 2 ft of soil. A spatula should be used to clean the litter pan, along with boiling water. Stray cats and rodents, flies, and cockroaches, which can mechanically carry oocysts, must be avoided or removed. Children's sandboxes should be covered and the area under back porches enclosed.

Treatment. In pigeons, treatment for *Toxoplasma* infection is generally not indicated. Levine (1973) has suggested the use of a combination sulfadiazine and pyrimethamine in laboratory animals and humans.

References

Bickford, A. A., and J. R. Saunders. 1966. Am J Vet Res 27:308–18.

Camargo, M. E., et al. 1978. Infect Immun 21:55–58.

Carini. 1911. Bull Soc Pathol Exot 4:518.

Dubey, J. P., and S. P. Sharnma. 1980. Am J Vet Res 41:794–95.

Dubey, J. P., N. L. Miller, and J. K. Frenkel. 1970. J Parasitol 56:447–56.

Frenkel, J. K. 1973a. Biol Sci 23:343–52.

______. 1973b. In *Parasite: Life Cycle, Pathology and Immunity,* ed. D. M. Hammond and P. L. Long, pp. 343–410. Baltimore: University Park.

______. 1981. J Parasitol 67:952–53.

Hutchinson, W. M., et al. 1971. Trans R Soc Trop Med Hyg 65:380–99.

Iannuzzi, L., and G. Renieri. 1971. Acta Med Vet (Napoli) 17:311–17.

Jacobs, L., and M. Melton. 1952. J Parasitol Suppl Abstr 38:12.

______. 1966. J Parasitol 52:1158–62.

Jones, S. R. 1973. J Am Vet Med Assoc 163:1038–42.

Levine, N. D. 1973. *Protozoan Parasites of Domestic Animals and of Man.* 2d ed. Minneapolis: Burgess.

______. 1985. *Veterinary Protozoology.* Ames: Iowa State University Press.

Lund, E. E. 1972. In *Diseases of Poultry,* 6th ed., ed. M. S. Hofstad, pp. 990–1046. Ames: Iowa State University Press.
Lyght, C. E., ed. 1966. *The Merck Manual.* 11th ed. Rahway, N.J.: Merck, pp. 903–5.
Manwell, R. D., and H. P. Drobeck. 1953. J Parasitol 39:577–84.
McKinney, H. R. 1973. Vet Med Small Anim Clin 68:493–95.
Miller, N. L., J. K. Frenkel, and J. P. Dubey. 1972. J Parasitol 58:928–37.
Nicolle and Manceaux. 1909. C R Acad Sci 147:763.
Olsen, O. W. 1974. *Animal Parasites.* 3d ed. Baltimore: University Park, p. 562.
Scheffield, H. G., and M. L. Melton. 1968. J Parasitol 54:209–26.
Siegmund, O. H., ed. 1979. *The Merck Veterinary Manual.* 5th ed. Rahway, N.J.: Merck.
Sparapini, G. C. 1950. Pediatria 58:411–14.
Springer, W. T. 1984. In *Diseases of Poultry,* 8th ed., ed. M. S. Hofstad et al., pp. 736–40. Ames: Iowa State University Press.
Voller, A., et al. 1976. J Clin Pathol 29:150–53.
Wenyon, C. M. 1926. *Protozoology.* Vol. 2. New York: William Wood.
Yilmoz, S. M., and S. H. Hopkins. 1972. J Parasitol 58:938–39.

Trichomoniasis

Definition. Trichomoniasis is a transmissible flagellated protozoan parasitic disease primarily affecting pigeons, doves, and turkeys. It is characterized by necrotic ulceration of the mouth, esophagus, crop, and proventriculus and by occasional invasion of the liver.

Synonyms for this disease are Canker, frounce, *cercomonas gallinae* infection (Rivolta 1878; Rivolta and Delprato 1880), *Cercomonas hepaticum* infection (Stabler 1954, quoting Rivolta 1878), *Trichomonas columbae* infection (Raillet 1885; Jacquette 1948a, quoting Neumann 1892). According to Jacquette (1948a), *columbae* was placed in the genus *Trichomonas* by Railliet (1885) and Neumann (1892). Stabler (1941a) studied *T. columbae* and *T. gallinae* and found them to be identical.

Classification. The parasite belongs in the phylum Sarcomastigophora, class Zoomastigophorasida (flagellata), order Trichomonadorida, family Trichomonadidae (Olsen 1974), genus *Trichomonas,* and species *gallinae* (Stabler, 1938b, quoting Rivolta 1878).

Nature

REPRODUCTION. The *Trichomonas* life cycle is simple.The parasite divides by longitudinal binary fission. There are no sexual stages and no cysts (Shorb 1964).

FORM. The single-celled parasite is characterized by a clear, narrow, longitudinal axial rod, the axostyle, which passes through the center of the body and emerges posteriorly. The undulating membrane which does not reach the end of the body, is bordered by a long flagellum directed posteriorly. This flagellum does not trail the parasite. Four anterior flagella arise from the blepharoplast, which is composed of several basal granules anterior to the single nucleus of the cell. An accessory filament is associated with the long flagellum. Another filament, the costa, originates at the blepharoplast and passes along two-thirds to three-fourths of the base of the undulating membrane.The sausage-shaped parabasal body adjacent to the nucleus also extends posteriorly from the blepharoplast and is provided with a parabasal filament. The anterior end of the axostyle is enlarged to provide a capitulum. Near it is a cytostome. The pelta is located on the anterior margin of the cell and just anterior to the blepharoplast (Tanabe 1926; Levine 1973). The organism is ellipsoid or roughly pear-shaped and varies in size, 6.2–18.9 μm by 2.3–8.5 μm (Stabler 1941a). (See Fig. 9.7.)

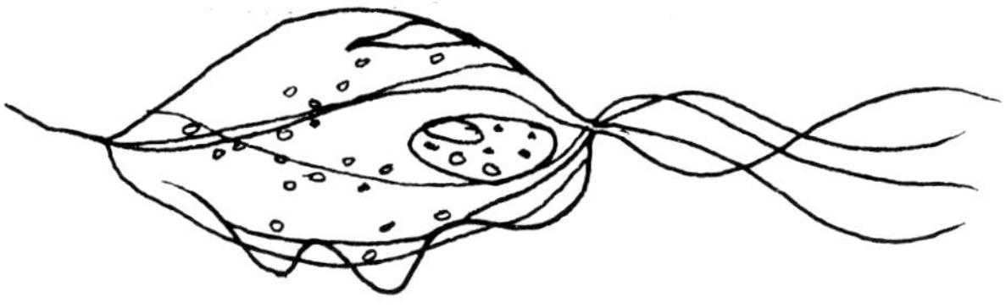

Fig. 9.7. *Trichomonas gallinae.* This is perhaps the most common pigeon parasite.

Distribution

GEOGRAPHY. The parasite is found wherever pigeons are raised. In 1878 Rivolta in Italy first described the parasite and named it *Cercomonas gallinae.* He also reported the organism in the small intestine and was the first to transfer the organism from pigeon to pigeon. Later, Rivolta and Delprato in 1880 described and sketched the organism. In 1890 Babes and Puscariu in Romania noted diphtheria in affected pigeons. In 1907 Jowett reported from South Africa the presence of the parasite in a pigeon liver. In Hungary, Ratz (1909) also described the organism. He was followed by Prowozek and Aragao in the same year in Germany. In 1919 Waterman found flagellates in pigeons in the West Indies. The next report came from van Heelsberger (1925) in Holland. Oguma then reported the disease in Japan in 1931. Waller (1934) in Delaware, first reported the occurrence of the parasites in pigeons in America. Yakimoff (1934) noted the parasite in chickens in Russia. In 1936 Cauthen in Washington, D.C., reported the infection in pigeons, ringdoves, and mourning doves in a cage colony on Long Island, N.Y. The next year Callender and Simmons (1937) gave an account of *T. gallinae* in Java sparrows in Panama. Hees (1938) recovered the organism from Shetland Island seagulls. Hart (1941) in Australia noted a similar infection in an Indian dove. Schorger (1952) quoted Moore in Great Britain in 1735 as having seen canker in pigeons. It would appear that the parasite has shared global recognition well before first written reports.

INCIDENCE. Virtually all pigeons carry *Trichomonas.*

SEASON. No seasonal variation occurs.

AGE, SEX, AND BREED. Pigeons at any age may be exposed and can contract trichomoniasis. Actually most squabs are infected within a short time after hatching and often show clinical disease. According to Florent (1938), squabs are particularly susceptible at the time of weaning and at the first molt. Adults usually are infected but show no sign of the disease.

Both sexes appear to be affected equally, and in the author's experience, no breed predisposition has been observed but Miessner and Hansen (1936) felt that roller and tumbler pigeons were more sensitive.

SPECIES. The following wide variety of birds are parasitized by *T. gallinae.* An asterisk (*) indicates they have been reported experimentally infected.

Pigeon (*Columba livia*): Rivolta (1878); Railliet (1885); Babes and Puscariu (1890); Jaquette (1948a), quoting Neumann (1892); Jowett (1907); Prowozek and Aragao (1909); Ratz (1909); Waterman (1919); van Heelsberger (1925); Oguma (1931); Cauthen (1934, 1936); Waller (1934); Wittfogel (1935); Miessner and Hansen (1936); Hurt (1940); Levine et al. (1941); Brocklehurst (1944); Rosenwald (1944); Hollander (1945); Stabler and Engley (1946); Jaquette (1948), Stabler (1948a,b, 1951 a,b, 1953a, 1954, 1957); Stabler and Kihara (1954); Perez Mesa et al. (1961); Stabler et al. 1964; McLoughlin 1966
Band-tailed pigeon (*Columba fasciata*): Stabler (1941b, 1951b); Stabler and Matteson (1950); Sileo (1969)
Wood pigeon (*Columba palumbus*): Jansen (1944)
Ringed dove (*Streptopelia risoria*): Cauthen (1934, 1936); Stabler (1941b)
Mourning dove (*Zenaidura macoura*): Cauthen (1934, 1936); Stabler (1941b, 1951a,b); Stabler and Matteson (1950); Stabler and Herman (1951); Haugen (1952); Kocan (1969)
White-winged dove (*Zenaida asiatica*): Levine (1973)
Indian dove (*Turtur suratensis*): Hart (1941)
Inca dove (*Scardafella inca*): Locke and James (1962)
Verraux's dove: Callender and Simmons (1937)*
Owl: Stabler (1941b)*
Golden eagle (*Aquila chrysaetos*): Stabler (1941a); Levine (1973)
Sea gull: Hees (1938)
Cooper hawk (*Accipiter cooperii*): Stabler (1941b, 1947); Levine (1973)
Red-tailed hawk (*Buteo iamaicensis*): Stabler (1947)
Duck hawk (*Falco peregrinus anatum*): Stabler (1941b, 1947); Levine (1973)
Red-shouldered hawk (*Buteo lineatus*): Stabler and Shelanski (1936); Stabler (1941b, 1947)
Sparrow hawk or kestrel (*Falco sparverius*): Stabler and Shelanski (1936); Stabler (1947); Stone and Jones (1969)
Swallow: Stabler (1953b)*
Java sparrow (*Muria oryzivora*): Callender and Simmons (1937)
English sparrow: Levine et al. (1941)
Song sparrow (*Melospiza melodia*): Stabler (1953b)*
Tovi parakeet: Callender and Simmons (1937)*
Cockatiel (*Nymphicus hollandicus*): Murtaugh and Jacobs (1984)
Orange-cheeked waxbill: Levine (1973)
Canary: Levine et al. (1941)*
Zebra finch (*Poephila castanotis*) Levine (1973)
Goldfinch (*Spinus tristis*): Stabler (1953b)*
Duckling: Levine et al. (1941)* (no lesions)
Bobwhite quail (*Colinus virginianus*): Jungherr (1929); Levine et al. (1941)*
Chicken (*Gallus gallus*): Schaaf and Scherle (1938); Levine and Brandly (1939); Levine et al. (1941); Levine (1973); Stabler (1941b)
Turkey (*Melagris gallopavo*): Jungherr (1929); Bushnell and Twiehaus (1940); Stabler (1938a, 1941b); Levine et al. (1941)*
Guinea pig: Bos (1933)*; Wittfogel (1935)*
Mice: Bos (1933)*; Wittfogel (1935)*; Hees (1938)*; Honigberg (1964)*
Rats and kittens: Levine (1973),* quoting Rakoff (1934)

NATURAL HOSTS AND CARRIERS. The domestic pigeon is the primary host and reservoir of

T. gallinae, but it also occurs in numerous other birds. They may serve as nonclinical, inapparent carriers for years.

Transmission. Trichomonads are transmitted from the adult birds to the squabs in the crop milk. Hawks become infected from eating their prey. Turkeys and chickens become infected as they drink from water sources contaminated by the mouth discharges of feral pigeons (Kocan 1969). Since these flagellates are very susceptible to drying and changes in temperature, they must be ingested soon after leaving the host if they are to be infectious. Droppings, however, are not considered a factor in transmission by Stabler (1954). Transmission by flies and other mechanical carriers must be assumed, but there is no congenital egg transmission, so squabs are not infected when hatched.

Epidemiology

DISTRIBUTION IN BODY. Trichomonads chiefly confine themselves to the upper digestive tract: mouth, esophagus, crop, and proventriculus. On occasion parasitism occurs in the liver, heart, lungs, pancreas, air sacs, and pericardium. Blood and abdominal fluid may contain parasites, and where conjunctivitis is present, parasites may be found in ocular discharges. According to Stabler (1954), parasites are rarely found in the spleen, kidney, trachea, upper respiratory tract, bone marrow, and ear linings.

ENTRY AND EXIT. Trichomonads enter by way of the mouth. They leave the body in the regurgitation of crop milk from the mouth and in ocular and nasal discharges.

PATHOGENICITY. In pigeons, clinical trichomoniasis is often found in young birds, but practically all adult birds are infected without showing clinical signs of disease. Regardless of the strain involved, the parasite produces the disease without the aid of intercurrent bacteria. Kocan (1971a) divided the strains into nonvirulent strains, oral canker strains, and visceral strains and found that dexamethasone increased the virulence of oral canker strains. He also showed (1971b) that the route of inoculation determined the course of disease; oral and pulmonary routes caused death by liver or lung infection whereas subcutaneous and intravenous routes produced local lesions.

Dwyer (1974) studied the antigenic nature of the parasite and concluded that virulence and antigenic composition were related. Stepkowsji and Honigberg (1972) provided an antigenic analysis of virulent and nonvirulent strains, and Kocan and Herman (1970) evaluated the serum protein values of immune and nonimmune pigeons.

IMMUNITY. Recovery from sublethal strains appears to produce immunity to more virulent strains (Stabler 1948b). This apparent immunity may also result from persistent low-grade infection. Kocan (1970) has shown that plasma from pigeons harboring mild strains protects other pigeons against a virulent strain.

Incubation. The clinical disease may develop in 2–3 days, but in most cases no evidence of infection will be noted unless a virulent strain is involved. Certainly infection is established upon transfer of parasites in sufficient numbers to overcome the immune response of the bird.

Course. The course of trichomoniasis starts when clinical infection is observed. The length of the disease in a bird varies, depending on the strain of the parasite, the age of the bird, and the treatment given. It can be rapidly fatal in young birds, producing death 4–18 days after infection or it can be a never-ending, long-standing, chronic infection.

Morbidity. Almost all birds carry *Trichomonas*. Most birds show no evidence of infection, but when a concurrent infection, such as pox is present, many of the flock may be visibly sick.

Mortality. Pigeons seldom die from trichomoniasis alone, but losses can reach almost 100% when birds are stressed or encounter a virulent strain of the organism.

Signs. Loss of weight, ruffled feathers, reluctance to move, huddling, inappetence, and stringy mucus in the mouth may be noted in the most heavily *Trichomonas*-infected birds. Very severely affected birds present, in addition, small, yellowish circumscribed areas on the mucous membrane of the mouth. These lesions increase in size and number, forming caseous masses that invade the roof of the mouth and sinuses. They may extend posteriorly, coating the pharynx and esophagus. As the masses thicken, they may crack and bleed. Cells die and a putrid odor develops. Thick, grayish, turbid mucus may coat the membranes of the mouth. A watery discharge may be observed from the conjunctival sac in isolated cases.

Necropsy. In trichomoniasis, mouth lesions may be observed in live or dead specimens. In severely affected birds, upper digestive tract lesions consist of raised, yellowish, purulent, independent or confluent caseous rough masses on the mucous membranes of the esophagus or crop that may ulcerate and extend into the soft tissue of the head and neck if the disease is untreated. A large amount of stringy, discolored mucus may accumulate in the crop and proventriculus. Other lesions resulting from systemic infections may be found in the liver and to a lesser extent in the lungs, air sacs, heart, and pancreas. A few white or yellowish solid, circumscribed nodules may form in the liver or lung substance, producing focal abscesses 1 cm or more in diameter.

Micropathology. Perez Mesa et al. (1961) described the mucosal cellular findings as a massive inflammatory response primarily involving heterophiles. In the liver the inflammatory reaction replaced destroyed liver cord cells with infiltrated heterophiles and mononuclear cells.

Diagnosis. A tentative diagnosis may be established by mouth scrapings suspended in saline on a warm slide. Typical localized, spermlike movement of the tiny translucent parasites may be observed under low-power light microscopy. The parasites do not leave the microscope field but circle in the area with jerky progression. Old infections may be suspected following a visual examination of the mouth for curdy lesions. Demonstration of the parasite is necessary to establish a diagnosis. Identification may be aided by growing the parasites on culture media followed by examination of Wright's-stained smears under oil with a light microscope. Darkfield or phase contrast illumination also aids in observing living specimens.

Differential Diagnosis. Trichomoniasis must be distinguished from vitamin A deficiency, pox, and thrush or moniliasis. The parasites themselves must be distinguished from other similar protozoa.

Biological Properties

STRAINS. Stabler (1948a) has reported *Trichomonas* strains with varying degrees of virulence. The virulent "Jones Barn" strain was passed through 119 *Trichomonas*-free pigeons, and 114 of these birds died (Stabler 1953a). This strain primarily affects the liver and only moderately affects the head and neck (Stabler et al. 1964). When a single organism of this strain was given to 10 clean pigeons, 5 birds died in 8–13 days (Stabler and Kihara 1954). Later 293 deaths were reported from 300 birds (Stabler 1957). The Mirza strain, on the other hand, produces extensive damage to the head and mucosa of the upper digestive tract and seldom affects the liver (Stabler 1954). Narcisi et al. (1991) reported on the virulent Eiberg strain.

STAINING. Most blood stains are effective in staining *Trichomonas* smears, and iron-haematoxylin may be used on tissue sections.

STABILITY. Trichomonads are readily destroyed outside the body by drying or by changes in temperature and osmotic pressure. It has no cystic form to protect it. Kocan (1969) found that the organism survived for up to 24 hr in 0.05–0.9% salt (NaCl) solution and in the water extract of soaked wheat, buckwheat, sorghum, and peas for 168 hr.

CULTURE. Cailleau (1935) and Matthews and Daly (1975) made extensive studies of the nutrition of *Trichomonas*. Diamond and Herman (1952) compared 28 culture media and introduced trypticase-yeast extract maltose-cysteine-serum media to cultivate the parasites. Liver infusion or nutrient agar slants overlaid with 5% horse serum in Drbohlav's modified Ringer's solution at a pH of about 5.5 to which 0.2% dextrose has been added is quite effective in holding cultures if nutrients and evaporated water are replaced. Stabler and Engley (1946) used modified Boeck-Drbohlav's media with success. Penev (1971) cultured trichomonads of chickens using modified "CPLM" media. (This media was described in 1943 by Johnson and Trussell. Ringer's solution was used in place of buffered normal saline.)

Prevention. All newly purchased birds and the entire loft should receive routine monthly preventative treatments. Trichomoniasis prevention is not the aim. Treatment to control trichomonads is essential whether clinical disease is present or not.

Control. Chemical suppression of trichomonads on a monthly basis generally maintains adequate control.

Treatment. All birds should be treated 1 day a month.Heavily infected birds should be treated 1 day every 2 wk until the disease is under control; then flock treatment should return to the monthly routine. Most drugs given to pigeons regardless of the type or nature will reduce the number of parasites. The aim is to control the number of parasites but not necessarily eradicate them. They are too common, and birds are too easily reinfected.

Dimetridazole (1,2-dimethyl-5-nitroimidazole) (Emtryl, Salsbury, Rhone-Poulenc) was the drug of choice at 3.6 g/gal drinking water for 1 day only, but the drug was discontinued in the United States. McLoughlin (1966) used the drug at 0.05% in the drinking water for 3–6 days with success. Murtaugh and Jacobs (1984) also reported its use. It is a neurotoxin and can cause temporary paralysis. Dosage levels must be strictly followed. Ipropran (Hoffmann–La Roche) was another effective drug, but it too was discontinued.

Bussieras et al. (1961) reported that the use of metronidazole (Flagyl, Searle) at 60 mg/kg orally for 5 consecutive days effectively controlled pigeon mortality. This was confirmed by Luthgen and Bernau (1967), but the drug was toxic at 25–50 mg/kg body weight. The author has used one 250-mg tablet mixed well in 1 gal water for 50–75 pigeons given 1 day a month. This is the only water provided until it is consumed and then replaced with fresh clean water.

Ronidazole (Ridzol-S, Merck), 4–6 mg/kg body weight, is reported to be less toxic for pigeons and is used in Belgium at 0.06 ppm in the drinking water for 6 days. It has been used for swine dysentery in Mexico and Canada, but it has not been cleared by the Food and Drug Administration in the United States.

Spartrix (Janssen in Belgium) tablets contain 10 mg carnidazole (0-methyl,2-2 methyl-5-nitro-1H-imidazol-1-y1 ethyl carbamothioate). They recommend one tablet per adult pigeon and half a tablet per newly weaned squab (20 mg/kg body weight). The drug must not be used on pigeons intended for food. Research indicates that the drug is effective and safe for pigeons and is recommended.

Enheptin-A (2-acetyl-amino-5-nitrothiazole) at the rate of 0.015–0.05% in the feed for 7–14 days or Enheptin soluble (2-amino-5-nitrothiazole) at the rate of 0.03–0.05% in the water for 7–14 days was recommended by the manufacturer for chicken use. It has been used with success in treating the parasite. Stabler and Melletin (1951, 1953) suggested using Enheptin-A at the rate of 20 mg/day per White King and indicated 50 mg/day would kill Homers. Stabler et al. (1958) used 6.3 g soluble Enheptin/gal water for 7–14 days but suggested that it not be used on breeding birds.

Copper sulfate was suggested by Jaquette (1948b) at a concentration of 35 mg/100 ml drinking water for 20 days or a dilution of about 1:3000 for breeders with squabs.The author has used 50 mg/1000 ml, but when copper sulfate is given in the water at levels above 50 mg/1000 ml, it will reduce water intake if no other water is available. Pigeons will finally resume drinking within 3–4 days if they are unable to get nonmedicated water. Obviously prolonged treatment is detrimental. One day at 1:3000 dilution will reduce parasite numbers, but this is not the best medication. It should not be repeated under 2 wk. Florent (1938) also indicated that pigeons would not tolerate the drug.

Hygromycin B (Elanco, Ind.), which was used to control Blackhead in chickens was tried on pigeon trichomonads by Conrad and Edwards (1970) and Kocan (1972) and was found to be harmful for pigeons.

References

Babes, V., and E. Puscariu. 1890. Z Hyg 8:376–403.
Bos, A. 1933. Zentralbl Bakteriol 130:220–27.
Brocklehurst, F. B. 1944. J R Army Vet Corp 16:130–32.
Bushnell, L. D., and M. J. Twiehaus. 1940. Vet Med 35:103–5.
Bussieras, J., R. Dams, and J. Euzeby. 1961. Bull Soc Sci Vet Lyon 63:307–12.
Cailleau, R. 1935. C R Soc Biol 119:853–56.
Callender, G. R., and J. S. Simmons. 1937. Am J Trop Med 17:579–85.
Cauthen, G. E. 1934. Proc Helminth Soc Wash D. C. 1:22.
______. 1936. Am J Hyg 23:132–42.
Conrad, R. D., and A. G. Edward. 1970. Avian Dis 14:599.
Diamond, L. S., and C. M. Herman. 1952. J Parasitol 38:11–12.
Dwyer, D. M. 1974. J Protozool 21:139–45.
Florent, A. 1938. Ann Med Vet 83:401–28.
Hart, L. 1941. Aust Vet J 17:20–21.
Haugen, A. O. 1952. J Wildl Manage 16:164–69.
Hees, E. 1938. J Egypt Med Assoc 21:813–37.
Hollander, W. F. 1945. Pigeon Loft 2:142–43.
Honigberg, B. M. 1964. J Protozool 11:7–20.
Hurt, L. M. 1940. Annu Rep Los Angeles Co Livest Dep 1939–40.
Jansen, J. 1944. Landbouwkd Tijdschr 33–34.
Jaquette, D. S. 1948a. Proc Helminthol Soc Wash 15:72–73.
______. 1948b. Am J Vet Res 9:206–9.
Jowett, W. 1907. J Comp Pathol Ther 20:122–25.
Jungherr, I. 1929. J Am Vet Med Assoc 71:636–40.
Kocan, R. M. 1969. Bull Wildl Dis Assoc 5:148–49.
______. 1970. J Protozool 17:551–53.
______. 1971a. J Protozool 18 Abstr 103:30.
______. 1971b. J Protozool 18 Abstr 39:15.
______. 1972. Avian Dis 16:714–17.
Kocan, R. M., and C. M. Herman. 1970. Bull Wildl Dis Assoc 6:43–47.
Levine, N. D. 1973. *Protozoan Parasites of Domestic Animals and of Man.* 2d ed. Minneapolis: Burgess, pp. 28–101.
Levine, N. D., and C. A. Brandly. 1939. J Am Vet Med Assoc 95:77.
Levine, N. D., L. E. Boley, and H. R. Hester. 1941. Am J Hyg 33:23.
Locke, L. N., and P. James. 1962. J Parasitol 48:497.
Luthgen, W., and U. Bernau. 1967. Dtsch Tieraerztl Wochenschr 74:301–5.
Matthews, H. M., and J. A. Daly. 1975. J Protozool 22:139–45.
McLoughlin, D. K. 1966. Avian Dis 10:288.
Miessner, H., and K. Hansen. 1936. Dtsch Tieraerztl Wochenschr 44:323–30.
Murtaugh, R. J., and R. M. Jacobs. 1984. J Am Vet Med Assoc 185: 441–42.
Narcisi, E. M., et al. 1991. Avian Dis 35:55–61.
Oguma, K. 1931. J Fac Sci Hokkaido Imp Univ 1:117–31.
Olsen, O. W. 1974. *Animal Parasites.* 3d ed. Baltimore: University Park, pp. 68–72.
Penev, P. 1971. Vet Med Nauki 8:71–74.

Perez Mesa, C., R. M. Stabler, and M. Berthrong. 1961. Avian Dis 5:48–60.
Prowozek, S., and H. B. Aragao. 1909. Muench Med Wochenschr 56:645–46.
Railliet, A. 1885. *Elements de Zoologie Med et Agricole.* Paris, p. 800.
Ratz, I. 1909. Allattani Kozl 8:90.
Rivolta, S. 1878. G Anat Fisiol Patol Anim 10:149–54.
Rivolta, S., and P. Delprato. 1880. Piza p. 500.
Rosenwald, A. S. 1944. J Am Vet Med Assoc 104:141–43.
Schaaf, J., and H. Scherle. 1938. Zentralbl Bakteriol Parasitol I Abst 141: 305–9.
Schorger, A. W. 1952. Auk 69:462–63.
Shorb, M. S. 1964. *Biochemistry and Physiology of Protozoa.* Vol. III. New York: Academic.
Sileo, L., and E. L. Fitzhugh. 1969. Bull Wildl Dis Assoc 5:146.
Stabler, R. M. 1938a. J Am Vet Med Assoc 93:33–34.
______. 1938b. J Parasitol 24:553–54.
______. 1941a. J Morphol 69:501–15.
______. 1941b. Auk 58:558–62.
______. 1947. J Parasitol 33:207–13.
______. 1948a. J Parasitol 34:147–49.
______. 1948b. J Parasitol 34:150–57.
______. 1951a. J Parasitol 37:473–78.
______. 1951b. J Parasitol 37:471–72.
______. 1953a. J Parasitol 39:12.
______. 1953b. J Colo-Wyo Acad Sci 4:58.
______. 1954. Exp Parasitol 3:368.
______. 1957. J Parasitol 43:40.
Stabler, R. M., and F. B. Engley. 1946. J Parasitol 32:225–32.
Stabler, R. M., and C. M. Herman. 1951. Trans North Am Wildl Conf 16:145.
Stabler, R. M., and J. T. Kihara. 1954. J Parasitol 40:706.
Stabler, R. M., and C. P. Matteson. 1950. J Parasitol 36:25–26.
Stabler, R. M., and R. W. Mellentin. 1951. Anat Rec III Abstr 584.
______. 1953. J Parasitol 39:637–42.
Stabler, R. M., and H. A. Shalanski. 1936. J Parasitol 22:539.
Stabler, R. M., S. M. Schnitter, and W. M. Harmon. 1958. Poult Sci 37:352.
Stabler, R. M., B. M. Honigberg, and V. M. King. 1964. J Parasitol 50:36.
Stepkowsji, S., and B. M. Honigberg. 1972. J Protozool 19:306.
Stone, W. B., and D. E. Jones. 1969. Bull Wildl Dis Assoc 5:147–49.
Tanabe, W. 1926. J Parasitol 12:120.
van Heelsberger, T. 1925. Tijdschr Diergeneeskd 52:1051.
Waller, E. F. 1934. J Am Vet Med Assoc 84:596.
Waterman, N. 1919. Tijdschr Vgl Geneeskd 4:40–47.
Wittfogel, H. 1935. Inaug Diss, Tieraerztliche Hochschule, Hannover, Germany.
Yakimoff, W. L. 1934. Bull Soc Pathol Exot 27:734–35.

Trypanosomiasis

Definition. Trypanosomiasis is an infection caused by trypanosomes, uncommon, single-celled parasites found in blood and tissue fluids of pigeons.

Classification. The organism is placed in phylum Sarcomastigophora, class Zoomastigophorasida, order Kinetoplastorida, family Trypanosomatidae, genus and species *Trypanosoma brucei* (Duke 1924, 1933), *T. hannai* (*T. avium*) (Hanna 1903; Pittaluga 1904) reported by Levine (1985).

Nature

LIFE CYCLE. The life cycle of *T. brucei* involves the bite of an infected male or female tsetse fly and the introduction into the pigeon of metacyclic trypanosomes. With *T. avium,* bloodsucking mosquitoes, blackflies, and hippoboscid flies are vectors. Birds can also be infected by eating infected insects (Levine 1973). No sexual process has been observed. All multiplication is by simple longitudinal binary fission with division starting at the kinetoplast followed by the flagellum, nucleus, and cytoplasm (Soulsby 1968).

Leishmanial rounded forms with no external flagellum and crithidial forms with the kinetoplast and basal granule just anterior to the nucleus both develop in the midgut of the fly. Flies infected with *T. brucei* become infectious after 15–35 days when these metacyclic forms are present in the salivary glands. Several thousand may be injected by one bite (Levine 1973). In the pigeon, replication of *T. brucei* possibly includes the formation of leishmanial forms in heart blood and cerebrospinal fluid. Baker (1956), however, indicates that no multiplication takes place in avian hosts. The parasites simply increase in size.

FORM. General statements concerning these parasites are difficult to make with accuracy. Each species has its own characteristics. The mature parasites are curved, flattened, bladelike in shape and have a single anterior flagellum attached to the body of the organism by an undulating membrane (Fig. 9.8). The kinetoplast and basal granule are located near the posterior extremity. The position varies between *T. brucei* and *T. avium.* The flagellum in *T. brucei* arises at the basal granule and forms the border of the undulating membrane as it

Fig. 9.8. Trypanosome. These blood parasites are seldom observed.

passes anteriorly to become a free flagellum as it leaves the body. A large nucleus is centrally located. Avium trypanosomes, including *T. hannai,* are pleomorphic. They may reach 26–60 μm in length and have a free flagellum (Levine 1973).

Distribution

GEOGRAPHY. *T. brucei* has been widely reported in Africa. *T. hannai* was first observed in India by Hanna (1903).

INCIDENCE. The *T. brucei* disease is found in areas where the tsetse fly is located in tropical Africa. The lack of reports in literature would suggest a low incidence in pigeons.

SPECIES. *T. hannai* was noted in pigeons (Hanna 1903; Pittaluga 1904). Levine (1973) considers this probably to be *T. avium. T. brucei* was collected by Sallazzo (1929, reported by Duke 1933), in fowl. Levine (1973) lists the latter species in horses, mules, a donkey, an ox, a zebra, sheep, a goat, a camel, a pig, a dog, and antelopes. He further indicates antelopes to be the natural host reservoir. Humans are not susceptible to *T. brucei* (Chandler 1949).

CARRIERS. The chief vector of *T. brucei* is the tsetse fly in the genus *Glossina.*

Transmission. Biting flies may mechanically transmit infection. Mosquitoes, blackflies, and hippoboscid flies may also serve as the vector for avian species (Baker 1956; Molyneaux 1977).

Epidemiology

DISTRIBUTION IN BODY. *T. brucei* may be found in blood, lymph, and cerebrospinal fluid (Levine 1973).

ENTRY AND EXIT. The parasite enters and leaves by the bite of a bloodsucking arthropod. In addition, when infective insects are ingested the organism probably penetrates the crop and enters the lymphatic system (Baker 1956).

IMMUNITY. Immunity has been reviewed by Brown (1967) and Gray (1967), among others. According to Levine (1973), *T. brucei* loses its antigenic identity when it enters the tsetse fly but regains the original formulas in the metacyclic stage in the salivary glands of the fly. The change in antigen inhibits attempts to immunize potential hosts.

PATHOGENICITY. Species observed in birds appear to be nonpathogenic (Soulsby 1968; Levine 1973).

Incubation and Course. Large forms of *T. avium* first appear in the blood 18–24 hr after infection (Baker 1956). Once infected the pigeon will probably remain infected.

Signs and Necropsy. Very little is mentioned in literature concerning these aspects. The signs must be somewhat similar to those of malaria even though avian species are considered nonpathogenic.

Diagnosis. Blood and bone marrow smears provide evidence of infection.

Biological Properties

STRAINS. Numerous strains can be demonstrated by antibody production (Levine 1973). Weitz (1963) showed the *T. brucei* antigens to be variable.

TISSUE AFFINITY. Femoral bone marrow is likely to harbor the parasite (Stabler and Holt 1963).

STAINING. Blood smears may be fixed with methyl alcohol and stained with Giemsa but other specialized stains are listed by Levine (1973).

STABILITY. The organism can be preserved and stored by slow freezing to −80°C to −196°C in the presence of 7.5% glycerol (Levine 1973).

CULTURE. The organism has been cultivated on Novy, MacNeal, and Nicolle medium and saline peptone blood medium (Diamond and Herman 1954).

Control and Prevention. Control is chiefly dependent on chemotherapy, but drug resistance is reported as an obstacle in large animals. Control must therefore be directed toward preventing the bite of the vector. In pigeons the louse fly is a hippoboscid fly that can be the vector of *T. avium.* It can be controlled by weekly dusting with insecticides. Control of the tsetse fly and *T. Brucei* in central Africa is much more difficult. Tsetse flies are repelled by white; thus lofts should be whitewashed and residual insecticides applied to the walls.

References

Baker, J. R. 1956. Parasitol 46:308–52.

Brown, K. N. 1967. Pan Am Health Org Sci Publ 150:21–24.

Chandler, A. C. 1949. *Introduction to Parasitology.* 8th ed. New York: Wiley, pp. 157–80.

Diamond, L. S., and C. M. Herman. 1954. J Parasitol 40:195–202.

Duke, H. L. 1924. Ann Trop Med Parasitol 18:4.

______. 1933. Parasitol 25:171–91.

Gray, A. R. 1967. Bull World Health Org 37:177–93.

Hanna, W. 1903. J Microsc Sci 47:433–38.

Levine, N. D. 1973. *Protozoan Parasites of Domestic Animals and of Man.* 2d ed. Minneapolis: Burgess, pp. 36–62.

______. 1985. *Veterinary Protozoology.* Ames: Iowa State University Press.

Molyneaux, D. H. 1977. Adv Parasitol 15:1–82.

Pittaluga, G. 1904. Rev Acad Ciencias (Madrid) 2:331–79.

Soulsby, E. J. L. 1968. *Helminths, Arthropods, and Protozoa of Domestic Animals.* Baltimore: Williams & Wilkins, pp. 535–65.

Stabler, R. M., and P. A. Holt. 1963. J Parasitol 49:320–22.

Weitz, B. 1963. Ann N Y Acad Sci 113:400–408.

10

External Parasite Infestations

Eight kinds of external parasites infest pigeons. They are classified as Insecta, having the body divided into head, thorax, and abdomen, with three pairs of walking appendages and two pairs of wings, and Arachnida, with the head and thorax fused and with four pairs of walking appendages and no wings.

Lice

Definition. Lice are common six-legged, wingless, crawling external parasites of birds, animals, and humans. All bird lice have chewing mouthparts.

Cause and Classification. Lice are placed in the phylum Arthropoda, class Insecta, and order Mallophaga. Pigeon species are included under two families: Philopteridae and Menoponidae. Pigeon species essentially infest only pigeons (Table 10.1).

Life Cycle. Mallophaga live their entire lives one generation after another on the host. The eggs are glued in clusters to the feather shafts. They hatch in 4–7 days and the nymphs require three molts and about 3 wk to mature. Lice live several months on the birds but survive only 5–6 days off the birds (Martin 1934; Loomis 1984).

Form. Lice are insects 1–2 mm in length that have a chitinous exoskeleton divided into three flattened body segments: head, thorax, and abdomen. They have three pairs of jointed legs attached to the central thorax. Each distal segment bears two tarsal claws. One pair of short four- to five-segmented antennae arise from the broad head. Tracheal tubes exit at spiracles that permit breathing. They have chitinized mandibles adapted for chewing that enable them to feed on skin scales, feather debris, and hair. They do not suck blood. Fig. 10.1 outlines the slender pigeon louse, *Columbicola columbae.*

Incidence. Lice are fairly common among pigeons in lofts where management procedures are unsatisfactory. If the owners are not knowledgeable or overlook bird inspection, lice can multiply quickly. One pair can produce over 100,000 progeny in a month.

Season and Age. Birds are often given less attention during the fall and winter months. This is when lice multiply the most. Any age may become infested, but younger birds may be irritated the most.

Transmission. Infested live pigeons serve as carriers and transmit pigeon lice species to other live pigeons. Lice will transfer from one bird species to another if they are in close contact, but they are unlikely to become established. Lice leave dead birds; however, they survive only 5–6 days off a warm host. Kiernans (1975) has shown that *Pseudolynchia canariensis,* the pigeon louse fly, can transport the slender pigeon louse *Columbricola columbae* from one bird to another.

Signs. Inspection of louse-infested birds reveals elongated, small, straw-colored parasites moving on the skin or among the feathers. Egg masses may be found adherent to the base of the feather and the feather shaft (Fig. 10.2).

Prevention. Purchased birds or birds returning from shows and/or races should be

Table 10.1 Lice reported to infest pigeons

Family, Genus, and Species	Reported by	Location	Pigeon Incidence[a]	Common Name	Body Location
Philopteridae					
Campanyulotes bidentatus compar[b,c,d,e]		North America	+ + +	Small	Body feathers
Columbicola columbae[b,c,d,e,f]	Nobrega	North America	+ +	Slender	Feathers
Coloceras damicorne[c,d,e]			+ +	Little	Feathers
Physconelloides zenaidurae[b,c,d,e]		Florida, Kansas	+ +		Feathers
Goniodes minor[f]	Piaget (1880)	Europe	+		
Menoponidae					
Colpocephalum turbinatum[b,c,d,e,f,g]		Panama, Florida, South Carolina	+	Narrow	Feathers
Hohorstiella lata[b,c,d,e,f]		California, Florida	+ +	Large	Feathers
Bonomiella columbae[b,c,d,f]		Kansas, Canada	+ +		
Menopon longicephalum[f,g]	Kellogg (1896)	Kansas, Pennsylvania	+		Feathers
M. latum[f,g]	Piaget (1880)	Europe	–		Feathers
M. biseriatum[f,h]	Piaget (1880)		–		Feathers
Menacanthus latus[f]		Florida, South Carolina	+	Largest	Tail, flight feathers

[a] – = unknown, + = incidental, + + = occasional, + + + = common.
[b] Becklund (1964).
[c] Emerson (1956).
[d] Emerson (1962).
[e] Loomis (1984).
[f] Levi (1941).
[g] Wood (1922).
[h] Lahaye (1928).

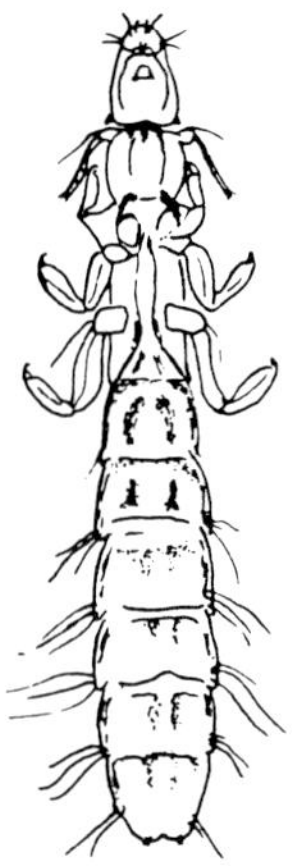

Fig. 10.1. Slender pigeon louse, *Columbicola columbae.*

dusted with an insecticide powder to prevent loft infestation with lice. Wild and domestic birds should also be excluded from lofts. Exercise yards should be enclosed with half-inch wire mesh to help prevent contact with feral pigeons and other birds. Even though other birds do not host pigeon lice, they may mechanically carry them.

Control. Lice can be controlled by any one of several insecticides, such as Malathion (American Cyanamid) 4% dust or carbaryl (Sevin, Union Carbide) 5% dust at 1 lb/100 birds. Sodium fluoride may also be used at one pinch per bird, which amounts to 1 lb/100 birds. The applicator should wear a mask and gloves when applying any dust. Birds may be dusted weekly or as needed on the neck and breast, below the vent, and under each wing; ½ tsp dust may also be placed in each nest box without eggs. Hatchability may be altered by prolonged egg contact with insecticides. Sprays are not as effective on birds because moisture does not penetrate to the skin. Lice like it warm and thus stay near the skin. Residual 1% water sprays may be applied in nests and on perches using emulsifiable concentrates (EC) of Malathion, 57% EC; Sevin, 50% EC; Ronnel, 25% EC (Dow); and Lindane, 20% EC (California Spray, discontinued). For roosting birds, nicotine sulfate 40% may be applied with an oilcan to the perches just before the birds roost at night. Windows should be open when nicotine is applied and the operator should back out of the loft. The nicotine gas that kills the lice can kill birds and people unless ventilation is provided.

References

Becklund, W. W. 1964. Am J Vet Res 25:1380–1416.
Emerson, C. K. 1956. J Kans Entomol Soc 29:63–79.

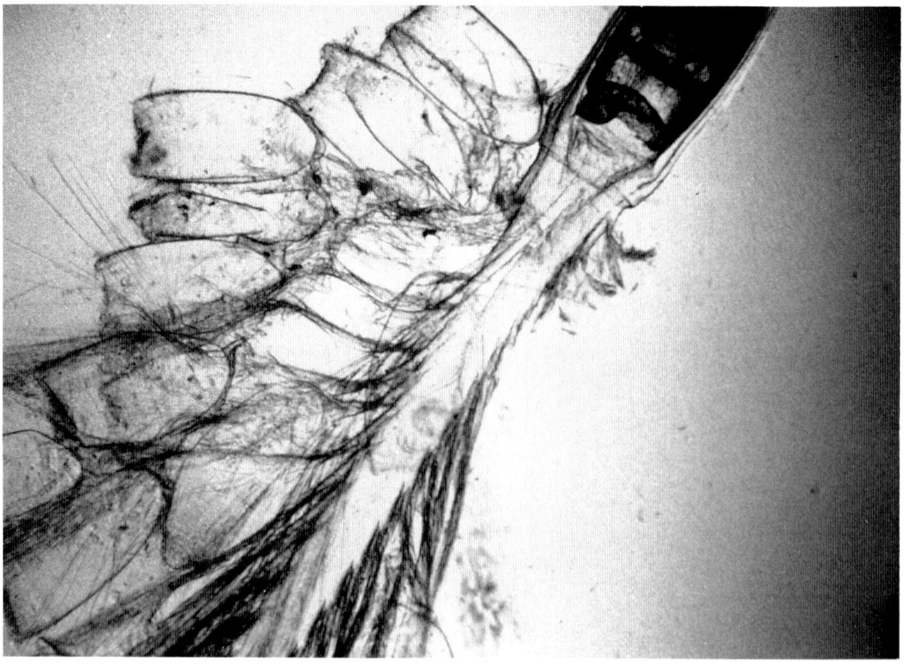

Fig. 10.2. Lice egg cases on a feather. Lice glue the egg cases to the feather.

______. 1962. 35:196–201.
Kiernans, J. E. 1975. J Med Entomol 12:71–76.
Lahaye, J. 1928. *Maladies des Pigeons.* 3d ed. Remouchamps, Belg.: Steinmetz-Haenen.
Levi, W. M. 1941. *The Pigeon.* Columbia, S.C.: Bryan.
Loomis, E. C. 1984. In *Diseases of Poultry,* 8th ed., ed. M. S. Hofstad et al. Ames: Iowa State University Press.
Martin, M. 1934. Can Entomol 66:6.
Wood, H. P. 1922. USDA Circ 213.

Flies

Definition. Flies are small, disease-carrying, winged insects with one pair of membranous wings and a second pair represented by knobbed halters. In addition, they are supported by three pairs of jointed legs.

Cause and Classification. Flies are placed in phylum Arthropoda, class Insecta, and order Diptera. (See Table 10.2.) The lowly quick-witted fly appears here to stay. It shares our plate and contemplates our every move. Each family or genus comes adapted for each occasion. The hippoboscid louse fly is perhaps the most important for pigeons, but midges, blackflies, flesh flies, blowflies, *Stomoxys* flies, and houseflies all demand our attention.

Life Cycle. Flies, in general, have complete metamorphosis but variations occur. Most flies lay eggs that transform into larvae, followed by pupation and the emergence of adults (Fig. 10.3).

Form. The body of the fly is divided into head, thorax, and abdomen. The head is attached to the thorax by a very flexible neck. The mouthparts are designed to conform with the feeding habits of the fly. Two antennae are jointed and modified in form, characteristic of individual families. The thorax has three fused component parts to which three pairs of jointed legs and one pair of membranous wings are attached. The halters located behind the base of the wings act as balancers as they vibrate. (See Fig. 10.4.)

Hippoboscid Fly: Life Cycle, Form, and Importance. The female pigeon louse fly gives birth one at a time to 4–5 fully developed, 3-

Table 10.2 Flies likely to affect pigeons

Family, Genus, and Species	Other Birds Affected[a]	Location	Pigeon Incidence[b]	Name	Body Location
Hippoboscidae					
Pseudolynchia canariensis[c,d]	Do	Worldwide	+ + +	Pigeon, louse fly	Skin
Microlynchia pusilla[e]		South America	–	Louse fly	Skin
Ornithomyia avicularia[e]		England	–	Louse fly	Skin
Muscidae					
Musca domestica[c,d]		Worldwide	+ + + +	Housefly	Body
Fannia canicularis[c,d]		United States	+ +	Little housefly	Latrine fly
Face					
F. femoralis[d]		United States	+ +	Coastal fly	Face
F. scalaris[c,d]		United States	+ +	Latrine fly	Face
Stomoxys calcitrans[c,d]		Worldwide	+ +	Stable fly	Face
Calliphoridae					
Callitroga americana[c,d]		United States	+	American screwworm fly	Flesh
Cochliomyia macellaria[c,d]		United States	+	Secondary screwworm fly	Flesh
Chrysomyia bezziana[c,d]		Asia, South Africa	+	Old world screwworm fly	Flesh
Chrysomyia megacephala[c]		Orient	+	Screwworm fly	Flesh
Lucilia illustris[c]		United States	+ +	Green bottle fly, blowlfy	Flesh
Phaenicia sericata[c]		United States	+ +	Surgical blowfly	Flesh
Phormia regina[d]		United States	+ +	Black blowfly	Flesh
Sarcophagidae					
Wohlfartia magnifica[d]		Europe	+	Screwworm fly	Flesh
Ceratopogonidae					
Culicoides sp.[d,f]	C, D	Worldwide	+ +	Midge fly	Face
Simuliidae					
Simulium sp.[d]	D, G, T	Worldwide	+ +	Blackfly, gnat	Face

[a]C = chickens, D = ducks, Do = doves, G = geese, T = turkeys.
[b]– = unknown, + = incidental, + + = occasional, + + + = common, + + + + = very common.
[c]Chandler (1949).
[d]Loomis (1984).
[e]Baker (1957).
[f]Fallis and Wood (1957).

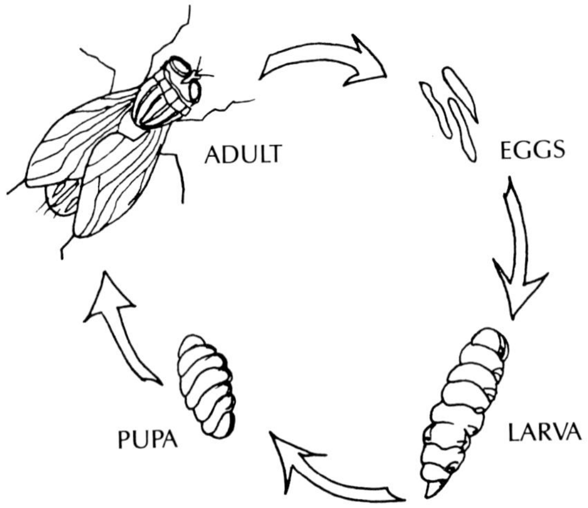

Fig. 10.3. Life cycle of the housefly, *Musca domestica.*

mm, whitish or cream-colored larvae. At the time of extrusion on the body of the bird, according to Drake and Jones (1930): "the mature larva is enclosed in an obovoid capsule or pupal case which is incapable of locomotion and is whitish in color with a black caplike structure at the caudal end. As the transformation continues, the purarium successively turns from white through yellow-orange, light to dark brown, and in the course of two to three hours becomes strongly indurated, somewhat shiny and jet black in color." Pupation on the floor or ground or in nest boxes requires 29–31 days under favorable temperature conditions. Emergence is delayed in colder weather, and only protected pupae hatch. Adults emerge to seek a blood meal from a pigeon or occasionally a dove.

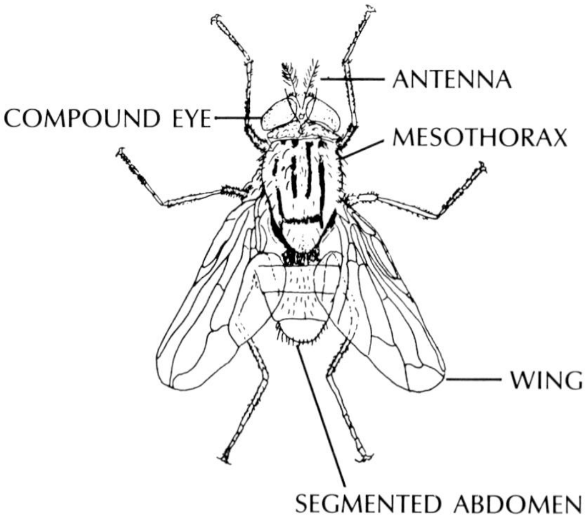

Fig. 10.4. Housefly structure. Lice have been known to hitchhike on flies.

The fly seldom leaves the host under normal conditions. It is adapted for quickly crawling sideways between the feathers, and the tarsal claws form strong hooks that enable the fly to cling to the feathers. When the bird is handled, the fly makes short, sudden, quick flights, frequently settling on nest boxes or walls for a brief period. Both sexes live about 45 days under the feathers of the pigeon (Benbrook 1965). Of interest is the fact that flies can copulate during flight (Prouty and Coatney 1934).

The hippoboscid fly is about 6 mm in length and has a flattened head; it is divided into head, thorax, and distinct abdomen. The head bears a short, stout beak, and elongated slender palpi form a sheath for the piercing proboscis. Two large prominent compound eyes adorn the head. There are three pairs of legs with jointed tarsi having five segments. Each distal segment has four strong tarsal claws. Two membranous, transparent wings are attached to the broad thorax and extend well beyond the body. The posterior fused segments of the rounded abdomen are adapted for reproduction. (See Fig. 10.5.)

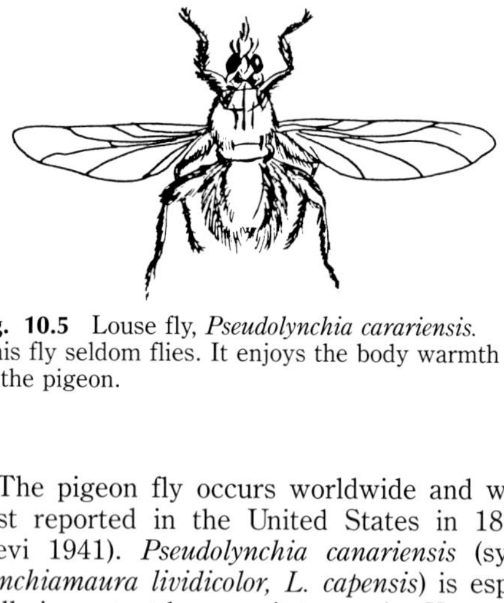

Fig. 10.5 Louse fly, *Pseudolynchia carariensis.* This fly seldom flies. It enjoys the body warmth of the pigeon.

The pigeon fly occurs worldwide and was first reported in the United States in 1896 (Levi 1941). *Pseudolynchia canariensis* (syn. *Lynchiamaura lividicolor, L. capensis*) is especially important because it transmits *Hemoproteus,* or pigeon malaria. Infested pigeons of all ages also suffer red blood cell destruction and considerable irritation from the fly. Such birds will often stomp their feet in an attempt to dislodge the flies.

Midge Fly: Life Cycle, Form, and Importance. The tiny biting midges, also called no-see-ums or gnats, are seldom over 1–2 mm in length. The eggs are deposited in masses secured to underwater objects in swamps or ponds. Saltwater or brackish water may be favored. Eggs hatch, forming slender wormlike larvae that seclude themselves in wet mud or sand. Gill-like tissues preclude the need for air. Pupae breath by means of surface tubes as they float in the water. Adults emerge to seek a blood meal. With favorable temperature the life cycle is completed in 2 wk. (From Chandler 1949.)

The midge fly body is divided into head, large thorax, and segmented abdomen. The fly has a short proboscis but quite long antennae. Wings are membranous and have five veins. The legs are long with jointed tarsi. (See Fig. 10.6.)

Fig. 10.6. Midge fly, *Phlebotomus* sp.

Only female midge flies are bloodsuckers, and these become active at dusk when the air is still. The bites produce intolerable irritation and itching. This *Culicoides* fly may be involved as an intermediate host in the transmission of filarial worms and *Hemoproteus* (Fallis and Wood 1957).

Blackfly: Life Cycle, Form, and Importance. Blackfly egg masses are laid under or virtually in running water. They hatch in 4–5 days, and the larvae attach to submerged objects by posterior hooklets. They breathe by means of tiny gills. After at least 2–3 wk, submerged pupal cocoons are formed and these breath by means of gill filaments. Adults are released to the surface in 3–7 or more days. The cycle may be completed in 1½–2 mo or more. (From Chandler 1949.)

The *Simuliidae,* or buffalo gnats, are small, humpbacked flies with short legs and broad wings in which the anterior veins are well developed. The short, stocky antennae have 11 segments but no bristles at the joints. The female proboscis is heavy and is designed for bloodsucking. The mouth parts include sharp, pointed mandibles and maxillae in addition to a hypo- and epipharynx. The male, however, does not suck blood. Northern species are black, but some may be yellowish or reddish brown and seldom over 4 mm in length. (See Fig. 10.7.)

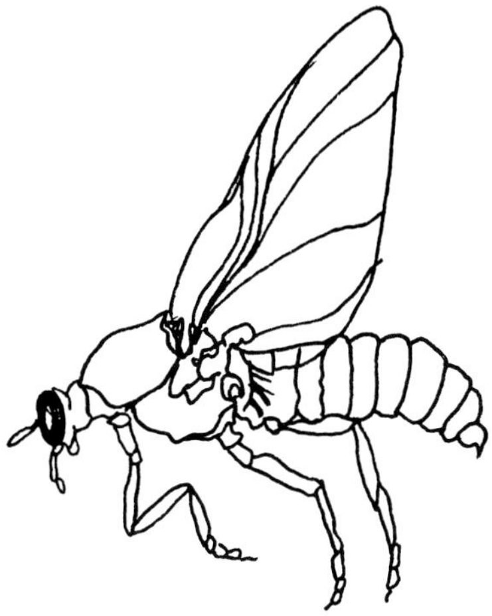

Fig. 10.7. Blackfly, *Simulium* sp.

Certain species of blackfly confine their attacks to birds and are serious bloodsuckers. They may be responsible for the transmission of microfilaria and *Leucocytozoon* infection.

Screwworm or Flesh Fly: Life Cycle, Form, and Importance. The cycle of *Callitroga americana,* a flesh fly, somewhat typifies this group of screwworm flies. Adults are not biting flies. Eggs are laid in flesh wounds and larvae hatch in about 12 hr. The voracious larvae eat flesh. When they mature or the bird dies, they leave the host within 48 hr and pupate in the ground for 7–9 days in summer. The cycle is completed in 18–22 days in warm weather. This fly is seldom observed in cooler areas.

The flesh fly has a compact, heavy-set, greenish blue body with an orange-red face. The wings are relatively small compared with the body. The larvae (maggots) that cause the tissue damage reach a length of 12–15 mm at maturity. They may be identified by the size of the spiracles and tracheal tubes. (From Chandler 1949.) (See Fig. 10.8.)

Living pigeons that incur wounds in warm climates may become occasional hosts for screwworm flesh flies such as *Callitroga americana, Chrysomyia bezziana,* and *Wohlfartia*

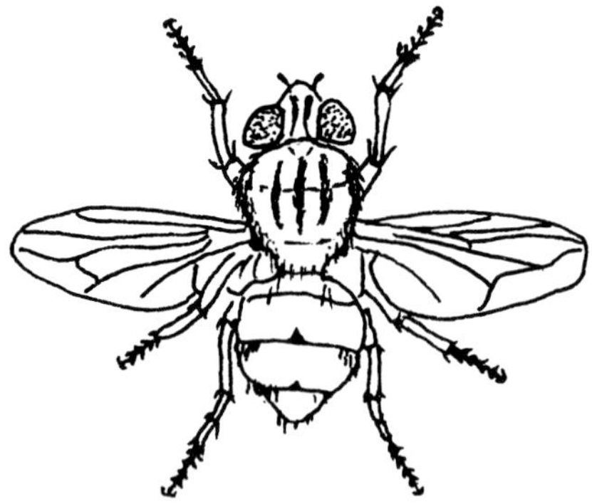

Fig. 10.8. Screwworm fly, *Callitroga americana.* Maggots of the fly ingest flesh of live or dead pigeons.

magnifica. Pigeons can be killed when these flies lay their eggs in wounds and maggots eat their way into the bird. Blowfly maggots may also enter the vent of manure-covered birds when flies lay their eggs in matted feathers.

Green or copper-colored blowflies, such as *Lucilia* spp. and *Phaenicia* spp. are more likely to lay eggs in dead carcasses. When conditions permit rapid putrefaction of a carcass, such maggots may encounter and ingest botulinus toxins A and C from *Clostridium botulinum* bacteria. If these maggots are themselves eaten by pigeons, limberneck poisoning may develop. According to Loomis (1984), larvae of *Cochliomyia macellaria* may also transmit the toxin. Carcasses of birds dead of tuberculosis may also serve to infect maggots and thus transmit this infection.

***Stomoxys* Fly: Life Cycle, Form, and Importance.** According to Chandler (1949), eggs of the *Stomoxys* fly are deposited in small masses in decaying vegetable material or manure. Larvae mature in 10–30 or more days and pupate in drier substrate. Adult flies form in 7 days in warm weather. (See Fig. 10.9.)

S. calcitrans, the stable or dog fly, is a stubborn, aggressive biting fly. It resembles the housefly, and it also mechanically transmits infections.

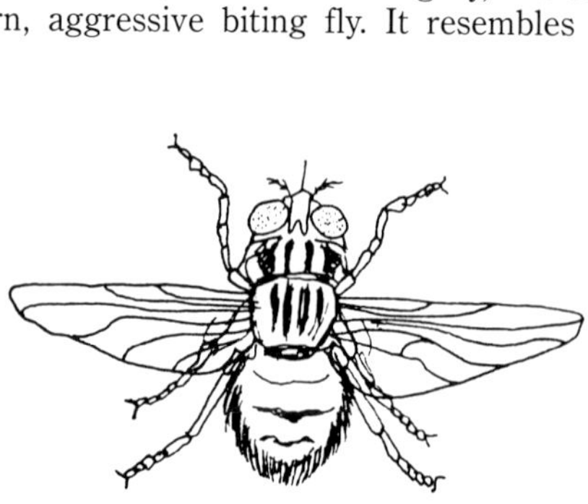

Fig. 10.9. Stable fly, *Stomoxys calcitrans.* These are biting flies.

Housefly: Life Cycle and Importance. The housefly, *Musca domestica,* usually lays its eggs in manure. These hatch in 12–24 hr, becoming larval maggots. Pupation occurs in a few days in a drier location, and adults emerge in 2–3 wk. The female lives 6–8 wk and produces about 2000 eggs. (From Chandler 1949.)

Muscidae houseflies, or latrine flies, are very prevalent common carriers of infection. They are filth feeders. They regurgitate and defecate where they feed or rest. Paratyphoid, tuberculosis, swine erysipelas, and other bacterial infections may be transported on the feet or in the discharges of such flies. In addition, the housefly serves as an intermediate host for the tapeworm *Choanotaenia infundibulum* (Loomis 1984). Also the Newcastle virus has been recovered from *M. domestica, Fannia canicularis,* and *F. femoralis* in California.

Control. Chemical insecticides should not be expected to completely control flies. Good sanitation is essential in removing breeding places. Houseflies and blowflies lay eggs in moist material, usually manure. To prevent fly breeding, it is necessary to repair leaks, drain wet areas, remove manure frequently, and collect dead birds or animals and place them in a fly-tight, scavenger-tight disposal unit.

For chemical control (Race and Robson n.d.), spraying should be begun early in the spring before flies become numerous. The following sprays are recommended (but not on the birds or with birds in the house): Pyrethrin 0.1–0.25% plus piperonyl butoxide 1–2% applied at 40 gal/hr according to the manufacturer's directions; fenthion (Baytex, Mobay, 45% emulsifiable concentrate [EC]), at 1 pt/5 gal water for 2500 ft^2; coumaphos (Coral, Mobay, 25% wettable powder) at 6 oz/5 gal water for 5000 ft^2; and tetrachlorvinphos (Rabon, Biotech, 24% EC) at 5–10 fl oz/5 gal water for 2500–5000 ft^3. Each spray may be repeated as needed. The addition of 4 oz cane sugar or 3 fl oz corn syrup/gal mix aids the kill. In addition, topical weekly applications of thiourea (McKisson Chemical) to chicken manure at 240 mg/hen is an effective larvacide (Lyons et al. 1988).

Permethrin, Ectiban, (ICI) 5.7% EC mixed at 13 fl oz/5 gal water may be applied to the loft at the rate of 1 gal/750 ft^2. Only Ectiban, at 6 oz/5 gal may be sprayed on birds, 1 gal/100 birds. Ectiban 0.25% dust may be applied on the birds at the rate of 1 lb/100 birds.

The louse fly may also be controlled by the use of a bird dip using 1 oz nicotine sulfate/6 gal water to which 4 oz laundry soap has been added (Drake and Jones 1930). Rubber gloves should be worn during the process. Pyrethrin and derris powders or any dog or cat flea powder containing 5% Sevin or 4% Malathion may be used once a week under the feathers of the wings, breast, and back. The use of a dust mask is indicated.

Label directions should be followed and protective clothing and a mask should be used during the handling and application of insecticides.

Presently screwworm fly control is being implemented with the release of sexually sterilized male flies. Eggs from such matings do not hatch, thus ending the cycle. Also, pheromones are being used in fly traps to attract flies. Beer and fishheads are good baits for blowflies. These baits placed under conical or V-shaped fly traps serve to reduce flies (Fig. 10.10). See Race and Robson (n.d.) for a further discussion of fly control.

References

Baker, J. R. 1957. J Protozool 4:204-8.

Benbrook, E. A. 1965. In *Diseases of Poultry,* 5th ed., ed. H. E. Biester and L. H. Schwarte, pp. 938-39. Ames: Iowa State University Press.

Fig. 10.10. Effective fly trap. Flies travel toward the light above after feeding on bait placed underneath.

Chandler, A. C. 1949. *Introduction to Parasitology.* 8th ed. London: Chapmann & Hall, pp. 613-59.

Drake, C. J., and R. M. Jones. 1930. J Sci 4:253-61.

Fallis, A. M., and D. M. Wood. 1957. Can J Zool 35:425.

Levi, W. M. 1941. *The Pigeon.* Columbia, S.C.: Bryan.

Loomis, E. C. 1984. In *Diseases of Poultry,* 8th ed., ed. M. S. Hofstad et al., pp. 586-602. Ames: Iowa State University Press.

Lyons, J. J., M. Vandepopuliere, and R. D. Hall. 1988. Poult Sci 67:407-12.

Prouty, M. J., and G. R. Coatney. 1934. Parasitol 26:249-58.

Race, S. R., and M. Robson. No date. Rutgers University Ext Leaf EO39.

Mosquitoes

Definition. Mosquitoes are small, six-legged, winged insects, the female of which sucks blood.

Cause and Classification. Mosquitoes are placed in phylum Arthropoda, class Insecta, order Diptera, and family Culicidae. The genera *Aedes* and *Culex* and rarely *Anopheles* bite birds (Table 10.3).

Life Cycle. The breeding habits of mosquitoes are varied. There are mosquito species adapted to breeding in fresh water, salt marsh, forest, prairie, and house and yard. They all have one feature in common: the eggs require moisture or water to hatch. The life cycles are much alike but differ in detail. The oval eggs of *Aedes* mosquitoes are laid singly out of water, *Culex* form floating rafts of upright eggs, and *Anopheles* lay them singly on water. In temperate zones the eggs may hatch in a day or in a few days. In colder climates the eggs may hatch in the spring, and in dry areas eggs may wait for the first drop of moisture or hatch intermittently as showers occur.

Larvae or wrigglers are always aquatic when they develop from the eggs. They molt four times during a period of 2 days to 2 wk to become pupae, but some pass the winter as larvae. Adult mosquitoes emerge from the pupae in a few hours to a week. The female mates then sucks blood, usually within 24 hr. Eggs are produced within several days to a week. The male with its bushy antennae feeds only on plant juice. (From Chandler 1949.)

Form. Most mosquitoes are about 5 mm in length. The adults have three pairs of jointed legs and one pair of veined, scaled, functional wings attached to the thorax. The proboscis, which is always longer than the head, is designed for piercing and blood suck-

Table 10.3. Mosquitoes (family Culicidae) reported likely to affect pigeons in North America

Genus and Species	Other Species Affected[a]	Pigeon Incidence[b]	Name
Aedes vexans[c,f]	B	+	
A. sollicitans[c]	M, B	+	Salt marsh
Culiseta melanura[c]	B	+ + + +	
C. morsitans[c]		+	
Culex pipiens[c,d,e]	B	+ + +	
C. salinarius[c,e]	B, M	+ +	
C. restuans[c,e]		+ +	
C. territans[c]	A, B		
Coquillittidia perturbans[c]	B, M	+	

[a]A = amphibians, B = birds, M = mammals.
[b]+ = incidental, + + = occasional + + + = common, + + + + = very common.
[c]Crans (1964).
[d]Downing and Crans (1977).
[e]Slaff and Crans (1981).
[f]Downe (1960).

ing in the female. The female proboscis has a dorsal groove that holds six stylets. When the mosquito bites, only the stylets pierce the skin. In the male the piercing stylets are usually absent. In addition, two segmented palpi, two antennae, and two eyes are prominent features of the head. A segmented abdomen terminates in especially adapted reproductive glands for each sex. Eggs are oval and hatch to form an elongated segmented larva with numerous segmental bristles and a breathing tube extending from the last segment of the abdomen. When larvae are in well-aerated water and are deprived of air, they can live for some time without breathing. (See Fig. 10.11.) (From Chandler 1949.)

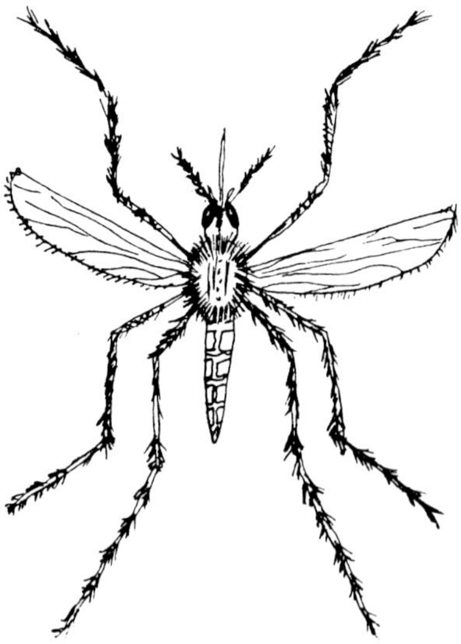

Fig. 10.11. Mosquito, *Aedes sollicitans.* These parasites transmit pox virus.

Geographical Distribution and Pathology. Mosquitoes are found worldwide. A number of the 140 species in North America are preferential bird biters. They in themselves are not generally a serious parasite of pigeons but they can and do create unrest and unthriftiness. It is the diseases they transmit that are cause for concern. Pigeon pox is an epitheliotropic virus that produces crusty scabs at the site of the infected mosquito bite. Arboviruses, including eastern and western encephalomyelitis and St. Louis viruses (Reeves and Hammon 1944; Downing and Crans 1977), are passed by mosquito bites but seldom produce clinical disease in pigeons. Likewise *Plasmodium* malaria has been carried by mosquitoes, and *Aedes, Culiseta,* and *Culex* species presumably serve as the definitive hosts for birds. *Culex pipiens* has been incriminated in the transfer of *P. cathermerium* malaria. Filariasis also occurs occasionally and can be transmitted by mosquitoes.

Control. Most insecticides are not only toxic for mosquitoes but are also poisonous for humans and other forms of life. The manufacturer's directions must be carefully followed in the application and storage of any product and in the disposal of such containers. In addition, as a general precaution, no chlorinated hydrocarbons should be used on water areas.

Control of mosquitoes involves destruction of adults and larvae and the elimination of breeding places. Larvae may be destroyed by (1) the application of larvacides; (2) draining water troughs, stagnant pools, ponds, and streams and swampland; (3) removing containers, such as cans, jars, tires, and stranded boats that are exposed to rain and screening cisterns and drinking water barrels; (4) clearing brush and sediment from streams and lily ponds to permit larval feeding by fish such as guppies, (*Lebistes reticulatus*) and killifish (*Fundulus* sp.); and (5) stocking attic or underground cisterns or open wells with *Gambusia patruellis* (Granett 1987).

Larvae may be destroyed by applying oil, such as mineral oil, kerosene, or clean number 2 fuel oil to water breeding locations and by using emulsified larvacides. Emulsifiers, such as nonionic Triton X-100 (Rohm and Haas) may be used as spreading agents.

The following larvicides are recommended (Sutherland 1986): (1) for catch basins, Baytex (Mobay) and temephos (Abate, Cyanamid) at 4 oz/basin and Altosid Briquets (Zoecon) per label instructions to disrupt larval molts and growth; (2) for floodwater, Abate at 0.016 lb actual insecticide (AI)/acre, Dursban emulsion (Dow) at 0.025 lb AI/acre, Altosid at 3-4 oz/acre, and *Bacillus thuringiensis israelensis* (Abbott Laboratories) for the ingestion by larvae according to manufacturer's directions; and (3) for impounded waters, Abate at 0.048 lb AI/acre, Baytex emulsion at 0.05 lb AI/acre, and Altosid 3-4 oz/acre.

Adult mosquitoes may be destroyed by fogging areas with 30% piperonyl butoxide in 70% mineral oil or soybean oil at the rate of 40 gal/hr (Vasvary 1987). Synthetic or natural pyrethrin spray (3-5%) at an ultralow volume of 4 oz/min is also effective. Synthetic pyrethrins are produced by Penick, Sumitomo, and ICI. Malathion (5%) in fuel oil is effective but expensive.

References

Chandler, A. C. 1949. *Introduction to Parasitology.* 8th ed. New York: Wiley, pp. 660-78.
Crans, W. J. 1964. Proc 51st Meeting of the New Jersey Mosq Extermination Assoc, pp.51-58.
Downe, A. E. R. 1960. Can J Zool 38:689-99.
Downing, J. D., and W. J. Crans. 1977. Mosq News 37:48-53.
Granett, P. 1987. Personal communication.
Reeves, W. C., and W. M. Hammon. 1944. Am J Trop Med 24:131-34.
Slaff, M., and W. J. Crans. 1981. Mosq News 41:443-47.
Sutherland, D. J. 1986. N J Agric Exp Stn Publ P 40400-01-86.
Vasvary, L. 1987. Personal communication.

Fleas

Definition. Fleas are reddish-brown, laterally compressed, wingless, jumping, bloodsucking, parasitic insects, the larvae of which are free-living.

Cause and Classification. Fleas are placed in the phylum Arthropoda, class Insecta, and order Siphonaptera. (See Table 10.4.)

Life Cycle of the Sticktight Fleas. The sticktight flea is one of several infesting poultry and pigeons in North America. It is discussed because its life cycle and form also characterize other fleas which can infest pigeons.

The life history includes a complete metamorphosis with egg, larval, pupal, and adult stages (Benbrook 1965). The adults are blood-feeding ectoparasites of birds and usually attach to the skin of the head for days or weeks. The female daily ejects one to four small, white, oval eggs among the feathers. These drop off and hatch on the floor or ground. The time required for eggs to hatch, usually 4-14 days, varies with the temperature. The most favorable temperature is 65-80°F with a rela-

Table 10.4. Fleas likely to infest pigeons

Family, Genus, and Species	Location	Other Species Affected[a]	Pigeon Incidence[b]	Name
Ceratophyllidae				
Ceratophyllus gallinae[c,d]	Europe, northeast and southeast United States	B, A, H, C, T, Ph, Q	+ +	European hen
C. columbae[d]	Canada		+ +	
C. niger[c,d]	Western North America	B, A, H	+	Western or black chicken
C. gibsoni[d]	Canada		+	
Pulicidae				
Pulex irritans[c,d]	Montana	H	+	Human
Echidnophaga gallinae[c,d]	Worldwide	A, B, H	+ +	Sticktight flea or tropical flea
Ctenocephalides felis[c]	Worldwide	Ca, D, A, H	+	Cat, flea
C. canis	Worldwide	Ca, D, A, H	+	Dog, flea

[a]A = animals, B = birds, C = chickens, Ca = cats, D = Dogs, H = humans, Ph = pheasants, Q = quail, T = turkeys.
[b]+ = incidental, + + = occasional.
[c]Loomis (1984).
[d]Reis and Nobrega (1936).

tive humidity of 70% or better. The larvae are tiny, slender, eyeless, legless, segmented, caterpillarlike organisms that feed on debris or blood. Pupae form in 14–31 days. Not all larvae of one brood pupate at one time; thus adults emerge at varied intervals, from 9–19 days. The cycle may be extended under cooler conditions. Fleas mature in 11–18 days. Adult fleas must have a blood meal to breed and produce eggs. At 60°F well-fed fleas may live 18 mo. At low temperatures in the absence of a host, fleas may survive 1–2 yr. Fleas usually feed daily or oftener, and unused blood passed in their feces is utilized by flea larvae. (From Benbrook 1965.)

Form. Adult fleas are 1.5–1.8 mm in length and have laterally compressed bodies. The adult head bears two ocelli (eyes) behind which are antennal grooves for the segmented antennae. The ventral edge of the anterior frons section of the head bears the spines of the genal comb. A ctenidium (pronotal comb) may be present on the rear dorsal margin of the first segment of the thorax. The piercing, sucking mouthparts project ventrally and consist of the maxillae and labial palpi. Wings are absent, and each of the three thoracic segments bears a pair of legs. The leg segments are formed by the coxa, trochanter, femur, tibia, and tarsus. The ventral sternite plates of the mesothorax are divided into areas called episternum, epimeron, and sternum. The abdomen has ten segments, and the last two are modified for reproduction. (See Fig. 10.12.) (From Chandler 1949.)

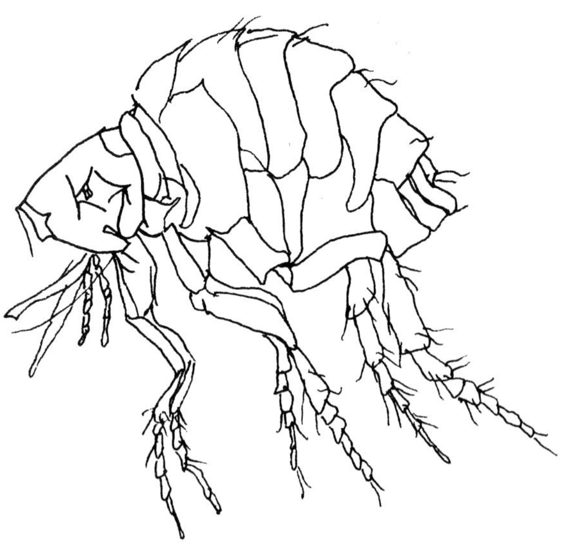

Fig. 10.12. Sticktight flea, *Echidnophaga gallinacea.*

Importance. The sticktight flea is more commonly found in warm areas of the United States and tropical and subtropical areas of the world. Irritation and blood loss are the chief causes for concern. Sticktight fleas are not generally considered as disease transmitters but they have the potential.

The European hen flea occurs more widely in the United States and in other parts of the world. It obtains a blood meal and then leaves the host. Other fleas, including the dog and cat flea, may occasionally attack pigeons.

Control. Control involves the use of residual insecticides on the birds to kill adult fleas and their application to the loft and premise to kill the larvae. Birds and nest boxes may be dusted lightly with 5% Sevin or 4% Malathion dust. This may be repeated in 1–2 wk as needed. Premise sprays include 2% Sevin or Malathion, 0.5% Lindane, discontinued 0.75% Ronnel, or 0.5% chlorpyrifos. These need not be repeated under 3 wk. In addition, Titchener (1982) used a pyrethroid permethrin spray in the hen house with success.

References

Benbrook, E. A. 1965. In *Diseases of Poultry,* 5th ed., ed. H. E. Biester and L. H. Schwarte, pp. 933–36. Ames: Iowa State University Press.

Chandler, A. C. 1949. *Introduction to Parasitology.* 8th ed. New York: Wiley, pp. 589–602.

Loomis, E. C. 1984. In *Diseases of Poultry,* 8th ed., ed. M. S. Hofstad et al., pp. 594–95. Ames: Iowa State University Press.

Reis, J., and P. Nobrega. 1936. *Tratado Doencas Das Aves.* Sao Paulo, Brazil: Sao Paulo Instituto Biologico.

Titchener, R. N. 1983. Poult Sci 62:608–11.

Bugs

Definition. Wingless bedbugs and winged assassin bugs are bloodsucking parasites of mammals and birds.

Cause and Classification. Bugs are placed in phylum Arthropoda, class Insecta, and order Hemiptera. (See Table 10.5.)

Life Cycle. The bugs included under the order Hemiptera have incomplete metamorphosis, with the adult form developing gradually by instars between successive molts of the nymphs.

Bedbug: Life Cycle and Form. The oval white, 1-mm eggs of bedbugs (*Cimex lectularius*)are laid in crevices. Females lay a total of 100–250 eggs singly or in batches. In warm weather these hatch in 6–10 days and form miniature bugs. After molting five times at in-

Table 10.5. Bugs likely to affect pigeons

Family, Genus, and Species	Location	Other Species Affected[b]	Pigeon Incidence[b]	Name
Cimicidae				Bedbugs
Cimex boueti[c]	Tropics	Po	+	
C. hemipterus[c]	Tropics	Po	+	
C. lectularius[c,d]	Southern United States	M, H	+ + +	Common
C. columbarius[c]	Europe		+ +	
Haemotosiphon nodorus[c]	Southwestern United States	C, O, T, H	+ + +	Adobe
	Central America			Mexican chicken bug
Ornithocoris toledoi[c]	South America		+ +	Brazilian chicken bug
O. pallidus[c]	South America, Florida, Georgia		+ +	South American chicken bug
Reduviidae				Assassin bugs
Triatoma sanguisuga[c,d]	Kansas, Texas, Florida, California, Maryland	Ph, Po	+ + +	
T. protracta[d]	Utah, California, Mexico		+ +	Western cone nose

[a]C = condors, H = humans, M = mammals, Ph = pheasant, Po = poultry, T = turkeys.
[b]+ = incidental, + + = occasional, + + + = common, + + + + = very common.
[c]Loomis (1984).
[d]Chandler (1949).

tervals of about 8 days, the adult form is achieved. A meal is essential between molts. Under favorable conditions the cycle is completed in 6–10 wk. Nymphs may live without food for about 70 days, but adults may survive up to 12 mo if the temperature is cool. They hide in cracks and crevices during the day and secure a blood meal at night. Their presence is characterized by an unpleasant odor from stink glands.

The bristle-covered body is divided into head, thorax, and abdomen. Adults are 2–5 mm long. Bedbugs have broad, flattened brown bodies and are devoid of wings. The head bears two eyes, two four-jointed antennae, and a three-jointed beak. The mouthparts are designed for piercing and bloodsucking and bear styletlike mandibles and maxillae. The thorax is divided into a larger forward prothorax and a small mesonotum with wing pads. Three pairs of segmented legs have three-jointed tarsi. Stink glands are located in the posterior thoracic section between the bases of the hind legs. (See Fig. 10.13.) (From Chandler 1949.)

Assassin Bug: Life Cycle and Form. The cycle of assassin bugs (*Triatoma* sp.) is similar to that of bedbugs except that it requires about 1–2 years to complete. After mating, the female produces a total of 100–300 eggs singly or in clusters. Eggs require 2–3 wk to hatch, depending on the temperature. Wingless nymphs emerge. They molt following each of five blood meals and finally reach adult forms. Adults invade the pigeon loft from the nearby countryside, usually at night. They feed on a great variety of hosts and consume 6–12 times their weight.

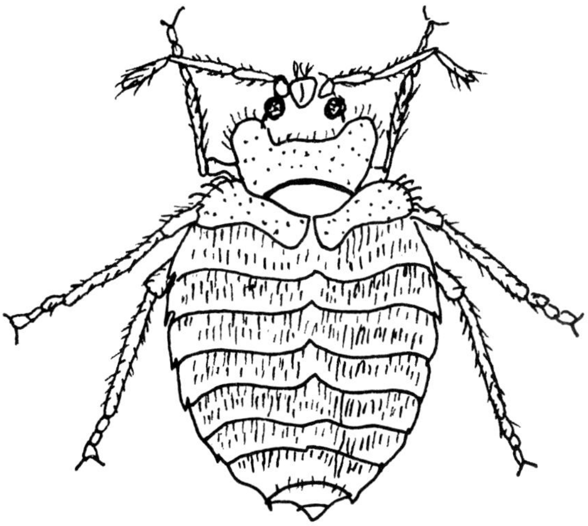

Fig. 10.13. Common bedbug, *Cimex lectularius.*

Assassin bugs have long narrow heads bearing prominent compound eyes, long four-jointed antennae, and usually two ocelli. The proboscis is long, slender, three-jointed, and straight, designed for piercing and bloodsucking. The prothorax bears leathery functional folded wings. The jointed legs are long and end in three-jointed tarsi. The abdomen is slender and segmented. The cylindrical body approaches 25 mm in length. (See Fig. 10.14.) (From Chandler 1949.)

Fig. 10.14. Assassin bug, *Triatoma* sp.

Importance and Pathology. Bedbugs are important bloodsucking parasites of birds. Pigeon lofts can be invaded, and pigeons may be repeatedly bitten. The saliva entering the wound produces swelling and itching. Affected birds become restless and unthrifty, and constant blood loss results in anemia. Pigeons are more often attacked where the loft is a part of the human domicile.

Assassin bugs are prevalent in the warm parts of the United States and the world. Their importance is emphasized by the fact that Kitselman and Grundmann (1940) recovered equine encephalomyelitis virus from *T. sanguisuga.* They have also been shown to carry *Trypanosoma cruzi,* which produces Chagas' disease in people (Packchanian 1942).

Control. Bug control involves spraying premises and equipment where the bugs hide. Recommended insecticides include 2% Malathion; 0.5% Lindane, which contains the gamma isomer of benzene hexachloride; 3% Sevin, and 0.5% chlorpyrifos. The manufacturer's directions should be closely followed during the handling and application of the insecticides.

References

Chandler, A. C. 1949. *Introduction to Parasitology.* New York: Wiley, pp. 553–67.

Kitselman, C. H., and A. W. Grundmann. 1940. Kans Agric Exp Stn Tech Bull 50.

Loomis, E. C. 1984. In *Diseases of Poultry,* 8th ed., ed. M. S. Hofstad, pp. 593–94. Ames: Iowa State University Press.

Packchanian, A. 1942. Am J Trop Med 22:623–31.

Beetles

Definition. Beetles are insects that seldom attack pigeons but serve as intermediate hosts for tapeworms.

Cause and Classification. Beetles are placed in phylum Arthropoda, class Insecta, and order Coleoptera. (See Table 10.6.)

Life Cycle and Form. Beetles have a complete life cycle with egg, larva, pupa, and adult stages. The larvae are wormlike grubs with three pairs of feeble legs.

Beetles are stoutly armored insects that have two pairs of wings. The horny, sheathlike forward surface wings, called the elytra, cover the membranous folded hind wings. Chitinous plates protect the ventral abdominal segments. Beetles have biting mouthparts with strong

Table 10.6. Beetles likely to affect pigeons

Family, Genus, and Species	Location	Pigeon Incidence[a]	Site of Infestation	Name
Tenebrionidae				
Alphitobius diaperinus	United States	+ + +	Grain	Darkling beetle
Tenebrio molitor	United States	+	Feet, skin	Yellow mealworm
Dermestidae				
Dermestes lardarius	United States	+	Squab skin	Larder beetle
Silphidae				
Silpha thoracica	United States	+	Squab skin, grain	Carrion or sexton beetle
S. opaca	United States	+		
Necrophorus vestigator	United States	+		

Source: Loomis (1984).
[a]+ = incidental, + + + = common.

mandibles. (See Fig. 10.15.) (From Benbrook 1965.)

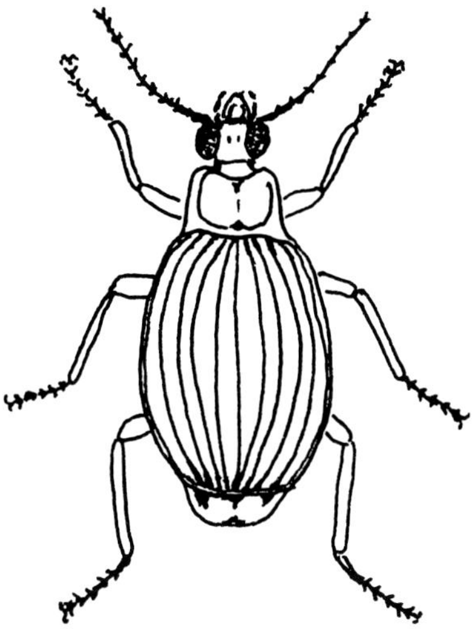

Fig. 10.15. Beetle, *Coleoptera* sp.

Importance. The darkling beetle (or lesser mealworm, *Alphitobius diaperinus*) may be found in the manure beneath the perches in warm moist conditions. Eggs hatch in 4–7 days, and when the larvae are ready to pupate they may burrow into the insulation and wood and can be very destructive (Loomis 1984). Adults 1/4 in. long emerge in 7–15 days and live at least a year. The cycle is completed in 2–3 mo (Race and Bezpa 1987).

Adult yellow mealworms (*Tenebrio molitor*), are shiny black beetles found in grain products. The grubs are smooth, hard, cylindrical, yellow, and wormlike. Grubs may attack the feet and skin of pigeons (Levi 1957).

Larder beetles (*Dermestes lardarius*) damage stored grain products, invade dead flesh, and frequent manure piles. The adult is black, about 7 mm long, with a brownish yellow wing cover marked by three black spots. The dark brown larvae are covered with brown hairs and may attack the skin of squabs (Loomis 1984).

Carrion or sexton beetles are scavengers that feed on dead flesh and are commonly found in manure. They may also attack the skin of squabs.

Control. When beetles are discovered the loft may be sprayed with Sevin 4F (Union Carbide) at 2.5 qt/5 gal water or with Rabon (Shell; Du Pont) at 6.5 oz 50% wettable powder/5 gal water. Both are applied at 1–2 gal of the spray mix/1000 ft^2. When squabs are produced for food, the Sevin spray must not be used closer than 7 days to slaughter (Race and Bezpa 1987).

References

Benbrook, E. A. 1965. In *Diseases of Poultry,* 5th ed., ed. H. E. Biester and L. H. Schwarte, p. 936. Ames: Iowa State University Press.

Levi, W. M. 1957. *The Pigeon.* 2d ed. Sumter, S.C.: Levi, p. 667.

Loomis, E. C. 1984. In *Diseases of Poultry,* 8th ed., ed. M. S. Hofstad et al., pp. 595–96. Ames: Iowa State University Press.

Race, S. R., and J. Bezpa. 1987. Darkling Beetle, Rutgers Univ Exp Stn Leafl, July.

Mites

Definition. Mites are common, tiny, eight-legged, wingless, crawling, internal and external parasites of birds, animals, and humans. The larval stages have six legs.

Cause and Classification. The mites are placed in phylum Arthropoda, class Arachnida, and order Acarina. The families, genera, and species likely to infest pigeons are listed in Table 10.7.

Life Cycle. Mites have four stages in their life cycle: egg, larva, nymph, and adult. Depending on the mite species, eggs may be laid under the skin, inside the host body, in cracks and crevices, or under the soil surface. Some mites, such as *Cytodites nudus, Laminosioptes cysticola,* and *Neoknemidocoptes laevis* are viviparous but usually six-legged larval mites hatch from eggs after a period of incubation. After a full meal, parasitic animal and bird mites molt and transform into nymphs with eight legs. Following a second feeding and one or more molts, adults develop. The life cycle for most pigeon mites require up to 1 wk for completion. Adults can survive off the host for 3–4 wk (Loomis 1984). Variations from this general pattern also occur.

Form. Mites are barely visible to the naked eye. The body is more or less round or oval and usually has a fused head, thorax, and abdomen. Many have the head and thorax fused as a cephalothorax. There are usually two pairs of mouthparts, composed of two chelicerae, two segmented pedipalps, and a hypostome adapted for piercing. There are four pairs of legs in the nymph and adult stages. Legs consist of six or seven segments, and they terminate in one or more tarsal claws. The digestive tract includes a stomach, a short intestine, and an anus. Malpighian tubules serve as excretory structures that empty into the intestine near the anus. Spiracles may be

Table 10.7. Mites likely to infest pigeons

Family, Genus, and Species[a]	Reported by	Location	Affected Species[b]	Pigeon Incidence[c]	Common Name	Site of Infestation
Analgidae						
Megninia columbae[d,e,f,g]	Buchholz 1980	South Carolina		++	Feather	Neck, body, tail
Cheyletidae						
Chyletiella yasguri[g]	Smiley 1965		B, M	+	Hyperparasite of Hippobosca	Feathers
C. heteropalpa[h]	Megnin 1878	Europe	D	+		Feathers
Sarcopterus nidulans[h]	Nitzsch 1818	Europe, United States		++	Feather	Feathers
Ornithocheyletia hallae[g]	Smiley 1977			++	Feather	Feathers
Springophilus columbae[d,e,f]	Hirst 1920	Tennessee, Texas		+	Quill	Feathers
Cytoditidae						
Cytodites nudus (V)[d,e,f,h]	Megnin 1879	Worldwide	C, T, Ph, Ca, Gr	++	Air sac	Air sac
Dermanyssidae						
Dermanyssus gallinae[d,e,f]	DeGeer	Hawaii, North America	C, T, Ca, B, H	+++	Red mite	Body
Ornithonyssus sylviarum[d,f]	Benbrook and Sloss	North America	C, T, Po, B, H, R	+++	Northern	Feathers
O. bursa[d,f]	Reis and Nobrega	Puerto Rico, South Atlantic, Panama	Po, B, H	++	Tropical	Feathers
Dermoglyphidae						
Pteroghagus strictus[g,h]	Robin and Megnin 1877			+		Feathers
Falculifer rostratus[d,e,f,g,i]	Buchholz 1869	United States, Canada, Europe		+++	Large wing	Wing
Ereynetidae						
Speleognathus striatus[d,f,g]	Crossley 1952	Texas		+	Nasal	Sinus
Knemidocoptidae						
Knemidocoptes mutans[d,e]	Robin and Lanquetin	North America	C, T, Ph, B	++	Scaly leg	Legs and feet
K. gallineae[d,e,f]		North America	C, Ph, G	++	Depluming	Skin
Neocnemidocoptes laevis (V)[g,h]	Railhiet 1887	Not North America		++	Itch mite	Skin
Laelaptidae						
Androlaelaps casalis[d,f,g]			C	+	Nasal	Sinus
Neonyssus columbae[d,f,g]	Crossley 1950	Texas		+	Nasal	Sinus
N. melloi[d,f,g]	Castro 1948	Brazil, Texas, Kansas				
N. zenaidurae[g]	Crossley			+	Nasal	Sinus
Laminosioptidae						
Laminosioptes cysticola (V)[d,f]	Kasparek 1907	Europe, United States	C, T, Ph, G	++		Subcutaneous flesh
Myialgesidae						
Myialges anchora[g]	Sergent and Trouessart			+	Hyperparasite	
Trombiculidae						
Trombicula alfreddugesi[f]	Ewing	United States	Re, M, B, H	+	Red bug chigger	Skin
T. batatas[f]	Michener 1946	Southern United States, Brazil	C, T, B, H, M	+	Chigger	Skin
Neoschongastia americana[f]		United States	C, T	+	Chigger	Skin

[a](V) = viviparous.
[b]B = birds, C = chickens, Ca = canaries, D = doves, G = geese, Gr = grouse, H = humans, M = mammals, P = pigeons, Ph = pheasants, Po = poultry, R = rats, Re = reptiles, T = turkeys.
[c]+ = incidental, ++ = occasional, +++ = common.
[d]Becklund (1964).
[e]Levi (1941).
[f]Loomis (1984).
[g]Smiley (1986).
[h]Hollander (1956).
[i]Lahaye (1928).

present for breathing but some absorb oxygen through the body wall. (See Fig. 10.16.) (From Chandler 1949.)

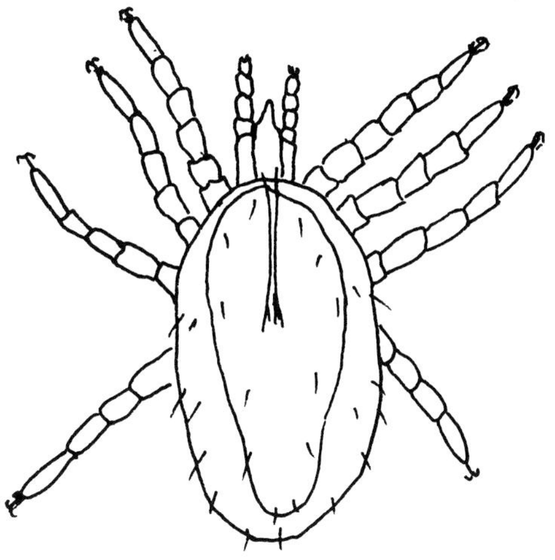

Fig. 10.16. Mite, *Dermanyssus gallinae.* Red mites are seldom found on the bird during the day.

Transmission. Infested birds of all types are carriers. When birds are housed or transported in close contact with each other, mites readily transfer from one host to another. Indirect transmission by recently contaminated equipment and supplies may also introduce mites to a loft.

Pathology and Characteristics. Each mite species produces its own characteristic damage. *Dermanyssus gallinae* are red mites that may be abundant at times. They cause skin irritation, scaliness, and anemia because they suck blood. Feather mites *Falculifer rostratus, Ornithonyssus sylviarum,* and *O. bursa,* which are more common, also cause skin changes and pruritis. Each of these parasites may be found on the skin and feathers and appear as black crawling dots after a blood meal. Direct contact between birds permits their transfer. *K. mutans,* the scaly leg mite (Fig. 10.17), tunnels under the skin and scales of the toes and feet of pigeons, causing the underside of the toes and balls of the feet to become roughened. This effect is produced by serum accumulation in the tissues. Leg shanks are seldom affected in pigeons. The depluming mite, *K. gallinae,* which is not common, is somewhat similar in form. It penetrates the base of the feather shaft, causing birds to pull or break off feathers. The quill mite, *Springophilus columbae,* enters the calamus and shaft of the feather and produces a fine, brown, powdery residue as it sucks blood. The latter three parasites complete their life cycle under or in the skin or feathers and are seldom observed on the surface of the body. *L. cysticola* remains totally within the body. Careful examination of connective tissues may reveal dead or live moving parasites. *C. nudus* has been found in the air sacs, lungs, and bone marrow of long bones of chickens. Kaupp and Surface (1939) reported this parasite in pigeons. It is possible for the eggs of both internal parasites *L. cysticola* and *C. nudus* to be coughed up with tracheal exudate and swallowed and thus spread to other hosts in droppings. Common nasal mites also increase respiratory discharges. These parasites are therefore conveniently transferred when infested birds contaminate feed or water. Larvae of chigger mites in the family Trombiculidae attach to the skin and inject an irritating substance that produces intense itching as it dissolves skin tissue. The mites cause damage as they feed on host tissue, not blood. Adults are free-living on the ground in scrub growth areas.

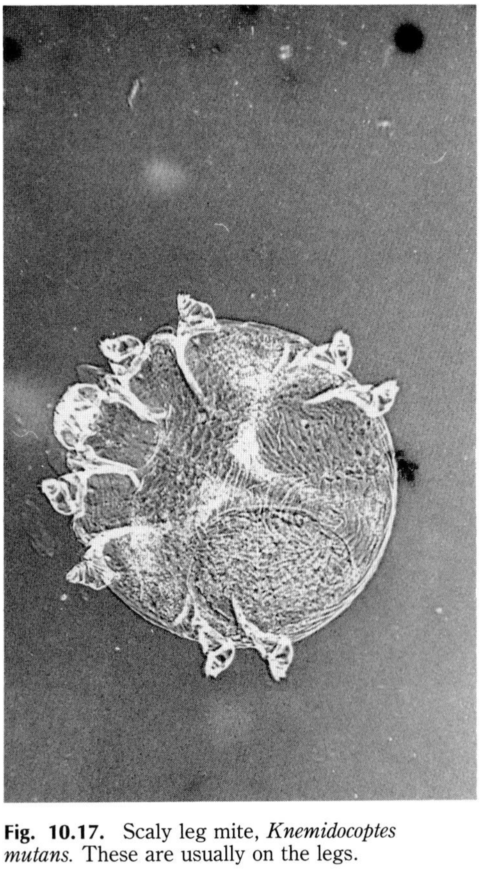

Fig. 10.17. Scaly leg mite, *Knemidocoptes mutans.* These are usually on the legs.

Aside from causing irritation and blood loss, mites have been responsible for transmitting poxvirus and for harboring western and eastern encephalitis viruses.

Signs and Necropsy. Mites appear as tiny, moving, grayish black dots. Red mites leave grayish deposits in cracks where they reside. They feed largely at night and thus are seldom observed on birds. Other mites live on the birds. In general, infested pigeons appear unthrifty and pale from the blood-sucking mites, because mites cause constant irritation and unrest. In addition, feather mites may damage feathers and quill mites may leave a brown powder in the shaft of the feather. Subcutaneous mites or air sac mites may be observed under a dissecting microscope. Dead mites calcify and may appear as white pinhead calcium deposits on tissue membranes. Identification of individual mites can be made by examination under a microscope. Xylol preserves mites for further study.

Control and Treatment. New birds or birds returning from shows or races should not be admitted to the loft unless they are treated for mites. Premise control as described for ticks may be followed. Vapor products are not recommended but loft and litter sprays have been helpful. Sprays include the use of carbaryl (Sevin, Union Carbide). Loomis (1984) also recommends pyrethoid compounds: fenvalerate (Pydrin, Du Pont) and permethrin (Ambush, Plant Protection Division, U.K.)

For red mite control, car drain oil or heavy vegetable oil is effective as a perch paint. Red mites live off the bird and will plug their spiracles as they cross oil-coated perches. Feather mites that live on the bird require individual bird treatment. A dip or spray of 1% carbaryl combined with 1% Malathion (American Cyanamid), repeated in about 10 days, has been effective but considerable mite resistance has been experienced. Sevin 5% dust applied to the bird at 1 lb/100 birds is also recommended for skin, body, and feather mites. This drug level should not be placed in the nest boxes when eggs are present. Scaly leg mites may be controlled by leg dips in mineral oil containing 0.5% Lindane, but the latter product has been removed from the market in the United States. Automobile drain oil may be effective if used repeatedly. Skin, subcutaneous, quill, nasal, and air sac mites may be treated with Ivermec 1% (Merck) at 1 ml/100 ml water, applied with a cotton swab to the bottoms of the feet 1 day/wk for 3 wk. The drug will be absorbed and thus reach the mites. Caution must be used in the application of this and other products. Rubber gloves must be worn to prevent absorption through the skin.

References

Becklund, W. W. 1964. Am J Vet Res 25:1380–1416.
Chandler, A. C. 1949. *Introduction to Parasitology.* 8th ed. New York: Wiley, pp. 498–524.
Hirst, S. 1920. Annu Mag Natl Hist 6:121–22.
Hollander, W. F. 1956. Trans Am Microsc Soc 75:461–80.
Kaupp, B. F.,. and A. C. Surface. 1939. *Diseases of Poultry.* Chicago: Kaupp and Surface.
Lahaye, J. 1928. *Maladies des Pigeons.* 3d ed. Remouchamps, Belg.: Steinmetz-Haenen.
Levi, W. M. 1941. *The Pigeon.* Columbia, S.C.: Bryan.
Loomis, E. C. 1984. In *Diseases of Poultry,* 8th ed., ed. M. S. Hofstad et al., pp.602–10. Ames: Iowa State University Press.
Smiley, R. L. 1986. Personal communication.

Ticks

Definition. Ticks are parasites similar to mites but considerably larger. They are common, eight-legged, crawling, wingless, blood-sucking, external parasites of birds, animals, and humans, with six-legged larval stages.

Cause and Classification. Ticks are placed in the phylum Arthropoda, class Arachnida, and order Acarina. Table 10.8 lists families, genera, and species of ticks likely to bite pigeons.

Life Cycle. Argasid soft-bodied ticks hide in nest boxes and cracks and crevices of lofts, where the fertilized female often lays her eggs. Mature females feed for short periods, usually at night, and after each of four or five blood meals, leaves the host to lay a cluster of eggs. The female dies after producing a total of 500 to 800 eggs over a period of weeks or months. Eggs laid in the fall may hatch in the spring. Six-legged larval seed ticks hatch in 6–10 days to 3 mo, depending on the temperature. The larval stage is followed by two eight-legged nymphal stages before adult male and female ticks develop. Each stage is fostered by a blood meal. Larval ticks engorge for 4–5 days before they molt and transform to the nymphal stage in 3–9 days. Larval ticks may live without feeding for several months, whereas nymphs may survive 15 mo and adults 48 mo without a blood meal. Warmth and dryness aid tick development and longevity (Benbrook 1965; Loomis 1984). In the laboratory Loomis (1961) determined the tick cycle to be completed in 7–8 wk.

Table 10.8. Ticks likely to infest pigeons

Family, Genus, and Species	Reported by	Location	Pigeon Incidence[a]	Affected species[b]
Argasidae				
Argas persicus[c,d,e,f]	Oken (1818)	United States, Europe, South America, Australia	++	Po, WB
A. radiatus[d,e,g]	Railliet (1893)	United States, Mexico Florida, Texas, Iowa	++	Po
A. reflexus[d,f,h]	Metz (1911)	Europe	+	
A. r. reflexus[i,d]	Fabricus (1794)	Europe, Asia United Kingdom	++	Po, WB M, H
A. r. hermanni[f,i,d]	Audouin (1827)	Africa	+	
A. abdussalami[f]		Pakistan	+	
A. miniatus[d,f]	Koch (1844)	South America	+	
A. africolumbae[d]		Africa	+	
A. vulgaris[f]		USSR	+	
A. sanchezi[d]	Duges	United States, Mexico, Texas	++	Po
A. neghmei[d]		Chile	+	H, Po
Ornithodoros coniceps[f]		Israel, Egypt, Crimea	+	
O. capensis[f]			+	
Ixodidae[f]				
Ixodes berlesei		USSR	+	
I. caledonicus		Hebrides	+	
I. c. sculpturatus		Germany	+	
I. arboricola		Belorussia	+	
I. crenulatus		Kirghizia	+	
I. redikorzevi		Azerbaijan	+	
I. persulcatus			+	
I. columbae			+	
I. ricinus		Germany	+	
Dermacentor marginatus		Kirghizia	+	
Amblyomma maculatum			+	
Haemophysalis punctata		Kirghizia	+	
Hyalomma plumbeum		Kirghizia	+	
H. marginatum		Kirghizia	+	
H. m. marginatum		Hissar Valley	+	
H. turanicum			+	
H. aegyptium		Sudan	+	
Rhipicephalus complanatus		Congo	+	
R. haemaphysaloides		India	+	
R. hurti		Kenya	+	
R. jeanneli		Kenya	+	
R. longus		Congo	+	
R. planus		Congo	+	
R. reichenouri		Kenya	+	
R. senegalensis		Cameroon	+	
R. simus		Kenya	+	
R. ziemanni		Congo	+	

[a] = incidental, ++ = occasional.
[b]H = humans, M = mammals, Po = poultry, WB = wild birds.
[c]Becklund (1964).
[d]Loomis (1984).
[e]Strickland (1986).
[f]USDA Index Catalogue.
[g]Kohls et al. (1970).
[h]Hollander (1956).
[i]Hoogstraal and Kohls (1960a,b).

The nymphal stage or stages and the adults of *Ixodid* hard-bodied ticks require several days for feeding but feed only once during each stage. After one feeding females lay 100 to 10,000 eggs over several days. The eggs are deposited on or just under the ground surface and hatch in 2–3 wk to several months, depending on the temperature. Eggs laid in the fall hatch in the spring. The male dies shortly after copulation. The six-legged larval seed ticks climb and attach to passing hosts to secure a blood meal. After a meal they molt and develop into eight-legged sexually immature nymphs. After the nymphs secure a blood meal they molt and become adults. Variations from this pattern occur (Chandler 1949; Georgic et al. 1985).

Form. The tick body is covered by a leathery cuticle that can expand as it engorges, becoming almost round and about 1 cm in length. Unengorged ticks are small and are flattened-ovoid in shape. The larval ticks are newly hatched and have three pairs of legs. The nymphs and adults have four pairs. Nymphs are sexually immature ticks. The head, thorax, and abdomen are fused. Antennae are absent. The mouthparts and the basis capituli form a false head, or capitulum. The mouthparts consist of two short palps, two mandibles or chelicerae, and a hypostome attached to the basis capituli. The hypostome, which pierces the skin, is armed with sharp recurved serrations. In Argasidae the capitulum is ventral whereas in Ixodidae the capitulum fits into an interior groove or camerostome. In Argasidae there is no scutum or dorsal shield whereas Ixodidae has a shield. A spiracle or breathing orifice is located on each side above and behind the third leg or coxa in Argasidae and behind the fourth leg in Ixodidae. The anus and genital pore are ventral. Both sexes are similar in the Argasidae family, but Ixodidae sexes are often dissimilar. (See Fig. 10.18.)

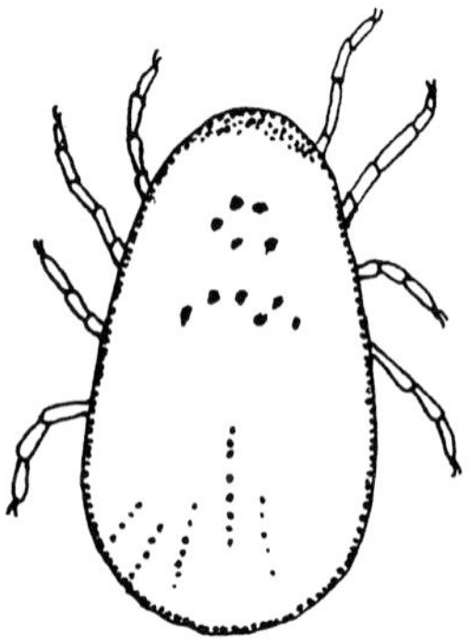

Fig. 10.18. Tick, *Argus persicus.*

Distribution. Ticks are found worldwide, but bird ticks are not common in northern states. Warm climates and summer temperatures encourage tick replication. Pigeons often share their ticks with other hosts, but ticks normally display a preference for certain birds and/or mammals.

Pathology. Ticks remain on the skin surface and extract a blood meal as the hypostome penetrates the skin. This portion of the tick may remain imbedded in the host if the tick is pulled from the bird. Tweezers should be used in their removal because ticks can transmit diseases. Spirochetosis caused by *Borrelia anserina,* egyptianellosis caused by *Aegyptianella pullorum,* and tick paralysis are known problems. Loomis (1984) also indicates that *Argas r. hermanni* in Egypt may transmit among pigeons the West Nile and Chenuda viruses and the Quaranfil virus group. These are arthropod-borne viruses isolated from argasid ticks in Egypt (Taylor et al. 1966).

Lyme borreliosis caused by *B. burgorferi,* a spiral-shaped bacterium, and transmitted largely by *I. dammini* in the eastern United States and by *I. pacificus* in California has become a major disease in people and animals. The states of California, Minnesota, Wisconsin, Pennsylvania, New York, Connecticut, Rhode Island, and New Jersey are infested with the tiny deer ticks. Pigeons have not been implicated in the transfer of the tick, but ticks are known to hitchhike, so fanciers should be aware of the problem.

Signs. Grayish brown, ovoid ticks may be observed adhering to the skin or feathers. Partially engorged ticks with their hypostomes imbedded may also be noted. Localized inflammation and crust formation may occur at the site of dislodgement.

Control. Individual ticks may be removed from the pigeon by touching the imbedded tick with the tip of a hot wire. Others suggest the application of nail polish remover or similar toxic chemicals to aid in the removal of the entire tick with the use of tweezers.

General tick control measures include removal of low-growing brush and weeds around loft buildings. Also, pressure spraying or painting of entire lofts, inside and out, without birds must be carefully done so as to penetrate cracks and crevices. Wood piles and equipment may be sprayed, but feeding and watering containers must be avoided or washed afterward.

All products used for tick control may be toxic for humans, especially children, animals, fish, and birds if contacted, ingested, or inhaled in sufficient quantities. The manufacturer's directions should be followed and care must be taken in their application. The applier should always back away from a spray application and wear a protective mask and clothing. If clothing becomes wet, it should be removed immediately and the skin washed with soap and water. After using any product, the applier's hands must be thoroughly washed.

Premise spray products that have been used to control ticks (Siegmund 1979; Loomis 1984) include 3.% carbaryl (Sevin, Union Carbide), 3.% Malathion (American Cyanamide), 0.5% chlorpyrifos (Dursban, Dow), 0.5% (Diazinon, Ciba-Geigy), 0.15% dioxathion (Deltic, Nor-Am), 0.75% ronnel (Korlan, Dow), 0.5% dichlorvos (Vapona, Shell), 0.5% toxaphene (Camphoclor, FMC), 1.% tetrachlorvinphos (Stirofos, Biotech), 0.5% Lindane (Calif. Spray), 2.% chlordane (Velsicol), 0.5% methoxychlor (Marlate, Du Pont), 0.3% naled (Dibrom, Amvac), and 1.% Ciodrin (Diamond Shamrock). Each product should be sprayed at the rate of 1 gal/1000 ft^2.

References

Becklund, W. W. 1964. Am J Vet Res 25:1380–1416.

Benbrook, E. A. 1965. In *Diseases of Poultry,* 5th ed., ed. H. E. Biester and L. H. Schwarte, pp. 952–54. Ames: Iowa State University Press.

Chandler, A. C. 1949. *Introduction to Parasitology.* New York: Wiley, pp. 660–706.

Georgic, J. R., V. J. Theodorides, and M. E. Georgi. 1985. *Parasitology for Veterinarians.* 4th ed. Philadelphia: Saunders, pp. 38–47.

Hollander, W. F. 1956. Trans Am Microsc Soc 75:461–80.

Hoogstraal, H., and G. M. Kohls. 1960a. Ann Entomol Soc Am 53:611–18.

______. 1960b. 53:743–55.

Kohls, G. M., et al. 1970. Ann Entomol Soc Am 63:590–606.

Loomis, E. C. 1961. J Parasitol 49:91.

______. 1984. In *Diseases of Poultry,* 8th ed., ed. M. S. Hofstad et al., pp. 610–12. Ames: Iowa State University Press.

Strickland, R. K. 1986. Personal communication.

Taylor, R. M., J. R. Henderson, and L. A. Thomas. 1966. Am J Trop Med Hyg 15:87–90.

U.S. Department of Agriculture. *Index Catalogue, Ticks and Tickborne Diseases. II Hosts,* Spec Publ 3, Part I A-F, pp. 307–928.

V

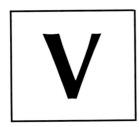

Management and Health

11

Disease-Control Program

Prevention is the aim of any disease-control program. Pigeon diseases can be transmitted in many different ways, depending to a large extent on the type of disease involved. Disease control and prevention must, therefore, be directed toward interrupting the mode of transmission. Diseases are spread by direct contact, for example, by bites of animals or insects, through breaks in the skin or mucous membranes, or through the egg. Transmission may also occur indirectly by inhalation of contaminated dust, by ingestion of infected food, or by contact with contaminated equipment.

Control of the various diseases is made more difficult because live birds from many sources are often placed in close contact with each other. Crowding encourages disease transmission, and in a loft the more timid birds are shoved around and given less chance to eat and drink. Crowding also results in injuries that interfere with performance of all birds. Disease control thus involves many management factors as well as the specific interruption of disease cycles.

Guarding pigeons against disease requires a planned program that is practical and economical. Shortcuts and dependence on drugs seldom provide a long-range approach to disease control and prevention. The program should include (1) selection of healthy birds, (2) essential quarantine, (3) good sanitation, (4) routine vector control, (5) proper nutrition, and (6) reliable diagnosis and appropriate treatment.

Selection of Healthy Birds

By selective breeding people have created the various types and species of pigeons enjoyed today; in so doing, however, they have permitted bad as well as good characteristics to exist. The gene pool, the sum of all the characteristics of all the pigeons, is limited and includes unfavorable qualities. Inbreeding (breeding within a family) quickly concentrates both good and bad characteristics and limits the size of the gene pool. A breeder must, therefore, be very selective of parents to secure progeny with desirable external and internal characteristics.

A breeder must maintain individual breeding records on each cock and hen and eliminate obvious defects. Obscured characteristics such as faulty or missing kidneys must be noted in families. Although many such birds will not live, some weak birds with physiological and/or anatomical demerits will survive and these birds must be sacrificed. When breeding for health, consideration must be given to stamina and vigor and resistance to disease. Other factors to consider include early natural maturity, general body conformation, and mating habits, including polygamy.

Purchased stock should be secured from reliable breeders who have demonstrated good performance and livability and whose stock is free from infectious disease. Nothing devastates a flock or demoralizes a fancier as much as disease.

Quarantine and Disease Control

Newly purchased birds and those returning from races or shows should be kept in isolation for up to 30 days or until they are tested and found to be free from infection. This means that separate lofts should be provided, ideally with 100 feet between units (although from a practical standpoint this is difficult to accomplish). The separate lofts should house young birds, old breeders, sick birds, and returning birds or new arrivals.

Quarantine also applies to personnel and animals. People, cats, dogs, and other birds or livestock should be excluded from lofts and yards.

Daily inspection of the flock is essential. If a disease outbreak occurs, sick birds should be isolated and cared for last. Chronically ill birds should be destroyed and burned or buried promptly. Ailing birds may be a threat to the entire flock. They must be removed to isolation facilities as soon as observed. Hospital pens or yards, however, may serve to spread infection, particularly if the sick birds from various pens are later returned to their respective quarters. Such isolation facilities should be separate from pens for newly purchased birds.

Sanitation

People and Pigeons. The owner, caretaker, or visitor should wear clean outer clothing when caring for and/or inspecting birds. A long lab coat, washable head cover, and rubber overshoes may be used over other clothing. The lab coat and hat may be disinfected in a galvanized or plastic, lid-covered container using 400 ppm quaternary ammonium disinfectant (Quatsyl 256, Lehn & Fink). After the used clothing is removed in a clean, dust-free location, the hands, coat, and hat must be immersed in the disinfectant and the coat is hung up to drip-dry. The solution may be used over and over again but should be replaced when visibly contaminated. The rubbers should be scrubbed off in a separate bucket containing 400 ppm of the same disinfectant.

Equipment. Borrowed or exchanged equipment should be carefully cleaned and disinfected before use. Waterpans should be scrubbed clean daily. A slotted metal cover or a wired incline to keep the birds out of the pan is essential. A wired platform under the waterpan may aid litter sanitation and keep the area drier. Feeders, including self-feeders, should likewise be constructed of metal or plastic and have a partial cover to exclude droppings. These waterers, feeders, and other equipment should be routinely cleaned and disinfected between flocks.

Contaminated crates returned to premises from races or the diagnostic laboratory may easily initiate infection and thus must be disinfected each time they are used. Vans or vehicles in which the birds are transported should be cleaned and disinfected between trips. Van cages or baskets should be mosquito-screened to prevent pigeon louse fly movement. Screen dust must be removed often to permit ventilation. Fiber- or wooden-framed crates and cloth covers on crates absorb contamination and should be avoided; metal frames and plastic covers are more readily cleaned and disinfected. Used feed bags or containers contribute to disease transmission. Crocks used for feed or water should be entirely glazed to prevent absorption of water and bacteria. The glaze must not contain lead. Most crocks are glazed only on the inside or top. All pottery and water bath pans must be routinely disinfected to avoid harboring infection. Disposable nests should not be reused, and crock nests must be cleaned and disinfected between service periods.

Buildings. The loft building should be constructed to enable easy cleaning and permit good sanitation. Smooth concrete floors sloping toward the door (1 in.:10 ft) can be easily washed down after each flock on a routine basis. Wooden floors tend to absorb contamination and are more difficult to clean and disinfect. Soil floors cannot be kept clean and should be avoided.

The exercise porch should have a wire floor. One-in. (25-mm)-mesh plastic-coated welded wire does not readily collect droppings and is preferable to smaller meshes, which are more difficult to keep clean. This prevents birds access to their droppings.

Loft windows and doors should also be screened and be designed to close securely to exclude birds and rodents. This also facilitates temperature and ventilation control, excludes rain and snow, and prevents predator losses. Half-in. welded mesh wire or plastic screen should cover windows, louvers between rafters, and screen doors to the loft and feed rooms.

The area beneath the porch and building should be enclosed to exclude feral birds and animals. Removable panels in this area aid periodic cleaning.

Good ventilation in a building is sometimes difficult to provide. In the construction of any loft, the aim is to supply fresh air to promote dryness and freedom from odors and at the same time avoid drafts. Fresh air helps to prevent airborne diseases, and in hot summer weather it aids in preventing heat prostration. Air stagnation often permits ammonia buildup and eye damage. Any condition offensive to the caretaker will be even more so for birds captive within their environment.

An A-roof must provide louvers at the peak of the roof to exhaust hot moist air. Shed roofs likewise require exhaust ports between the rafters at the highest point. Without removal of moist air at these points, water in the air condenses and falls to the floor, causing damp or wet floors.

Insulation is provided to help prevent rapid changes in temperature and to control condensation. A well-fed adult pigeon can take virtually any amount of uniform dry cold without distress, but young birds need protection. Insulation is of particular value in the roof and should be applied directly to the wooden roofers or sheet metal roof. Rats and vermin find a haven in loose insulation, especially if a false ceiling or wall is formed by the insulating material. Wall insulation must be covered with some impervious material to prevent ingestion by the birds. When soft fiberboard is used as the exterior wall, birds will often fill up on the indigestible wallboard and will not eat feed.

Exterior walls should be constructed to be impervious to water from the outside. Wet walls encourage fungus growth and systemic bird and feather infections.

The building itself should face the sun. Sunlight not only aids dryness but is beneficial to the birds by stimulating laying activity. Clean loft windows or skylights permit entrance of ultraviolet rays, which activate the pro–vitamin D within the skin of the birds, thus promoting health. Glass skylights filter out some of the ultraviolet rays, but plastic skylights exclude more of the rays.

A loft normally houses 20 pairs of pigeons in a pen with one to three pairs per square yard or meter. Each breeding pen should contain 40 nest boxes (2 per pair), one self-feeder, one drinker, and a perch area.

Hoppers and waterpans should be designed to exclude the feet of birds and bird droppings. The feeder or drinker located near the floor may be placed on a square 6- to 8-in. (15- to 20-cm)-high box frame covered with ½-in. (1.2-cm) wire mesh. Feed or water that is tossed out will then drop through the wire mesh where it cannot be reached by thc birds, preventing them from eating contaminated feed or water.

Properly designed wall perches will remain clean. They may be rounded projecting supports or box perches (Fig. 11.1). If the supports are properly spaced and stacked vertically the birds will not be able to bother each other and droppings will fall to the floor.

Lofts should be situated to provide adequate water drainage away from the loft. Birds often pick up disease from the ground. Water draining away should be directed toward underground sewage lines or catch basins. Exercise yards associated with the loft should be well drained and surface water should not be permitted to accumulate.

Fig. 11.1. Rounded corners on perches, to prevent injury. Injured birds do not win races.

Manure and Litter Disposal. Loft sanitation involves routine removal of droppings and litter disposal. Droppings are a potential threat to the health of birds. Diseases and parasites are transmitted by them. If a nonlitter management method is practiced, the droppings should be removed weekly or more often from nesting areas, perches, and floors before a light coating of new dry sand or kitty litter is added. When in use the nests should be cleaned and nesting material replaced every 2 wk. Yearly complete loft cleaning and disinfection is recommended. If capillary worms are present, the pens should be cleaned more often and certainly after worming the birds, because worm eggs are not killed by the action of the wormer.

Old droppings may harbor *Cryptococcus* or *Histoplasma* organisms, which can be fatal for the person cleaning the house. A dust mask should always be used for personal protection. A light coating of water helps to settle the dust prior to cleaning.

If litter is used, a light covering of clean, dry, nondusty, nonmoldy litter such as bran or sugarcane fiber is suggested. Parasites and disease germs cannot thrive in the absence of moisture. Caked or wet litter should be removed. Crowding measurably increases litter moisture and the possibility of infection. Improperly functioning automatic water fountains or leaky waterpans increase moisture problems. Droppings and contaminated litter may be held for disposal in a garbage can with a secure lid to reduce fly breeding. Droppings and litter may be composted in an isolated area. Care must be taken to avoid spillage in loading or transporting droppings. The equipment used in cleaning and transporting the manure should be cleaned and disinfected immediately after use. A residual 5% phenolic or cresol disinfectant is effective for this purpose. During routine cleaning operations separate clothing and rubber footwear are suggested.

Cleaning and Disinfecting. Good sanitation requires that buildings and equipment be cleaned and disinfected after each group of birds is removed. The following steps are recommended:

1. Settle the dust by spraying lightly with water or oil. If the litter is to be burned, oil is recommended.
2. Take all movable equipment out of the building before starting the cleaning operation.
3. Remove the coarse material, sweep down the ceiling and walls, and collect manure spilled outside the doorway. (Wear a mask during the cleaning operation.)
4. Hose down the floor and walls after scrubbing clean.
5. Apply a quaternary ammonium disinfectant at 400 ppm to the buildings and equipment. Other residual disinfectants may also be used. Feed and water containers may be disinfected with nonresidual disinfectants.

Ordinary 5.25% sodium hypochlorite (household bleach) is a good nonresidual disinfectant for clean surfaces. The recommended use dilution is 2 tbsp/qt water. It may be rinsed off after its application on waterpans or feed hoppers.

The area around the building should also be cleaned. Refuse, weeds, rotten wood, discarded equipment, and manure should be removed. Walkways, feed rooms, door areas, and roadways may be liberally dusted with dry wettable powdered sulfur at the rate of 1 lb/100 ft^2. This provides a long-term effect. In addition, dry sodium acid (bi) sulfate crystals should be used at the same rate for immediate effect. Application of dry crystals to a bare wood floor used by birds is not recommended. The dry combination may be raked into litter or kitty litter while the birds are in the loft. It may be applied on a weekly or biweekly basis to keep the floor acid if paratyphoid is a problem. The sulfate combines with manure moisture to form sulfuric acid. A liquid solution may be applied with a broom at the rate of 1 lb/3 gal water, avoiding metal equipment to prevent corrosion.

Birds may consume a small quantity of the acid salt without harm, and their feet will likewise be unharmed. Caution is advised. Protective goggles and rubber gloves and boots should be used when handling sodium acid sulfate. Clothing dampened by the solution should be immediately removed, and any skin surface in contact with the material should be washed off as soon as possible.

The aim of the procedure is to maintain a pH of 3.5 or below. Pigeon paratyphoid organisms normally grow and reproduce under moderately alkaline conditions and cannot survive at a pH of 3.5 in an acid environment. The normal pH of body fluids of birds or animals approaches 7.5, and it is at this pH that most disease-producing organisms live and thrive. Color-sensitive test Hydrion paper may be obtained at a pharmacy to determine the pH of the soil or litter; 1 tbsp test material is mixed in ½ cup water, the test paper is dipped in the mixture and the color read. The pH of the water should be tested first because minerals in the water may alter the reading.

Vector Control

Rats and mice spread infection. Elimination of their source of food and shelter by rat-proofing bins and storage rooms and by keeping buildings clean is fundamental in any rat-control program. Constant surveillance and routine use of poison baits and traps are essential in keeping rodent populations in check. Poison baits must be so placed as to prevent access by pigeons and/or other animals. The directions must be followed closely.

Insect pests, particularly louse flies and mosquitoes, are disease carriers for pigeons. Certain other insects may serve as intermediate hosts in the life cycle of tapeworms. Infectious agents such as pox may be mechanically carried by insects. Mosquito screens may therefore be necessary in areas where they are prevalent. Elimination of insect breeding places is essential in their control; thus shady wet areas should be avoided. Lice, mites, and flies may largely be controlled by insecticides.

Dogs, cats, other animals, free-flying birds, and pet birds should be excluded from the loft. Visitors likewise are a cause for concern. They should be restricted unless they wear clean clothing and footwear.

Nutrition

The keystone of flock health is proper nutrition. A balance of nutrients, including vitamins and minerals, is required. Any shortage or excess may lower a bird's resistance and cause a dietary disease. If a bird has a full breast it is usually a healthy bird. A bird lacking in weight is cause for concern.

Feed should not be moldy, wet, or contaminated and sweepings should be avoided. Stale, dusty or powdery feed, broken kernels, and grains high in tannin are usually deficient in nutrients and should be discarded. Drugs, including vitamins, do not replace good nutrition. They should not be given unless a specific problem is diagnosed and then only as recommended.

Plenty of fresh, clean water is necessary at all times. Pigeons drink an unusual amount of water, particularly when they are feeding their young. They require water containers that are about 2 in. deep so they can immerse the entire beak. Waterpans should be kept clean. If the caretaker is willing to drink from the pan, it is clean enough for the pigeons.

Clean feed, fed in clean containers, is desirable. Birds readily pick up parasites if feed is thrown on the ground or floor. Birds tend to "bill out" unwanted feed and discard it unless self-feeders are designed to prevent feed waste.

Reliable Diagnosis and Treatment

If birds become ill, a reliable diagnosis is needed from a veterinarian, diagnostic laboratory, or clinic. Without a diagnosis, treatment is of questionable value. If birds are taken to a clinic, three or four typical dead or live sick specimens should be selected and carried in a disposable cardboard box. The pathologist should be shown the cause for concern and/or the condition described, providing a clear, accurate, and complete history of the problem.

In warm weather, specimens should be wrapped in newspaper and refrigerated but not frozen. The birds should not be cut open. Mailed dead birds seldom arrive in good condition unless they are placed in a plastic bag and refrigerated in transit. Fecal samples can be mailed if they are placed in aluminum foil or wax paper enclosed within a self-sealing small plastic bag. Such samples should be examined every 2 to 3 mo as a routine disease-control measure. Air-dried blood smears may also be mailed for identification of blood parasites.

Once a diagnosis has been established, any indicated specific treatments should be employed. Suggested medication, directions, and procedures should be followed closely. Accurate dosage is very important. If the medication calls for grams, a gram scale must be employed because a teaspoon will not accurately measure grams, and too little may be ineffective and an excess often toxic. Outdated medications can also be a source of trouble. If vaccination is indicated, follow the instructions closely. Deviations from the proper level of medication, or from the specified method of application, may definitely alter the course of the disease or condition.

12

Breeding Techniques and Problems

Mating and Hatching

Mating. Individual mating cages provide the best means of progeny control. They enable unrestricted mating and permit nest building without interference from other birds. They also assure better brooding and hatching conditions than are possible in open breeding lofts. Breeding cages or nest boxes for a breeding pair should be provided with two nest bowls so the eggs of the second round can remain clean. Breeding pairs should be full fleshed but not overly fat. They should also be healthy and free from parasites. A mature, driving, noncannibalistic cock and a mature, robust hen, free from defects, are essential to good fertility and hatchability.

A hen 5 mo old may lay its first egg, but hens are seldom mated until a year of age and it is then that egg laying normally starts. After the first mating, an egg is laid in about 10 days. The second egg is laid two days later. An egg must be fertilized in the oviduct about 24 hr prior to being laid. The egg remains in the oviduct 16–24 hr while the three white albumin layers, the two shell membranes, and the shell are added. Semen is active in the oviduct for up to 8 days after mating. About 25 days after the first mating, when the squabs are weaned, a pair may mate again, and this time the egg is usually laid in 5 days. If a pair is separated for 2 wk and the hen is remated with a second cock, the semen from the second cock will fertilize the eggs.

After the second egg is laid, the hen and the cock commence serious incubation of the eggs. Incubation takes 19–20 days from the second egg. The hen sets during the night and is relieved by the cock by midmorning and until she returns in the afternoon. Variations in the pattern occur. In fact brooding birds may desert the nest at any time and for various reasons. If the eggs become cold, the embryos may die. Eggs deserted for 24–36 hr may be saved by a foster parent or by the use of an incubator.

If artificial incubation is desired, the eggs must be stored in a humid area, preferably a basement with a relative humidity of 70–75% and a temperature below 68°F and above 55°F. Germ development starts at 68°F. The eggs may be held under these conditions for up to 2 wk without significant loss of fertile eggs if they are turned from end to end twice daily.

Eggs. The eggs should be clean. They may be carefully cleaned with sandpaper and not washed or they may be quickly dip-washed in 200 ppm quaternary ammonium detergent disinfectant solution at a temperature of 115–130°F. The eggs are permitted to drain after a quick rinse with a clean solution of the same disinfectant at a temperature of 115°F. The cooled eggs are then stored as previously outlined or they are incubated.

Incubation. Another disinfectant procedure involves fumigation of dry, clean eggs for 20 min prior to hatching in a tightly closed cabinet. The eggs are placed in a wire basket in the cabinet and 24 ml Formalin is added quickly to 23 g potassium permanganate held in a separate container within the cabinet. These amounts are for 20 ft^3 cabinet space. Following fumigation the eggs may be stored or incubated.

Incubation in a still-air incubator requires a uniform egg temperature of 99–99.5°F. The temperature at the bottom of the machine should approximate 105°F, and the temperature at the egg level should approach 101–103°F. Humidity should be kept as high as possible. The application of a wet towel to the top of the eggs may be helpful. A wet bulb reading of 87°F or 60% relative humidity is suggested for the first 14 days, followed by 97°F or 90% humidity for the end of the hatching period. In addition the eggs should be turned or rolled three or four times a day to keep the embryos from sticking to the shell.

Prior to artificial incubation, the empty machine should be cleaned, washed, and fumigated; 30 ml Formalin and 15 g potassium permanganate/20 ft^3 for 1 hr with closed vents and high humidity is recommended. Also 30 g paraformaldehyde/100 ft^3 for 20 min with closed vents is effective. In both instances the machine should be in operation without eggs.

Hatchability. Many factors can alter hatchability. Fertility is the initial problem; this is determined by candling eggs under a bright light after 3–5 days of incubation. Clear eggs or the lack of development can be observed quite readily, and no movement is obvious. Infertility of breeders arises from poor health or nutrition, crowding in open lofts, cold weather, heavy feather development near the vent area (which must be cut away), holding eggs too long, overweight pairs, fighting among cocks, and early matings with immature cocks. Blood rings or slight development arise from irregular heating and chilling or overheating of eggs prior to setting. Dead germs developing between 3 and 15 days result from inadequate turning, improper ventilation, or carbon dioxide buildup within the incubator from closed air vents, holding eggs too long, cracked eggshells, improper nutrition of breeders, low voltage and thus heat variation, diseases such as paratyphoid, and inherited low hatchability.

Pipped eggs that do not hatch may be due to insufficient moisture or too high a temperature. If eggs hatch too early, the temperature may have been too high. Conversely if they hatch too slowly the temperature may have been low.

Inadequate nutrition causes dead germs. Low vitamin G and B_{12} and poor-quality protein increase second-week mortality. Low vitamin E exerts its effect during the first week. Low manganese levels can cause embryo deaths at any time. The effects of decreased calcium are noted during the last 3 days of hatch. Adequate vitamin D is also essential at this time. A deficiency of pantothenic acid in the breeder ration may also increase the percentage of weak squabs and squeakers that die early. Whenever a problem occurs, unhatched eggs and dead squabs should be examined and cultured by a veterinarian at a laboratory to exclude the possibility of an infectious disease.

Crop Milk

At about 30 days of age a squab is called a *squeaker,* and at about 6–7 wk it is termed a *youngster.* All this time both parents feed the nestlings until they are mature enough to eat on their own. The hormone prolactin from the pituitary gland triggers broodiness and crop milk secretion. The crop wall forms longitudinal folds and is lined with epithelial cells, which begin to thicken by the eighth day of broodiness. Considerable thickening and proliferation may be noted by the thirteenth day, and by the following day the holocrine crop gland begins the production of curdy, cream-colored crop milk. Sheets of gravid, lipid-containing mucosal epithelial cells are desquamated to disintegrate and become part of the secretion (Weber 1962). This reaches a peak about the eighteenth day, when hatching occurs. Concentrated crop milk ceases about 10 days posthatching, but it may continue to a lesser degree for 25 days after the hatching date (Vandeputte-Poma and Desmeth 1973).

An analysis of fresh, wholesome pigeon milk as reported by Ferrando et al. (1971) yielded 75–77% water, 11–13% protein, 5–7% fat, and 1.2–1.8% minerals. The minerals were 0.14–1.17% phosphorus, 0.12–0.31% calcium, 0.11–0.15% sodium, and 0.13–0.15% potassium. There were virtually no carbohydrates. Reed et al. (1932) reported 64.3% water, 18.8% protein, 12.7% fat, and 1.6% ash. In addition, Carr and James (1931) identified rennet, amylase, and saccharase enzymes in the crop contents. They further quoted a graduate thesis that reported an analysis of crop milk from corn-fed pigeons as 20–30% solids, 14–16% protein, 30–39% ether extract, 1.9–2.7% P205, 1.1–1.9% CaO, and 5–6% total ash. Leash et al. (1971) studied the crop contents of White Carneau breeders over a period of 27 days. On day 1, the hatching day, they found 70% water, 27% fat, 46% protein, and 21% carbohydrate. On day 7 the crop milk composition abruptly changed to 5% fat and 27% protein and continued to change until day 27, when they recorded 27% water, 3% fat, 17% protein, and 74% carbohydrate. Water was calculated as a percentage of wet weight, the other ingredients were a percentage of dry matter. The carbohydrate content resulted from hard grains being mixed with the crop milk. Desmeth and Vandeputte-Poma (1980) evaluated the lipid content of crop milk and found that triglycerides form 80% of the fat and phospholipids another 12%. In addition, free fatty acids, glycerides, and cholesterol and its esters were present. The most common fatty acid was oleic; others were palmitoleic, stearic, linoleic, and palmitic.

Unfortunately the quality and/or quantity of crop milk can be altered by diseases such as: trichomoniasis, herpes, psittacosis and mycosis. Therefore, breeding birds should be examined prior to mating to insure freedom from any infection that can be transferred to the young squabs by the parents.

References

Carr, R. H., and C. M. James. 1931. Am J Physiol 97:227–31.
Desmeth, M., and J. Vandeputte-Poma. 1980. Comp Biochem Physiol 66B:129–33.
Ferrando, R., et al. 1971. Ann Nutr Alim 25:241–51.
Leash, A. M., et al. 1971. Lab Anim Sci 21:86–90.
Reed, L. L., et al. 1932. Am J Physiol 102:285–92.
Vandeputte-Poma, J., and M. Desmeth. 1973. Vlaam Dr Diergeneeskd Tijdschr 47:231–35.
Weber, W. 1962. Z Zellforsch Mikrosk Anat 56:247–76.

Genetic Problems

Numerous inherited conditions in pigeons cause structural and physiological changes. Some of these are fatal genetic defects.

Kidney Structure. There appears to be an increase in birds with *faulty kidney structure.* One or more lobes on one or both kidneys may be missing, and those present are often deficient in tissue substance. Normally birds have three lobes per kidney, which fill the lumbosacral compartments. Dilated ureters often accompany the condition. Jeffrey et al. (1937) observed the condition in inbred Single Comb White Leghorn chickens in New Jersey.

Eyes. *Albino pink eyes* were reported by Hollander (1951) in White Carneaus, White Kings, and Tumblers. Albinos fail to see well, their down is short, and they grow slowly; the character is inherited as a single-factor recessive.

In 1938 Hollander described an inherited *clumsy pigeon syndrome,* a flight problem that appears to be due to defective vision involving the retina and is a simple recessive trait.

Another eye disturbance is that of *microphthalmia,* or *small eyes.* The author has observed a few such cases in which retinal abnormalities were also present. Hollander (1948) also recorded the hereditary defect in blind squabs of Racing Homers. The eyeballs were about half the normal size, with a thickened choroid coat. The retina was defective and the eyelids were not functional.

Levi (1941) reported a female White Carneau that hatched with *no eyes.* Hollander (1970) received a bird from Indianapolis in 1959 that could not see to pick up feed grains. He has shown this to be caused by an inherited recessive gene. The trait continues to appear in isolated flocks.

Recently *blind birds* in Maryland were brought to the author's attention in which histological findings revealed multiple foci of retinal degeneration. The cause remains undetermined. Cuenod et al. (1970) found ballooning of synaptic vesicles to be an indication of optic nerve degeneration.

Cataracts were reported by Hollander (1958). The cause is unknown and no treatment is indicated. In the author's experience, older pigeons and chickens have the characteristic white opaque lens capsule as observed through the pupil.

In Europe, Descheemaecker (quoted by Ferreira 1989) noted a *growthlike eye condition* in young pigeons in which a raised, rough, dry, millimeter-sized, inflammatory, circular lesion appeared on the conjunctiva or cornea of the eye. The cause has not been determined.

Dyslexia in children is characterized by difficulty in distinguishing mirror-image letters such as *b* and *d.* In research work at the University of Maryland, Miller (1982) observed that pigeons likewise have a similar visual problem. Nine of 48 pigeons tested could not distinguish between lateral mirror images. When one eye was covered, definite improvement in performance was noted. Mello (1966) studied the condition and found that the interocular reversal of left-right mirror image has not been reported in any other species. He further indicated that retinotectal fibers were completely crossed at the optic chiasma.

Toes. Levi and Hollander (1942) observed *polydactyly,* a sublethal double-hind-toe problem. An anonymous writer (1916) illustrated this same condition observed in an Australian Racing Homer. Fontaine (1922) also saw extra toes in a Syrian Spot pigeon breed.

In 1945 Hollander studied an *amputated toe* condition in which White Kings had the toe tip of one or all toes missing. Some of these survivors had an extremely short lower beak. The condition appeared to be dependent on a single recessive mutant. Outcrossing covered up the trait (Hollander 1971).

In 1974 Hollander reported *webbed toes* as a

single-gene recessive condition, typically affecting only the middle-outer toe connection in Racing Homers. In 1982 Hollander and Miller discovered another "web lethal" inherited trait in which the rear toes were webbed to the inner toes and the outer toes to the middle toes. This mutant type is generally lethal, causing death before maturity.

Beak. Hanebrink (1970) reported on a *cross-beak* recessive deformity in German Beauty Homers. Parona first described the condition in Italy in 1880. The condition is fairly common. The author has observed the problem on several occasions.

Skeleton. In 1945 Hollander called attention to *lethal achondroplasia* in White Kings, in which the skeleton was shortened and the embryos were unable to hatch.

Hermaphrodite Intersexuality. Early in 1936 Riddle and Schooley first noted *hermaphrodite intersexuality* of pigeons. This was reported by Riddle et al. (1945), who observed that 80% of the male pigeons had male and female genitalia. This they considered was due to one or more genetic factors together with estrogen.

Plumage. Another hereditary malformation recorded by Riddle and Hollander (1943) described *scraggly plumage* that prevented flight, as the barbules failed to interlock. In addition, these birds lacked coordination of body movements from defects in the spinal cord and brain and crusty skin growth. This was generally considered a recessive trait but at times was dominant.

In 1918 Riddle reported on a similar case of *hereditary ataxia* in pigeons. Later Hollander and Miller (1978) published an account of sideburns, or face feathers projecting forward, a dominant mutant trait accompanied by an undesirable head tremor or palsylike shaking. In 1938 a condition in Racing Homers was studied in which the birds were *featherless,* resulting in their inability to mate (Levi 1941).

Earlier Hollander and Levi (1940) reported defective wings and tail in a White Carneau hen. They also observed a White Carneau hen from the same place with *no tail vertebrae, uropygial gland, or large tail feathers.*

Nervous System. Perhaps related to the ataxia problem is the *tumbler* or *roller syndrome* so well reviewed by Quinn (1971). He outlined the hereditary nervous convulsive disorders investigated in chickens and turkeys and presented the following conclusions of pigeon studies. Rolling in flight and on the ground are transmitted by allelic genes. Rolling is not due to abnormalities of the cristae or otoliths of the inner ear. Rolling is not dependent on visual stimulation. It is not a form of epilepsy. Rolling in flight appears to be the manifestation of a recessive gene expressed as hypersensitivity of roller neck muscle fibers to acetylcholine.

Connective Tissue. Another genetic problem involves *atherosclerosis,* a fatty degenerative change in the connective tissue of arterial walls. White Carneaus and Silver Kings have a high genetic incidence of aortic atherosclerosis, while Racing Homers and Show Racers have a low incidence (Lofland and Clarkson 1959; Bullock 1974). Santerre et al. (1972) reported a difference in severity between White Carneaus and Show Racers by 1 yr of age. Patton et al. (1975) observed that Homers were not prone to aortic atherosclerosis but developed coronary lesions equal to those of the White Carneaus.

Other Genetic and Fertility Problems. Genetic problems must include *double-yolk* eggs. Twin embryos form from double-yolk eggs and live almost to the point of hatching. Abnormal embryo development results in monstrosities, reported by Hollander and Levi (1940). Another fertilization problem is the over 190 cases of *mosiacism* involving sex-linked color genes. This was reported by Hollander (1949, 1975) to be caused by bipaternity or the survival of supernumerary sperm.

Of further interest is the work of Riddle and Johnson (1939) and Cole and Hollander (1950) dealing with the *crossing of the common pigeon with the ring dove,* in which only male or virtually only male offspring resulted. Surviving female hybrids produced no eggs. Backcrosses to pigeons resulted in practically no fertility but backcrosses to doves gave 2% fertility.

References

Anonymous. 1916. J Hered 7:320–24.

Bullock, B. C. 1974. Adv Cardiol 13:134–40.

Cole, L. J., and W. F. Hollander. 1950. Am Nat 84:275–308.

Cuenod, M., C. Sandri, and K. Akert. 1970. J Cell Sci 6:605.

Ferreira, J. 1989. Personal communication.

Fontaine, R. 1922. Les Races de Pigeons et Leur Elevage. Paris: Charles Amat.

Hanebrink, E. L. 1970. Am Pigeon J 59:105.

Hollander, W. F. 1938. J Hered 29:55–56.

______. 1945. J Hered 36:297–300.

______. 1948. J Hered 39:289–92.

______. 1949. J Hered 40:271–77.

______. 1951. All Pets Mag 22:70–72.

______. 1958. Am Pigeon J 47:248.

______. 1970. Am Racing Pigeon News, Dec., p. 37.

______. 1971. In Pigeon Genet Newsl 58:7.

______. 1974. Am Racing Pigeon News, Oct., p.15.

______. 1975. J Hered 66:197–202.

Hollander, W. F., and W. M. Levi. 1940. Auk 57:326–29.

Hollander, W. F., and W. J. Miller. 1978. Behav Genet 8:101.
______. 1982. Am Racing Pigeon News 98:50–51.
Jeffrey, F. P., F. R. Beaudette, and C. B. Hudson. 1937. J Hered 28:335–38.
Levi. W. 1941. *The Pigeon.* Columbia, S.C.: Bryan.
Levi, W. M., and W. F. Hollander. 1942. J Hered 33:385–91.
Lofland, H. B., and T. B. Clarkson. 1959. Circ Res 7:234–37.
Mello, N. K. 1966. J Exp Anal Behav 9:11.
Miller, J. A. 1982. Sci News 122:332–36.
Parona. 1880. Atti Soc Ital Sci Nat 23:127–33.
Patton, N. M., R. V. Brown, and C. C. Middletown. 1975. Atherosclerosis 21:147–54.
Quinn, J. W. 1971. Pigeon Genet Newsl 59 (July):5–20.
Riddle, O. 1918. Proc Soc Exp Biol Med 15:56–58.
Riddle, O., and W. F. Hollander. 1943. J Hered 34:167–72.
Riddle, O., and M. W. Johnson. 1939. Anat Rec 75:509–27.
Riddle, O., and J. P. Schooley. 1936. Carnegie Inst Wash Yearb 35:55.
Riddle, O., W. F. Hollander, and J. P. Schooley. 1945. Anat Rec 92:401–23.
Santerre, R. F., et al. 1972. Am J Pathol 67:1–22.

Insemination

Artificial insemination is seldom conducted on pigeons because it is time consuming, it requires considerable patience, and the results are often disappointing. The male constantly fights to be released and is seldom cooperative. The technique is done to permit more extensive use of proven breeding cocks and to overcome breeding difficulties in research projects.

Bonadonna (1939), Dowling (1974), Owen (1941), and Raises (1968) have reported methods of pigeon insemination. Successful insemination of fowl has been reported by Quinn and Burrows (1936), Burrows and Quinn (1939), Van Wambeke (1972), Sexton (1978), and Lake and Ravie (1979, 1980). Their procedures have been adapted for the artificial insemination of pigeons.

Females to be inseminated must be separated from the male, but isolated pigeons do not lay. To overcome this problem, a male and a female should be placed in adjacent cages so they can hear and see each other and the male can touch her through the bars of the cage. The female will then lay regularly and will usually incubate her eggs and rear the squabs. A female may also be stimulated to lay by placing two hens together in a cage and separating them just before the eggs are due (Raises 1968).

A single insemination maintains fertility for 8 days. The most fertile insemination period is from the third to the sixth day before the first egg is laid. Since the ovarian cycle of the pigeon is predictable, it is possible to determine when an egg will be laid and thus when to inseminate. Pigeons normally lay two eggs per mating. Eight days after mating, the first egg is laid in the late afternoon. The second egg is passed 2 days later in early afternoon (Raises 1968).

To collect semen from the male pigeon, two people are required. Cloacal massage must be done on a regular basis for about a week to accustom the bird to the attention and reduce the constant fighting to be released. Massage should be conducted as soon as the bird is caught and should not be done more that once a day for best results. Every effort must be directed toward avoiding distraction, and it may be helpful to conduct the procedure near the mate. It is advisable to pluck the feathers from a 1-in. area around the vent several days before semen is to be collected. Dirt should likewise be removed. The operator should hold the pigeon in the left hand with the pigeon's head directed toward the left elbow. The fingers of the left hand are directed along the back toward the tail with the little finger and third finger securing the right wing. The index finger and the thumb are held against opposing sides of the upper portion of the tail. The right hand is free to milk the cloacal area. This is done with the thumb and second finger of the right hand directed forward under the tail on either side of the vent. The index finger applies deep pressure between the pubic bones and immediately below and ahead of the cloaca. Coordinated massage by the index finger and thumb of the left hand toward the thumb and second finger of the right hand will expose the ventral cloacal folds. Usually a small drop, (0.01–0.02 ml) of clean, white semen will be ejaculated on the surface of the cloaca after a short period of massage. When this appears the assistant should endeavor to collect the semen in a small clean eyedropper or tuberculin syringe. It is not uncommon to have a sample contaminated with urates or feces; such a sample should be discarded. If blood appears, massage should be discontinued to avoid further injury to the cock.

After the semen is collected 0.01 ml of semen may be diluted with 0.03 ml lukewarm physiological saline in a small vial (Owen 1941). While the hen is held on its back by the operator, the vent may be gently everted by the assistant and the tip of the syringe inserted. The oviduct connects with the left dorsal side of the cloaca. By gentle probing with the syringe the semen may be delivered in or near the oviduct opening. Both the vial and the syringe must be clean and be sterilized in boiling water before use. The syringe must be warmed to human body temperature before it is filled to keep the semen active.

References

Bonadonna, T. 1939. Proc 7th World Poultry Congress, pp. 79–82.
Burrows, W. H., and J. P. Quinn. 1939. USDA Circ 525.
Dowling, J. 1974. Proc Rutgers Univ Pigeon Health and Disease Conf.
Lake, P. E., and O. Ravie. 1979. J Reprod Fertil 57:149–55.
______. 1980. 9th Int Congr Anim Reprod Artif Insemination (Madrid) 5:113–16.
Owen, R. D. 1941. Poult Sci 20:428.
Quinn, J. P., and W. H. Burrows. 1936. J. Hered 27:31–37.
Raises, M. B. 1968. Aust Vet J 44:486.
Sexton, T. J. 1978. Poult Sci 57:285–99.
Van Wambeke, F. 1972. Br Poult Sci 13:179–83.

Pigeon Breeder's Lung

The allergic pulmonary disease occurs in up to 10% of pigeon fanciers, who have a hypersensitivity to pigeon proteins found in feathers, droppings, and discharges. Symptoms of the human disease includes coughing, fever, chills, shortness of breath, muscle soreness, and loss of appetite and energy. It appears first as a persistent cold that develops into hypersensitivity pneumonitis following several months of exposure. The disease initially responds to antibiotics and steroids, but as the disease progresses from continued exposure, interstitial fibrosis occurs within the lung tissue. Patients then usually experience permanent lung damage. Afflicted individuals can no longer breath normally and intercurrent respiratory diseases pose a serious threat to life. Sensitive fanciers must dispose of their birds and avoid pigeons.

APPENDIX A

Disease Conditions

Pathology deals with the essential nature of disease and especially the functional and structural changes caused by disease. Following are pathological and anatomical terms for some of the many common disease problems that occur in organs and tissues of pigeons.

Pathological Conditions

These terms represent conditions that are not otherwise designated as infectious or noninfectious and do not describe an anatomical change.

abscess. A localized collection of pus following tissue degeneration and liquefaction. Pus is composed of organisms, destroyed cells, and tissue fluid that cause pain, fever, and reddening. Abscesses may be found in any part of the body but are observed most often in the footpad, wing joint, sinuses, breast, beneath the skin, about the eyes and in tonsil and larynx mucus glands. *Staphylococcus* and *Pseudomonas* organisms are often involved.

adhesions. The abnormal joining of body structures by fibrous tissue bands, which most commonly occurs between the air sacs, oviduct, intestine, and liver. Perforation or infection of the abdominal cavity usually involves these organs in birds, causing them to adhere.

anemia. A state in which the blood is deficient in either quality, as in a deficiency of hemoglobin, or in numbers of red cells and volume, as from hemorrhage. This results in paleness of the skin and membranes and weakness. Wounds, toxins, parasites, and nutritional factors may cause anemia.

aneurysm. Spontaneous rupture of the aorta. The condition is unlikely in pigeons.

anorexia. A loss or decrease in appetite.

anoxia. Without oxygen or having insufficient oxygen.

arthritis. The inflammation of a joint, marked by pain, redness, heat, and swelling. *Mycoplasma* spp. and reoviruses are important infectious causes in birds.

ascites. The accumulation of straw-colored fluid in the peritoneal cavity (also called *dropsy*). Faulty heart function and the ingestion of toxic substances such as toxic fat containing polychlorinated derivatives of dibenzo p-dioxin (Flick et al. 1973) and crotolaria seed may cause fluid accumulation.

ataxia. Lack of muscular coordination or control, unsteadiness, and weakness, as in nerve changes or degeneration from injury, toxins, or infection. This may occur with paramyxovirus or herpesvirus infection.

atherosclerosis. Arteriosclerosis with fatty connective tissue degeneration of arterial walls.

avitaminosis. Without vitamins or lacking these nutrients.

AV block. Artrioventricular block, blockage of electrical impulses initiated by the AV mode. This is one cause of cardiac arrhythmias in pigeons (Miller 1978).

blindness. Loss of sight. Mechanical, chemical, or thermal damage to the cornea of the eye, the lens, or the retina may result in blindness. Cataracts (lens opacity) occurs in pigeons, but the cause is undetermined.

bruise. An impact injury without laceration to

surface tissues. Diffuse subcutaneous hemorrhages with tissue contusion characterize many injuries.

burns. Pathological lesions that, depending on the degree, produce redness, blistering, and skin and feather destruction followed by charring of tissues. Burns are classified by cause as thermal, chemical, electrical, radiant energy, friction, and extreme cold. Thermal heat results from fire, directly or indirectly, and hot liquid or steam. Chemical burns are caused by alkali, such as lye; acids, such as sulfuric acid; and caustics, such as disinfectants and carbolineum roost paint. Electrical burns are less common but may occur during the winter from shorted electrical waterheaters. Radiant energy given off from infrared heatlamps used to warm squabs during the winter may cause burns if not adjusted properly. Friction is unlikely to cause burns in pigeons. Extreme cold can cause skin dehydration or desiccation with much the same tissue destruction as in electrical burns. Frozen or frosted ceres seldom occur in pigeons.

cancer. Malignant or spreading types of growths.

carditis. An inflammation of the heart. Bacteria, including *Mycoplasma,* are often involved in heart disease.

cataract. An opacity of the eye lens or its capsule. The condition is caused by long-term action of light and oxygen on the protein of the eye's lens. It is sometimes observed in older birds. Research at Tufts University indicates that vitamin C is 30 times more plentiful in the eye than in the blood. This would suggest a possible relationship with cataracts.

cirrhosis. A condition resulting from liver disease in which fibrotic stroma replaces functional cells. This stroma contracts distorting the liver structure. Any parasitic, bacterial, or viral infection that destroys liver cells may cause scar tissue formational.

concussion. A condition resulting from a violent blow to the head. Traumatic injury during flight can be the cause.

congestion. The abnormal accumulation of blood in body structures. Dilation of blood vessels with stagnation of circulation is observed in many systemic infections or toxicities.

conjunctivitis. An inflammation of the membrane on the undersurface of eyelids and the forward surface of the eye. Lime dust, other chemicals, chlamydia, and trichomonads may produce the inflammation.

constipation. Infrequent or difficult evacuation, often with retention of feces. Disease of the intestinal tract may cause reflex inhibition. Lead poisoning may induce spasmodic constriction of portions of the intestine and constipation.

coryza. The term describes an acute thick mucus discharge from the nasal membranes. Most respiratory diseases cause a nasal discharge.

cyanosis. A bluish discoloration of tissues from lack of oxygen.

cyst. Any saclike structure in or on the body that contains air, fluid, or semisolid material. Cysts are most often observed in the kidney, ovary and oviduct.

dermatitis. A skin inflammation. The condition seldom occurs in pigeons, but when present, fungi are often recovered.

diarrhea. An abnormal, profuse, rectal discharge. Any intestinal inflammation regardless of cause and/or excessive dietary salt may be the cause.

dyspnea. Difficulty in breathing. Vagal paralysis may alter the rib movement and thus breathing. Tracheal inflammation or injury and lung and air sac infection may cause dyspnea.

edema. An abnormal fluid accumulation in tissues such as the liver and kidney.

emaciation. The loss of body flesh. The key to the health and nutrition of a bird lies in the fullness of the flesh covering the keel.

embolism. The plugging of a blood vessel by a clot or other obstruction. Embolism is observed in anemic infarction of sections of the liver or lungs.

emphysema. A pathological accumulation of air in tissues or organs. Whenever air enters interstitial connective tissue, alveolar walls tend to break down, permitting enlargement of some alveoli and compression of adjacent lung tissue.

encephalitis. An inflammation of the brain caused by viral, bacterial, or parasitic agents. Helfer and Dickinson (1976) described a roundworm encephalitis in pigeons.

endocarditis. An inflammation of the interior lining of the heart.

enteritis. An inflammation of the intestine, classified as catarrhal, hemorrhagic, ulcerative, or necrotic. Various infectious and parasitic organisms produce enteritis.

gout. A condition resulting from abnormal purine metabolism and the excess precipitation of urates in and on body tissues. Thallium poisoning is one cause of gout.

heat exhaustion. The inability to respond to stimuli following prolonged exposure to heat. The pigeons' heat-regulatory mechanism may be overcome by high environmental temperature, high humidity, and/or inadequate ventilation. Confined birds unable to avoid the direct rays of the sun may develop a prolonged high body temperature, which is often fatal. Panting occurs when the rectal body temperature reaches 108°F, and death ensues from heat exhaustion at 113°F. If the ambient air temperature exceeds 105°F in the shade for more than 1 hr, losses occur. When these conditions are present, obesity, excitement, and exercise predispose birds to exhaustion. Whenever the air temperature reaches 90°F, birds should be given cool ice water to aid in reducing their body temperature. A fan installed in the loft to provide ventilation also helps to remove body heat and moisture released from the lungs of the birds. Exhaled water vapor is the chief means of lowering body temperature; thus humidity is a big factor in heat exhaustion.

hematoma. A circumscribed effusion of blood. This condition is observed following the rupture of blood vessels, such as the pooling of blood under the liver capsule.

hepatitis. An inflammation of the liver. The cause may be infectious or noninfectious.

hydropericardium. The accumulation of straw-colored fluid within the pericardial sac. Any inflammation of the heart or toxin ingestion may cause this condition.

icterus. See **jaundice.**

infection. The invasion and replication of pathogenic organisms within the body. Common sites of infection include air sacs, crop, eyes, footpads, gizzard, head, heart, intestine, kidney, lungs, ovary, oviduct, sinus, skin, spleen, subcutaneous tissue, and wings.

infiltration. An accumulation of cells or substances not normal to the tissues. Fat collects in damaged liver cells following circulatory stagnation and after a high-calorie diet. Lymphoid cells and polymorphonuclear leukocytes migrate and collect within various tissues in response to chemical stimuli.

inflammation. The reaction of a tissue to infection or irritation, resulting in pain, swelling, redness, fever, and loss of function.

ingluvitis. An inflammation of the crop. It occurs when *Candida albicans* infection is present.

iritis. An inflammation of the iris of the eye. Eye worms, which are uncommon, may cause the problem.

jaundice. A condition in which the concentrated greenish yellow bile pigment appears in the skin and mucous membranes. Pigeons seldom present evidence of yellow icterus. Following red blood cell destruction from toxins, infection, or blood parasites, the released hemoglobin that normally accumulates as bile in the gallbladder spills over into the tissues.

keratitis. An inflammation of the cornea of the eye. It may be caused by various irritants or eye infections.

keratosis. A horny growth of epithelial tissue. Excessive exposure to sulfur in the litter will result in skin thickening and scabby crusts. Filthy wet floors can also stimulate footpad thickening and cracking. Lack of specific vitamins in the diet, such as vitamin A and pantothenic acid, can cause crusts to form about the eyelids and mouth.

laryngitis. An inflammation of the larynx. This occurs when poxvirus grows in these tissues.

necrosis. Focal tissue death. Cellular death in pigeons is most often observed in the eyes, breast muscle, gizzard, liver, spleen, and toe. Zenker's degeneration of muscle tissue is noted in suffocation.

nephritis. An inflammation of the kidney. Swelling, congestion, and brownish discoloration are frequent indications of kidney disease. Klumpp and Wagner (1986) described enlarged pale kidneys in pigeons with hypercellularity and hypertrophy of the glomerular tuft with adhesions to the parietal epithelium of Bowman's capsule. Siller (1981) considered avian interstitial nephritis to be frequently caused by a virus.

nephrosis. A degenerative kidney disease that is characterized by edema, swelling, necrosis of renal tubules, and albuminuria.

obesity. An excessive accumulation of fat. Abnormal fat storage is usually related to a high-calorie diet.

one-eye cold. A condition with a watery discharge from one or both eyes. The condition is a sign of trouble and not a single entity. A deficiency of vitamin A often results in ocular inflammation and tearing. Trichomonads may pass by the tear duct to the inner canthus of the eye where they cause irritation and eye discharge. *Chlamydia* or *Mycoplasma* spp. have also been known to cause pus-filled, sticky eyes and eyelids. Excess ammonia in a loft may also contribute to the problem.

osteomalacia. Bone softening (osteoporosis) resulting from resorption of calcium and phosphorus in adults. It is caused by inadequate dietary vitamin D and/or sunlight and/or a lack or imbalance of calcium and phosphorus.

osteomyelitis. An inflammation of bone.

Pyogenic organisms, such as *Staphylococcus* spp., may grow in the joints and bone marrow, producing lameness, fractures, and death.

paralysis. A loss of movement from impaired nerve or muscular function. This occurs in Newcastle, paramyxovirus, or herpesvirus encephalitis infections.

paresis. Partial paralysis. The condition may involve one or more appendages with restricted movement.

peritonitis. An inflammation of the peritoneum, a serous membrane that lines the abdominal cavity. Perforation of the gizzard wall by a foreign body or the infection of the body cavity with migrating pathogens, such as *E. coli,* can cause peritonitis.

pneumonia. An inflammation of the lungs. Paratyphoid organisms and other bacteria may invade lung tissue, producing fluid and cellular infiltrations and consolidation of alveoli.

pseudomembraneous stomatitis. A false membrane on the oral and pharyngeal mucosa. This condition, reported by Klumpp and Wagner (1986), involves white Carneaus and Show Racers under 6 mo of age. No cause was determined.

salpingitis. An inflammation of the oviduct. Coliforms or mycoplasma may create an infection of the oviduct with residual caseous plugs.

septicemia. Pathogenic bacteria and their toxins in the bloodstream.

sinusitis. An inflammation of sinus membranes. Trichomonads, chlamydiae or mycoplasmal organisms may cause an excessive mucus discharge typical of the condition.

sour crop. A acid, fetid odor to the crop contents. *Candida albicans* infection of the crop mucosa is often the cause.

splenitis. An inflammation of the spleen. Most bacterial infections will involve the spleen, causing edema, enlargement, and often focal necrosis.

suffocation. Killing by obstructing normal breathing.

synovitis. An inflammation of the synovial joint membranes. *Mycoplasma* spp. may infect joints. Reoviruses have been recovered in chickens with synovitis and arthritis. Pigeons may be affected.

trauma. Any injury or wound inflicted to the body. Gunshot and predators contribute to mechanical damage.

ulcer. An open skin or mucous membrane degenerative lesion other than a wound. Ulcers most often occur in the crop, intestine, gizzard, and proventriculus.

uremia. An accumulation of urinary components in the blood. Acute congestion of muscles and viscera with deposition of uric acid in the kidneys and viscera often characterize the condition.

vomiting. The forcible ejection of the contents of the crop or proventriculus.

Anatomical Conditions

Following are terms for noninfectious conditions that describe anatomical changes.

atrophy. The reduction in size of cells, tissues, or organs, resulting from inadequate nutrition or enervation. This pertains to the regression of ovarian follicles, sciatic or brachial nerve injury with pectoral or leg muscle shrinkage, and liver and spleen contraction.

deformity. Anatomical defects appearing after hatching for the purposes of this text. These include distortion of the sternum, legs, hocks, neck, back, and beak.

dilation. Distention or stretching of a part, such as the crop, gallbladder, heart, and intestine.

eversion. Turning inside out, for example, when the cloaca and oviduct protrudes but does not prolapse.

fracture. A break in a bone. Wings, ribs, and leg bones are frequently broken.

hypertrophy. A compensatory enlargement of an organ or part. Racing pigeons often develop an increased muscular thickness of the heart.

impaction. Firmly lodged material causing blockage. This occurs in the crop, gizzard, intestine, oviduct, and uterus. Undigested feed, grit, and foreign material may block the crop or intestine. Broken eggs may cause obstruction of the oviduct.

internal laying. Laying eggs within the body cavity. Reverse peristalsis of the oviduct delivers the egg into the peritoneal cavity.

introsusception. Telescoping of the intestine or proventriculus within itself. It occurs following ingestion of irritant materials which cause intense peristaltic waves in the intestine.

malformation. Genetic defects evident upon hatching for the purposes of this text. Birds hatch with one eye, pink or small eyes, single or partial kidneys, hernias, crossed or hooked beaks, faulty feather formation, and extra or missing toes.

obstruction. Blockage or partial closure, impeding movement. This may be the result of a stricture or ulcer of the intestine or esophagus or of thickening of the trachea by roundworms.

perforation. Puncture, as from the ingestion of a foreign body. For example, a wire may pierce an organ or part.

prolapsis. This means turning inside out as when an oviduct completely everts from the vent.

rupture. This is the tearing of a part. Cervical air sacs along the neck may become distended and break releasing air under the skin. The spleen following injury may split permitting fatal hemorrhage.

stricture. An abnormal reduction in the size of a duct from contraction or from tissue deposition is implied. This may occur as a result of intestinal ulcer, scar contraction or from an oviduct infection.

References

Flick, D. F., et al. 1973. Poult Sci 52:1637–41.

Helfer, D. H., and E. O. Dickinson. 1976. Avian Dis 20:209.

Klumpp, S. A., and W. D. Wagner. 1986. Avian Dis 30:740–50.

Miller, M. S. 1987. Proc Northeastern Avian Disease Conference, p.51.

Siller, W. G. 1981. Avian Pathol 10:187–262.

APPENDIX **B**

Glossary of Medical Terms

absorption. A process by which drugs and foods pass through a barrier, such as the intestinal wall or the skin, and enter the bloodstream.
adsorption. The attachment of one substance to the surface of another.
acid. A broad category of chemical substances marked by a sour taste and the ability to react with alkaline substances (bases) to form salts. Most body functions depend on the maintenance of balance between acids and bases in cells, blood, and other body fluids.
acidification. Producing an acid condition below a pH of 7.
acute. Having a short, relatively severe development that may cause death.
administration. The method of introducing a drug or medication to the body. This includes dosage (how much), schedule (how often), and route (by mouth, injection, etc.).
affinity. Natural attraction.
agglutinin. Any antibody causing clumping of bacteria or cells.
air sacs. Membrane sacs that join the lungs. The pigeon has 11 of these air passages: 2 cervical, 1 interclavicular, 2 anterior thoracic, 2 posterior thoracic, 2 abdominal, and 2 humeral or axillary.
alkaline Having a pH above 7.
amino acid Basic chemical units into which food proteins are broken down during digestion and from which body proteins are formed in cells and organs. Each is genetically coded by three nucleotides.
anaphylaxis. An acute allergenic response followed by possible fatal shock, resulting from prior sensitization to an allergen.
anorexia. Lack of appetite.
anterior. Toward the front or head.
antibiotic. Soluble chemical substances produced by various microorganisms and fungi that kill or inhibit other microorganisms.
antibody. A specific biological response of the body to the introduction of an antigen.
antigen. A biological substance, protein in nature (e.g., bacterium, fungus, or virus) that, when introduced into the body, induces an immune response and binds to a corresponding antibody.
antiseptic. A chemical germicide used on the body surface.
antitoxin. An antibody that neutralizes bacterial toxin.
arbovirus. A virus carried and reproduced in an arthropod and transmitted by bite to a vertebrate host in which it also reproduces.
asexual. Without sexual reproduction.
asymptomatic. Without signs of disease.
attenuated. Reduced in pathogenicity or virulence.
bacteria. Minute, one-celled organisms that multiply by dividing.
bacteriostatic. Inhibiting or retarding bacterial growth.
bronchi. Bifurcation of trachea into two smaller air passages that lead to the lungs.
capsid. Rigid protein cover about the nucleic acid viral filaments, which are in a tightly coiled ball.
capsomer. Units forming the viral capsid coat.
carrier. An individual who harbors an infectious agent and sheds it without clinical evidence or signs.
cervical. Pertaining to or of the neck.
chorioallantoic membrane. The membrane sac within the embryonated egg that surrounds the embryo and is applied to the inner shell membrane.
chronic. An infection characterized by continued presence; long-lasting.
clinical disease. Disease with visible evidence of its presence.
cloaca. The lower end of the avian digestive

tract, which collects and releases undigested feed residues and urinary excretions.
clone. A population of cell or viral particles obtained from a single particle or cell that has essentially the same genome.
coccidiostat. A chemical used in water or feed to control oocyst production in the intestine.
concurrent. Referring to a disease existing at the same time as another.
congenital. Referring to an inherited genetic disease.
congestion. An increased amount of blood in tissues to the point of stagnation.
connective tissue. Tissue that binds and supports body structures.
contagious. Referring to a disease agent capable of spreading from one individual to another.
course. The period from clinical disease to recovery or death.
crop. The storage sac and milk gland receiving feed from the esophagus.
culture. Taking of a sample; the growth of microorganisms or of living tissue cells in special media.
cytoplasm. The semifluid substance of a cell, exclusive of the nucleus.
dehydration. Loss of body fluid.
distal. Farthest from the center or point of attachment.
DNA. Deoxyribonucleic acid, which forms a principal constituent of genes.
dorsal. Referring to the upper surface.
double-stranded. Referring to nucleic acid that forms a helix with two strands.
ectoderm. The outermost layer of skin or membrane.
eczema. Inflammatory skin disease with watery discharge.
edema. The presence of abnormal amounts of fluid in tissue spaces.
ELISA. Enzyme linked immunosorbent assay.
embryonated. Having an embryo or development; refers to hatching eggs in which the germ has started activity.
envelope. The lipoprotein outer coat of a virion, derived from host cell membranes, which contains virus-specific proteins.
enzootic. A disease that persists in a specific bird or animal population.
enzyme. An organic compound, often a protein, capable of catalytic activity in the body.
epidemiology. The science or study of factors affecting epidemics.
esophagus. The tubular structure that connects the pharynx to the crop.
exotic disease. An infection not normally present in a country.
exudate. A substance discharged by the tissue following disease or a vital process.
flaccid. Flabby, limp, without firm shape.
flagella. Whiplike structures that enable motility.
focal. Referring to localized centers of change.
fomites. Contaminated nonliving objects that transmit infection.
fungus. Any of a group of microscopic or larger plants that do not contain chlorophyll and reproduce by spores.
genome. The total complement of genetic material; all of the genes of an individual.
glycogen. A carbohydrate produced by the liver for conversion to dextrose.
hemagglutination. The clumping of red blood cells.
hemolytic. Causing release of hemoglobin from red blood cells.
hormone. The "messenger" chemicals of glands that help to coordinate the actions of various tissues.
host. Susceptible species of animals or birds.
humoral immunity. Acquired immunity that is antibody mediated.
icosahedral symmetry. A nucleocapsid formed by protein capsomers arranged into a symmetrical polyhedron with 20 equilateral triangular surfaces.
immunity. The degree of resistance possessed by an individual to a specific infection or poison.
inapparent infection. A disease without clinical signs.
incidence. The rate at which a disease occurs within a population.
inclusion body. Irregularly shaped bodies within the nucleus and/or cytoplasm of the cell with altered staining properties, caused by virus infection.
incubation period. The interval between exposure to infection and the onset of clinical disease.
inflammation. Tissue reaction to injury.
latent infection. Persistent concealed, underlying infection, with seeming inactivity.
life cycle. Stages in the development of the life of an organism between successive recurrences of a primary stage.
mechanical carrier. An individual that transmits an agent without participating in its replication.
metabolite. A product of metabolism.
microorganism. Any microscopic or ultramicroscopic organism.
monoclonal. Referring to one (mono) specific type of antibody molecule cloned to produce a population of identical molecules.
morbidity. The rate of individuals becoming sick from a disease within a specific period.

mortality. The rate of individuals dying from a disease within a specific period.
mucous membrane. Internal membranes that secrete mucus and line the digestive tract and trachea.
mucus. The viscid secretion that covers mucous membranes.
navel. The point at which the yolk enters the body of the squab prior to hatching.
pathogen. Any disease-producing organism.
pathogenicity. The ability to cause infection and thus disease.
pH. Acidity or alkalinity, represented by a number below or above 7, the neutral point.
plaque. A localized area of cell lysis caused by cell-to-cell spread of virus in a cell culture monolayer.
potency. The ability to be effective.
regurgitate. To vomit crop milk and partially digested food to feed squabs.
reservoir. The species or material serving as the main source of infection.
RNA. Ribonucleic acid, present in the nuclei and cytoplasm of cells.
secretion. The process whereby products are separated from blood and transformed into new substances.
serous. Pertaining to or resembling serum.
serous membrane. Membranes, such as the air sacs or peritoneum, that are bathed by serum.
single-stranded. Referring to nucleic acid that forms a helix with a single strand.
strain. A group of organisms within a species or variety with a distinct hereditary character.
subacute. Referring to diseases that seldom cause death and are less serious than acute diseases.
susceptibility. The degree of infectivity of a host cell or host.
systemic. Involving all body systems.
tissue culture. The growth of specific cells in sterile fluid media.
toxic. Having a poisonous effect.
trophozoite. A protozoan in an early stage when it occupies an epithelial cell or blood cell.
vaccine. A product designed to induce an immune reaction within the body.
vector. An intermediate host, such as a mosquito, that transmits the causative agent of disease.
ventral. Pertaining to or situated on the lower surface of the structure.
vestigial. Referring to a rudimentary or degenerative part that remains undeveloped in the adult.
viremia. Virus in the bloodstream.
virion. A complete virus particle.
virus. Any group of submicroscopic infective agents that are not capable of multiplying outside living cells.
viscera. Organs within the body cavity.

APPENDIX C

Abbreviations and Equivalent Measurements

1 meter (m)	100 cm	39.37 inches (in.)
1 centimeter (cm)	0.01 m	1/100 m
1 millimeter (mm)	0.001 m	1/1000 m, or 0.03937 in.
1 micrometer (μm) or micron (μ)	0.000,001 (1 millionth) m	1/1000 mm
1 angstrom (Å)	0.000,000,000,1 (1 10-billionth) m	0.0001 (1/10,000), μm
1 nanometer (nm), or millimicron (mμ)	0.000,000,001 (1 billionth) m	
1 teaspoon (tsp)	4 ml	
1 tablespoon (tbsp)	16 ml	
1 ounce (oz)	29.573 ml	
1 pint (pt)	473.16 ml	16 fl oz
1 quart (qt)	946.33 ml	
1 gallon (gal)	3.785 l	
1 liter (l)	1000 ml	33.81 fl oz, or 1.0567 qt
1 milliliter (ml)	0.001	1/1000 l
1 microgram (μg)	0.000,001 (1 millionth) g	0.001 (1/1000) mg
1 milligram (mg)	0.001 g	1/1000 g
1 centigram (cg)	0.01 g	
1 gram (g)		
1 kilogram (kg)	1000.0 g	
1 pound (lb)	454.5 g	
1 fluid ounce (fl oz)	29.573 ml	

cubic foot (ft^3)
hour (hr)
week (wk)
month (mo)
year (yr)
parts per million (ppm)

Index